OLD AND NEW UNSOLVED PROBLEMS IN PLANE GEOMETRY AND NUMBER THEORY

VICTOR KLEE AND STAN WAGON

THE
DOLCIANI MATHEMATICAL EXPOSITIONS

Published by
THE MATHEMATICAL ASSOCIATION OF AMERICA

The Dolciani Mathematical Expositions

NUMBER ELEVEN

OLD AND NEW UNSOLVED PROBLEMS IN PLANE GEOMETRY AND NUMBER THEORY

VICTOR KLEE
University of Washington

STAN WAGON
Macalester College

Published and Distributed by
THE MATHEMATICAL ASSOCIATION OF AMERICA

Library of Congress Catalog Card Number 91-61591

Complete Set ISBN 0-88385-300-0
Vol. 11 ISBN 0-88385-315-9

Printed in the United States of America

Current printing (last digit):
10 9 8 7 6 5 4 3 2 1

The DOLCIANI MATHEMATICAL EXPOSITIONS series of the Mathematical Association of America was established through a generous gift to the Association from Mary P. Dolciani, Professor of Mathematics at Hunter College of the City University of New York. In making the gift, Professor Dolciani, herself an exceptionally talented and successful expositor of mathematics, had the purpose of furthering the ideal of excellence in mathematical exposition.

The Association, for its part, was delighted to accept the gracious gesture iniating the revolving fund for this series from one who has served the Association with distinction, both as a member of the Committee on Publications and as a member of the Board of Governors. It was with genuine pleasure that the Board chose to name the series in her honor.

The books in the series are selected for their lucid expository style and stimulating mathematical content. Typically, they contain an ample supply of exercises, many with accompanying solutions. They are intended to be sufficiently elementary for the undergraduate and even the mathematically inclined high-school student to understand and enjoy, but also to be interesting and sometimes challenging to the more advanced mathematician.

DOLCIANI MATHEMATICAL EXPOSITIONS

1. *Mathematical Gems,* Ross Honsberger
2. *Mathematical Gems II,* Ross Honsberger
3. *Mathematical Morsels,* Ross Honsberger
4. *Mathematical Plums,* Ross Honsberger (ed.)
5. *Great Moments in Mathematics (Before 1650),* Howard Eves
6. *Maxima and Minima without Calculus,* Ivan Niven
7. *Great Moments in Mathematics (After 1650),* Howard Eves
8. *Map Coloring, Polyhedra, and the Four-Color Problem,* David Barnette
9. *Mathematical Gems III,* Ross Honsberger
10. *More Mathematical Morsels,* Ross Honsberger
11. *Old and New Unsolved Problems in Plane Geometry and Number Theory,* Victor Klee and Stan Wagon

CONTENTS

Preface xi

Chapter 1. Two-Dimensional Geometry

Introduction 1

1. Illuminating a Polygon 3–11, 71–79

   Is each reflecting polygonal region illuminable?

2. Equichordal Points 12–15, 80–85

   Can a plane convex body have two equichordal points?

3. Pushing Disks Together 16–20, 86–90

   When congruent disks are pushed closer together, can the area of their union increase?

4. Universal Covers 21–24, 91–94

   If a convex body C contains a translate of each plane set of unit diameter, how small can C's area be?

5. Forming Convex Polygons 25–28, 95–102

   How many points are needed to guarantee a convex n-gon?

6. Points on Lines 29–35, 103–110

   If n points are not collinear, must one of them lie on at least $\frac{1}{3}n$ connecting lines?

7. Tiling the Plane 36–44, 111–119

   Is there a polygon that tiles the plane but cannot do so periodically?

8. Painting the Plane 45–49, 120–127

   What is the minimum number of colors for painting the plane so that no two points at unit distance receive the same color?

9. Squaring the Circle 50–53, 128–131

   Can a circle be decomposed into finitely many sets that can be rearranged to form a square?

10. Approximation by Rational Sets 54–57, 132–136

    Does the plane contain a dense rational set?

11. Inscribed Squares 58–65, 137–144

    Does every simple closed curve in the plane contain all four vertices of some square?

12. Fixed Points 66–70, 145–150

    Does each nonseparating plane continuum have the fixed-point property?

References 151

Chapter 2. Number Theory

Introduction ... 167

13. Fermat's Last Theorem ... 168–172, 199–202

Do there exist positive integers x, y, and z and an integer $n \geq 3$ such that $x^n + y^n = z^n$?

14. A Perfect Box ... 173–174, 203–205

Does there exist a box with integer sides such that the three face diagonals and the main diagonal all have integer lengths?

15. Egyptian Fractions ... 175–177, 206–208

Does the greedy algorithm always succeed in expressing a fraction with odd denominator as a sum of unit fractions with odd denominator?

16. Perfect Numbers ... 178–181, 209–214

Is there an odd perfect number? Are there infinitely many even perfect numbers?

17. The Riemann Hypothesis ... 182–185, 215–220

Do the nontrivial zeros of the Riemann zeta function all have real part $\frac{1}{2}$?

18. Prime Factorization ... 186–190, 221–224

Is there a polynomial-time algorithm for obtaining a number's prime factorization?

19. The $3n + 1$ Problem ... 191–194, 225–229

Is every positive integer eventually taken to the value 1 by the $3n + 1$ function?

20. Diophantine Equations and Computers ... 195–198, 230–233

Is there an algorithm to decide whether a polynomial with integer coefficients has a rational root?

References ... 234

Chapter 3. Interesting Real Numbers

Introduction ... 239

21. Patterns in Pi ... 240–242, 251–254

Are the digits in the decimal expansion of π devoid of any pattern?

22. Connections between π and e ... 243, 255–257

Are π and e algebraically independent? Is their ratio rational?

23. Computing Algebraic Numbers ... 244–247, 258–260

If an irrational number is real-time computable, is it necessarily transcendental? Is $\sqrt{2}$ real-time computable?

24. Summing Reciprocals of Powers ... 248–250, 261–264

Is $1 + \frac{1}{2^5} + \frac{1}{3^5} + \frac{1}{4^5} + \cdots$ irrational?

References ... 265

Hints and Solutions: Two-Dimensional Geometry 269
Hints and Solutions: Number Thoery 300
Hints and Solutions: Interesting Real Numbers 311
Glossary 315
Index of Names 325
Subject Index 331

PREFACE

THE ROLE OF UNSOLVED PROBLEMS

As mathematics becomes ever more complex, more structured, and more specialized, it is easy to lose sight of the fact that the research frontiers still contain many unsolved problems that are of immediate intuitive appeal and can be understood very easily, at least in the sense of understanding what a problem asks. Parts of plane geometry and number theory are especially good sources of such intuitively appealing problems, and those are the areas emphasized in this book. For each of the presented problems, we look at its background, try to explain why some mathematicians have found the problem of special interest, and tell about techniques that have been used to obtain partial results. Some of the exercises will help out readers establish partial results for themselves, and references are provided for those who want to learn more about the problems.

All mathematical discoveries may be said to consist of the solution of unsolved problems. Sometimes a question is answered almost as soon as it is formulated. Then the question and its answer are published together, and the question never becomes known as an unsolved problem. In other cases, a mathematical question may be formulated long before anyone is able to answer it. If the question becomes widely known, many researchers may direct their efforts toward answering it and the unsolved problem may have a significant influence on the development of mathematics.

Some unsolved problems are conceptually very simple and specific—for example, "Is the answer 'yes' or 'no'?", "What is the value of this function at a specified point of its domain?" Others are conceptually sophisticated—for example, "How can such and such a theory or argument be extended so as to apply to a certain more general class of objects?" Both the simple and the sophisticated problems have played an important role in the development of mathematics, and often they are inextricably mingled. Some of the "simplest" unsolved problems, such as Fermat's last "theorem" (see Section 13) have led directly to some of the most sophisticated mathematical developments and, in turn, to sophisticated unsolved problems. On the other hand, the solution of a sophisticated problem often requires answering a number of simple, specific questions. Thus it is probably unreasonable to try to decide whether the simple or the sophisticated problems have been more important in the development of mathematics. In any case, this book is devoted exclusively to problems of the simple sort—ones whose statements are short and easy to understand.

Some of the more modern areas of mathematics are poor sources for the sort of unsolved problem that is emphasized here. If one thinks of mathematical disciplines as buildings, some could be represented by buildings that are very tall but also quite narrow, with a few important research problems sprouting out of the roof. In order to reach these unsolved problems, one must climb through the many lower floors of the building, and on each floor must make one's way through numerous definitions and theorems that may be necessary for even understanding what the unsolved problems are about. By contrast, we are concerned here with some parts of mathematics that can be represented as broad, low buildings, with a high ratio of roof area (research problems) to volume. Any particular problem can be reached, at least in the sense of understanding what it asks, by entering the building through the proper door and climbing only one or two flights of stairs. It is striking that plane geometry and elementary number theory, the oldest branches of mathematics, still offer such a large supply of easily understandable, intuitively appealing, unsolved problems. This book contains some of our favorite examples from each subject.

Before proceeding to the problems themselves, we should admit that our picture of branches of mathematics as buildings is misleading in some important respects. For example, we have pictured elementary number theory as a broad, low building, and many mathematicians would think of the subject of algebraic geometry as one of the tallest, most highly structured buildings. However, algebraic geometry has provided some of the most important tools for dealing with some of the most "elementary" problems of number theory, such as Fermat's last theorem. This illustrates the intermingling of simple problems and sophisticated problems, and the important distinction between understanding a problem in the sense of having an intuitive grasp of what it asks, and the much deeper understanding that requires knowing enough to have some chance of solving the problem or at least discovering new results that are relevant to it. For each problem in this book, we hope to have provided enough information to enable the reader to understand the problem in at least the first sense. It remains to be seen what is required for understanding in the second sense. However, we don't believe that all of the problems presented here will require sophisticated methods for their solution. Probably some will be solved by means of a single really clever idea. There's no monopoly on clever ideas, and your chance of solving some of the problems may be as good as anyone's!

Part of the appeal of elementary problems is the feeling that they ought to have solutions that are equally elementary. But elementary problems can have exceedingly complex solutions. A notorious example is the four color theorem (every map in the plane can be colored with four colors so that adjacent countries are colored differently); its statement is simple but its proof, discovered in 1976 by K. Appel and W. Haken, is long, subtle, and involves an extensive computer calculation. Another example is a less well-known result due to P. Monsky: It is impossible to dissect a square into an odd number of triangles having equal area (see Section 7 for a discussion of this result and its extensions). The only known proofs use a sophisticated concept from field theory. Mathematics is full of simple-sounding statements that have been proved, but whose proofs are complicated, using the most sophisticated tools mathematicians know. But that's not the worst of it. It may be that for certain elementary conjectures, *no proof of either the conjecture or its negation exists.* Most interesting questions deal with infinitely many instances; indeed, much of the power and beauty of mathematics consists of finding a finite proof that answers infinitely many instances of a question. As observed by Haken[1] in an interesting paper that contains both details of the four color theorem's proof and some philosophical observations, it may be that "infinitely many single statements of a conjecture have infinitely many individual and significant differences so that a proof of the conjecture is impossible since it would require infinitely many case distinctions." (For more on the possibility that simple problems are undecidable, see Section 20.) Thus amateur and professional mathematicians are in comparable situations. An amateur might spend years searching for an elementary solution to an elementary unsolved problem, confident that such a solution exists. A professional, armed with supercomputers and the cumulative technical knowledge of generations of brilliant mathematicians, might attack a problem confident that a solution, perhaps a complicated one, exists. But both run the danger of being stymied by the nonexistence of the objects of their search.

Although we believe that most of our problems do have solutions, we would not be surprised if, for most of them, the solutions seem much less elementary than the problem statements. However, some problems that we expect to prove intractable may turn out to be solvable after all. For example, among the geometric problems that we originally planned for inclusion, both of us might well have cho-

---

[1] "An attempt to understand the four color problem," *Journal of Graph Theory,* 1 (1977) 193–206.

sen the main problem of Section 9—the modern form of the problem of squaring the circle—as the one least likely to be solved in the near future. But when the book was nearly completed, that problem was settled by M. Laczkovich, who proved that a circle (with interior) is equidecomposable to a square. His proof is quite complicated, using tools that on first glance would seem to have little to do with the problem.

ORGANIZATION OF THE BOOK

Each section of the book is devoted to a single problem or a related group of problems, and each section is divided into two parts. The first part is elementary. Its main goal is to supply the necessary definitions, convey some intuitive feeling for the problem, and present some easy proofs of relevant results. The second part contains more advanced arguments and mentions references. We hope that all of each first part and some of each second part will be accessible to undergraduate mathematics majors, and that everything in the book will be accessible to all graduate students of mathematics. On the other hand, since the presented problems are unsolved, and since fairly extensive references are supplied, there may be some interest in the research community as well.

The sections are arranged in three chapters. In each chapter, the first parts are grouped together, followed by the second parts. Thus the reader may peruse an entire collection of first parts without being distracted by technical details or references, and may then consult the second parts and the references for more information about whichever problems seem most appealing.

ACKNOWLEDGEMENTS

For helpful mathematical or editorial comments, or for supplying useful material, we are indebted to Donald Albers, Vojtech Bálint, Miroslav Benda, Joe Buhler, Vasilis Capoyleas, Kiran Chilakamarri, Herbert Edelsbrunner, Paul Erdős, John Ewing, Klaus Fischer, Richard Gardner, Chris Godsil, Eli Goodman, Peter Gruber, Branko Grünbaum, Richard Guy, Charles Hagopian, Joan Hutchinson, John Isbell, Aleksander Ivić, Peter Johnson, James Jones, Martin Kneser, Wlodzimierz Kuperberg, Milos Laczkovich, David Larman, Dragoslav Ljubic, Erwin Lutwak, Howard Masur, Mark Meyerson, Giacomo Michelacci, Paul Monsky, William Moser, Andrew Odlyzko, János Pach, Roger Penrose, Richard Pollack, Carl Pomerance, George Purdy, Basil Rennie, Peter Renz, Boyd Roberts, Jack Robertson, Raphael Robinson, Joel Rogers, Moshe Rosenfeld, Reinhard Schäfke, John Smillie, Sherman Stein, Walter Stromquist, Bernd Sturmfels, Sam Wagstaff, Gerd Wegner, and Joseph Zaks. Of course, any errors are our own, and we hope that readers will write to tell us about them.

Finally, we want to record our special thanks to the MAA's Beverly Ruedi, who took our jumbles of computer files and turned them into a book. It was a pleasure to work with her.

Victor Klee
University of Washington

and

Stan Wagon
Macalester College

CHAPTER 1

TWO-DIMENSIONAL GEOMETRY

INTRODUCTION

Of the ideas that lead beyond the geometry of Euclid, the notion of a convex set may be the one that would have been most easily understood by the ancient Greeks. In fact, a close relative of this notion was defined by Archimedes. A set is *convex* if it includes, with each pair of its points, the entire line-segment joining them. (Intuitively, a convex set is one that has neither holes nor dents.) As a source of appealing, easily stated unsolved problems, the study of low-dimensional convex sets rivals number theory and graph theory. Thus it is not surprising that convex sets play a central role in many of the sections of this chapter. However, there are also problems that don't involve convexity, and some of them involve modern (though elementary) set-theoretical and topological notions.

In the title of this book, the term "plane geometry" was used for brevity. However, that may suggest an axiomatic approach, and our approach here is certainly not axiomatic. Rather, our viewpoint is that the geometric objects really exist, and we want to study them in an efficient way. For that purpose, the most appropriate setting seems to be the cartesian plane $\mathbb{R}^2$, complete with its usual topology and basic vector operations. The title of this chapter, Two-Dimensional Geometry, is intended to suggest that, although all of the main discussion is in $\mathbb{R}^2$ and the various Parts One are confined to $\mathbb{R}^2$, many of the problems have interesting higher-dimensional analogues that are discussed briefly in Part Two.

In each section, the "title problem" is the one that begins the section and is stated in the Table of Contents. However, most of the sections include a variety of other, equally interesting problems that are related in spirit to the title problem. The variety is especially marked in Section 1, which concerns the role of reflections in illumination and billiards.

Problem 6 is conceptually the simplest. It does not involve distance or angle or area, only the incidence of points and lines. Problem 5 involves convex polygons as well, and Problem 3 involves areas of unions of disks, but these problems would also have been easily understood by the ancient Greeks. Problems 2 and 4 are simple, but they do involve more general convex sets than the other problems just mentioned. Problems 7 and 8 concern subdivisions of the entire plane, rather than just part of it. Problem 9 deals with a striking interaction between the simplest convex sets (circles and squares) and sets that probably no one would agree to call "geometric figures." Problem 10 involves some relationships between geometry and number theory, while Problems 11 and 12 involve topological notions.

Several of the problems are striking for the great disparity between what is known and what may be true. For some of them, this disparity can be expressed in a quantitative way. Problem 3 originates from the conjecture that when several congruent disks are pushed closer together, the area of their union can't increase. However, the best that has been proved is that it can't increase by a factor of more than 9. Problem 6 was motivated by a conjecture that when n points in the plane are not all collinear, and when all the lines determined by pairs of these points are added to the configuration, then one of the points is on at least $\frac{1}{2}n$ of the lines. Some small counterexamples show that $\frac{1}{2}n$ doesn't always work, but perhaps $\frac{1}{3}n$ does. It was proved initially (but not easily) that $10^{-1087}n$ always works, and that has been improved to $10^{-32}n$—still a long way from $\frac{1}{3}n$. A more technical result in Section 1 involves a smoothness assumption for the boundary curvature of a convex billiard table. The original theorem assumed the curvature to be 553 times differentiable. The 553 has been reduced to 6, but a further reduction is probably possible.

The most easily stated problems are 8 and 10. Problem 8 asks, "What is the minimum number of colors for painting the plane so that no two points at unit distance receive the same color?" It has been known since 1960, when the problem first appeared in print, that the number is 4, 5, 6, or 7, but no one has been able to narrow the possibilities further. Problem 10 asks, "Does every simple closed curve in the plane contain all four vertices of some square?" An affirmative answer has been known since 1913 for all sufficiently smooth convex curves, and more recent results have avoided the convexity assumption. However, there is a sense in which most simple closed curves are not smooth at all, and thus it is possible that the answer is negative for most curves. Despite their ease of statement (or perhaps because of it!), Problems 8 and 10 have, like many of the other problems considered here, been the subject of incorrect published results. As we mentioned in the Preface, it is tempting to believe that an easily stated problem should have an easy solution. The reader is cautioned against believing that for most of the problems stated here.

There are other good sources of information about easily understood unsolved geometric problems. We mention especially the book on tiling by B. Grünbaum and G. Shephard [GS4] and the books on lattice point problems by J. Hammer [Ham] and P. Erdős, P. Gruber, and Hammer [EGH]. In addition, there are three excellent collections of unsolved problems that will soon become books—those of Erdős and G. Purdy [EP3], of H. Croft, K. Falconer, and R. Guy [CFG], and of W. Moser and J. Pach [MP] (developed from an earlier collection of L. Moser). We should also mention the numerous articles of Erdős, Erdős and Purdy, L. Fejes Tóth, Grünbaum, and H. Hadwiger, some of which are included in our list of references. Another good source of geometric problems is the "Unsolved Problems" section in the *American Mathematical Monthly*. And there are many other sources—too many to mention them all. It seems clear that the oldest part of mathematics will continue to provide challenges (and surprises) for many years to come.

1. ILLUMINATING A POLYGON

Problem 1

Is each reflecting polygonal region illuminable?

Imagine life in a two-dimensional world where electricity and skilled construction are very expensive but reflecting materials are cheap. One might live in a room whose shape resembles that of Figure 1.1 and whose walls are mirrors, and one might hope to illuminate the entire room by a single light source. Can this always be done, regardless of the room's shape? If it can be done, how should one discover where to place the light source?

Light rays issue in all directions from the source, and each ray continues on its way according to the usual rule: "angle of reflection equals angle of incidence." The room is *illuminated* by a given source if each point of the room lies on at least one of the rays. For each position of the light source s, the points that are illuminated *directly* from s (without using reflections) form a set that is geometrically simple, for it's a union of straight-line segments issuing from s. (Such a set is said to be *star-shaped* from s). However, this simple picture rapidly becomes complicated as more and more reflections are taken into account, and when the number of reflections is

FIGURE 1.1
Can a light source be placed so as to illuminate the entire room, assuming each wall is a mirror?

unlimited the analysis becomes very difficult. In fact, both of the following questions are open, and they may have different answers.

Problem 1.1

Can each reflecting polygonal room be illuminated from at least one of its points?

Problem 1.2

Can each reflecting polygonal room be illuminated from each of its points?

Perhaps there is a polygonal room that is illuminated for at least one placement (or even for "almost all" placements) of the light source, but for which at least one placement does not result in total illumination. If there is such a room, it provides an affirmative answer to 1.1 but a negative answer to 1.2.

We shall also be concerned with rooms whose walls are curved rather than polygonal. When a ray strikes the boundary curve at a point where the curve has a tangent, the angles of incidence and reflection are defined as those that the ray makes with the tangent. And we assume that when the ray strikes a boundary point where there is no tangent (e.g., a corner of a polygonal room), the ray is absorbed and does not continue (see Exercises 1 and 2).

Unless the contrary is stated, we assume that the room's boundary consists of a single simple closed curve (also called a *Jordan curve*)—a continuous path in the plane that doesn't cross itself and ends where it started. In the polygonal case, this path is formed from a finite sequence of line-segments. The answers to Problems 1.1 and 1.2 may change if the room is permitted to have "holes," and may also be affected by a requirement that the light source should be interior to the room, or

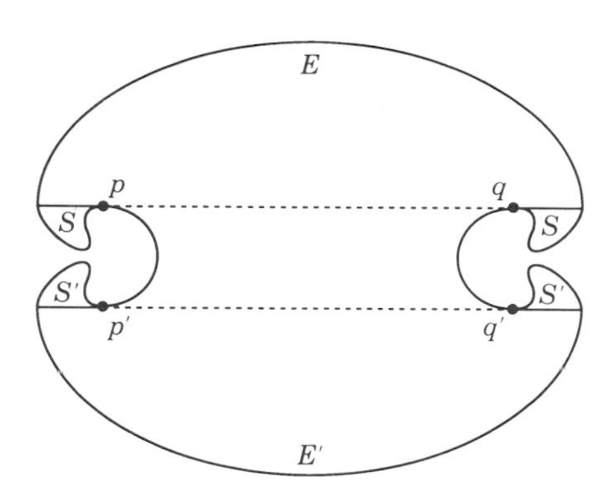

FIGURE 1.2
This region with smooth boundary is not illuminable from any point.

on the room's boundary. (Unless the contrary is stated, both sorts of placement are permitted here.)

Now let's consider smooth rather than polygonal boundaries. In Figure 1.2, the boundary is formed by an upper half-ellipse E with foci p and q, a lower half-ellipse E' with foci p' and q', and two curves that connect these to form a single smooth Jordan curve. Recall that the foci of an ellipse are so called because whenever a light ray issues from one of the foci and strikes the ellipse, it is immediately reflected back through the other focus. From this it follows easily that when a ray crosses the segment pq and strikes E, it is immediately reflected back across pq (Exercise 3). Hence the room is not illuminable from any point, for if the light source is below the line pq then the areas labeled S are not illuminated and if the source is above the line $p'q'$ then the areas labeled S' are not illuminated.

For each positive integer n, n copies of Figure 1.2 can be joined as in Figure 1.3 to produce a smoothly reflecting region that is not illuminable from any n (or even from any $2n-1$) sources—that is, even when $2n-1$ light sources are permitted, it is impossible to place the sources so as to illuminate the entire room. By extending this construction still further, using a shrinking sequence of copies of Figure 1.2 and adding a single limit point p, we obtain a reflecting region that is smooth except at p and cannot be illuminated from any finite number of light sources.

The easy analysis of the preceding smooth examples suggests an intriguing question: Can we obtain an unilluminable polygonal region by taking a sufficiently close polygonal approximation of an unilluminable smooth region?

If good workmanship should become less expensive, we might be able to afford a room that is convex. Then we could illuminate the entire room directly by an arbitrary placement of a single light source that emits rays in all directions. But suppose now that the light source is a laser, emitting only a single ray. Can we, with

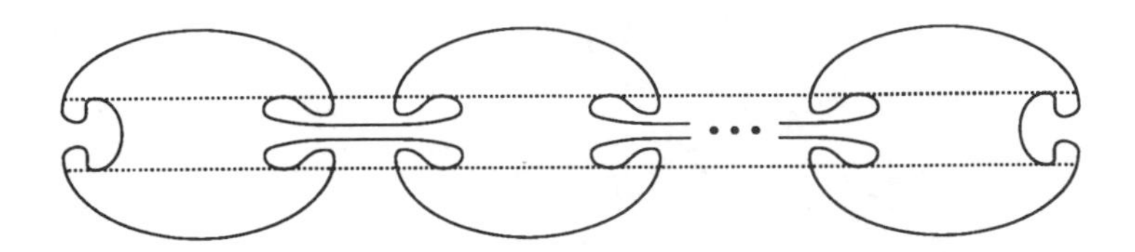

FIGURE 1.3
For each n there is a smoothly bounded reflecting region such that even n light sources do not suffice for total illumination.

the aid of reflections, hope to illuminate the entire interior of the room? The answer is negative, regardless of the shape of the room's boundary and even if an arbitrarily large finite number of single-ray light sources is permitted. This follows from the fact that no circle can be completely covered by a sequence of lines (Exercise 4).

Though we can't illuminate the entire room by a single reflected ray, can we at least come close to doing so? As an aid to intuition, let's regard the room as the surface of a billiard table and refer to the rays as *billiard paths.* In our special game of billiards, an opponent invisibly marks a small disk somewhere on the table. We're free to put the ball wherever we like, and to strike it in any direction we choose. We win the game if our billiard path passes over the opponent's disk (equivalently, if our laser ray illuminates some point of the disk). We don't know where the disk has been placed or how small it may be, so we want to be sure that our path passes over every disk on the surface of the table. In mathematical terms, the path should be *dense* in the table. The shape of the table determines whether such a path exists, and this question of existence is still open for many shapes. In particular, the following question arises.

Problem 1.3

Does each polygonal table admit a dense billiard path? How about triangular tables?

Let us say that a polygonal table is *rational* if each of its angles is a rational multiple of π. The following known result is too deep to prove here, but the special case of a square table is discussed in Part Two.

Theorem 1.1. *On each rational polygonal table there is a dense billiard path.*

A billiard path may be described by the sequence $p_0p_1p_2\ldots$ that lists the path's starting point p_0 and its successive contacts with the boundary. The path is said to be *periodic with period* k if $p_{i+k} = p_i$ for each $i \geq 1$. Such a path is very far from being dense in the table, for it consists of only a finite number of segments. It's easy to see that when the table is an acute triangle, there is a billiard path of period 3 (Exercise 5) and it is known also that each right triangle admits a billiard path of period 6. However, the following problem is open.

Problem 1.4

Does each polygonal table admit a periodic billiard path? How about obtuse triangular tables?

As in the case of Problem 1.3, special methods are available for rational polygons.

Theorem 1.2. *On each rational polygonal table there is a periodic billiard path.*

Now let's take a look at smooth tables. It's easy to see that when the table T is circular, no billiard path is dense in T. For if the initial segment p_0p_1 is collinear with the center c, then p_1p_2 is a diameter of T and the ball merely bounces back and forth along this diameter. If p_0p_1 is not collinear with c, then p_1p_2 is tangent to a smaller circle C centered at c, and all the later segments are also tangent to C (Figure 1.4). (The circle C is said to be a *caustic* for the table T. Caustics are discussed in Part Two.) Thus it's clear that no matter which path we produce, we can't be sure it will pass over the opponent's invisible disk. However, we can be sure of winning if we know the size of the disk in advance (Exercise 6).

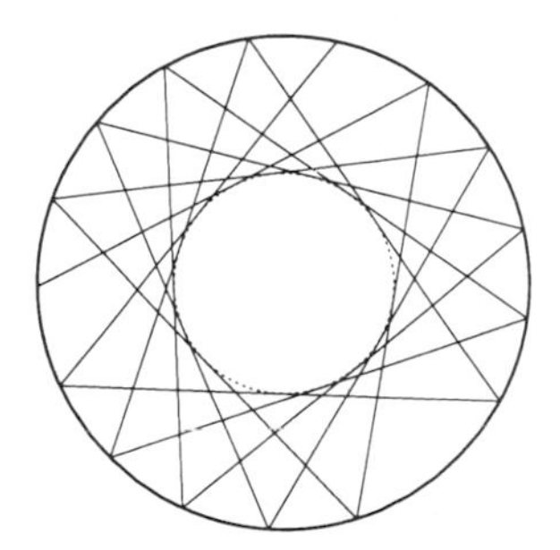

FIGURE 1.4
If T is a circular table, C is a smaller concentric circle, and P is a billiard path in which one segment is tangent to C, then all segments of P are tangent to C.

How about periodic paths on a smooth table? On a circular table, each diametral path is of period 2, and for each $k \geq 3$ an inscribed regular k-gon provides a billiard path of period k. Here's a more general result.

Theorem 1.3. *Suppose that the table T is smooth and convex, and that $k \geq 2$. Let P be a convex k-gon of maximum perimeter inscribed in T. Then P provides a billiard path of period k.*

Outline of Proof. A "2-gon of maximum perimeter" is obtained by taking two boundary points p and q at maximum distance in T. Using the maximality and the fact that T is convex, it's easy to see that T's tangents at p and q are both perpendicular to the segment pq. Hence a billiard path that starts along the segment pq continues to run back and forth along that segment.

Suppose now that $k \geq 3$, and consider the successive vertices $v_0, v_1, \ldots, v_k = v_0$ of a convex k-gon P of maximum perimeter inscribed in T. For each i, let L_i denote the line that is tangent to T at the point v_i. To show that P provides a billiard path of period k, it suffices to show that for each i, the segments $v_{i-1}v_i$ and $v_{i+1}v_i$ make equal angles with the line L_i. Suppose, for example, that this fails when $i = 1$, and that the segment v_0v_1 makes a smaller angle with L_1 than does the segment v_2v_1. Then, from the fact that L_1 is tangent to T at v_1, it can be shown that moving the point v_1 along the boundary slightly in the direction of v_0 will produce an inscribed k-gon of greater perimeter. (See Figure 1.5 and Exercise 7.) □

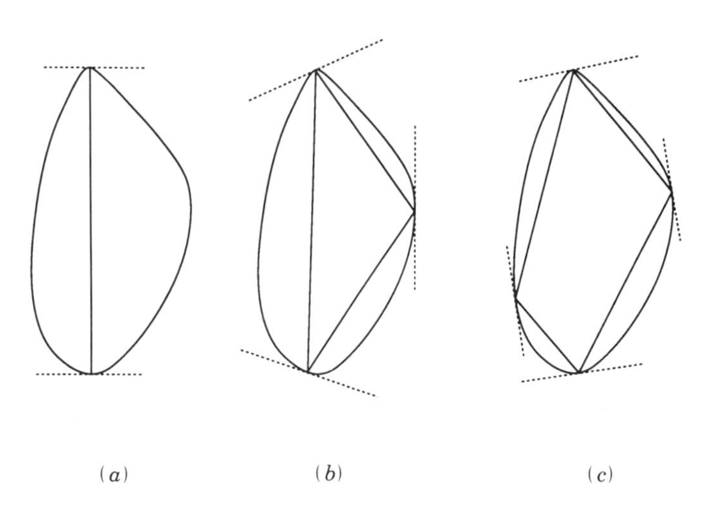

FIGURE 1.5
If an inscribed convex k-gon is of maximum perimeter, it produces equal angles of incidence and reflection and hence provides a billiard path of period k.

Finally, let us consider the problem of illuminating a polygonal room when the costs of careful construction and reflecting materials are prohibitive. We can no longer afford to build with mirrors, so we must rely on direct illumination. We cannot afford a convex (or even a star-shaped) room, so we cannot hope to illuminate it with a single light source. Nevertheless, we would like to control the cost of electricity by using the smallest possible number of light sources. The number of sources needed is related to the number of edges of the polygonal boundary.

Theorem 1.4. *For the direct illumination of a polygonal room with n edges, $\lfloor n/3 \rfloor$ light sources are occasionally necessary and always sufficient.*

Outline of Proof. Remember that $\lfloor x \rfloor$ denotes the greatest integer not exceeding x. Consideration of Figure 1.6 and its extensions shows that as many as $\lfloor n/3 \rfloor$ sources may be needed. To show that direct illumination can always be accomplished by placing the sources at $\lfloor n/3 \rfloor$ or fewer carefully chosen vertices of the polygon, we use the following fact.

(A) If a Jordan polygon P has at least four vertices, there is a pair of vertices for which the connecting open segment is interior to P.

To prove (A), let the vertices (in order of traversal of the polygon) be $v_0, v_1, \ldots, v_n$ $(= v_0)$, and let v_i be such that the polygon's interior angle $v_{i-1}v_iv_{i+1}$ is less than π. (Since the sum of the interior angles of a simple n-gon is $(n-2)\pi$, such an i must exist.) Assume for notational simplicity that $i = 1$. If the open segment v_0v_2 is not interior to P, then the interior of the triangle $v_0v_1v_2$ contains at least one vertex of P, and among such vertices there is a vertex w whose distance from the line v_0v_2 is a maximum. But then the open segment v_1w is interior to P.

Now the following can be proved by induction on the number of vertices, using the observation (A).

(B) If P is a Jordan polygon, then certain triples of P's vertices can be used to form triangles whose union consists of P and the points inside P, the triangles being such that whenever two of them intersect, the intersection is an edge of each or a vertex of each. Also, the vertices can be colored in three colors so that all three colors are present in each triangle. (See Exercise 8.)

Using (B), the proof of Theorem 1.5 is immediate. When the n vertices are colored in three colors, some color (say red) is used at most $\lfloor n/3 \rfloor$ times. Each of the triangles has exactly one vertex that is colored red, so we may simply place the light sources at the red vertices. □

Theorem 1.5 is sometimes called an "art gallery theorem" or a "watchman theorem," for it has the following interpretation: if the walls of an art gallery are made up of n straight segments, then the entire gallery can be supervised by $\lfloor n/3 \rfloor$

watchmen placed at appropriate corners of the gallery. There are other art gallery theorems and related unsolved problems, some of which will be encountered in Part Two.

FIGURE 1.6
When there are n edges in the boundary of this sort of room, direct illumination cannot be accomplished with fewer than $\lfloor n/3 \rfloor$ light sources.

Exercises

1. We assume here that light rays are absorbed when they strike a corner of a polygonal room. However, if a ray R strikes a corner at which there is an interior right angle, it is reasonable (on grounds of continuity) to assume that R is reflected back on itself. Explain why this is so. (It may be helpful to consider Exercise 2 in conjunction with this one.)

2. Suppose that half-lines H and L form an angle A of radian measure α, with $2\pi/5 < \alpha < 2\pi/3$. With B denoting the half-line that bisects A, suppose that two light rays R_H and R_L run parallel to B but are on opposite sides of B, with R_H striking H first and R_L striking L first. Show that after R_H is reflected from H, it strikes L and then continues along a half-line S_H that lies entirely in A. Similarly, after R_L is reflected from L, it strikes H and then continues along a half-line S_L that lies entirely in A. Show that the half-lines S_H and S_L are parallel if and only if $\alpha = \pi/2$.

3. Consider a reflecting elliptical room with foci p and q. Show that if a segment of a light ray crosses the segment pq, then the next segment of the ray also crosses pq.

4. Show that if C is a circle and $L_1, L_2, \ldots$ is a sequence of lines, then there is a point of C that does not belong to any L_i.

5. (a) Show that for each triangle W there is an acute triangle T that admits a billiard path of period 3 similar to W.

(b) Show that each acute triangle T admits a billiard path of period 3.

6. For the special game of billiards that we have discussed, suppose that the table is a circle of radius 1 and we know our opponent's invisible disk is of radius at

least ϵ. How can we be sure of winning the game by producing a billiard path that crosses the disk?

7. For the case in which $k \geq 3$, complete the proof of Theorem 1.3. (Hint: In order to see clearly how the cosines of the angles of incidence and reflection are relevant, carry out the argument first under the assumption that in a neighborhood of the point v_i, the boundary curve T actually coincides with the tangent L_i. Then discuss the general case by choosing a coordinate system in which v_i is the origin and L_i is the x-axis.)

8. Prove the statement (B) by induction on the number of vertices. (Hint: If the Jordan polygon P has n edges, with $n \geq 4$, then P has an interior diagonal D of the sort described in (A). This divides P into two Jordan polygons, Q and R, whose intersection is D. Since each of Q and R has fewer than n edges, each is subject to the inductive hypothesis.)

9. For the simple closed polygonal rooms shown in Figures 1.1 and 1.6, produce triangulations of the sort described in (A) of the proof of Theorem 1.4. Then color the vertex-set in the manner described in (B).

2. EQUICHORDAL POINTS

Problem 2

Can a plane convex body have two equichordal points?

A *chord* of a region R is a segment that joins two boundary points of R. Each chord xy through an interior point p of R is divided by p into two subsegments px and py. The point p is an *equichordal point* if all chords through p are of the same length (i.e., the sum of lengths $px+py$ is constant), an *equireciprocal point* if the sum of reciprocals $1/(px)+1/(py)$ is constant, and an *equiproduct point* if the product $(px)(py)$ is constant.

The center of a disk is an equichordal point, an equireciprocal point, and an equiproduct point. It's natural to wonder whether points of these and related sorts are found in noncircular regions, and how many such points a region might have. Most of the basic questions have been answered, but the simplest of all—whether a plane convex region can have two equichordal points—is still open. It was posed in 1916.

If the answer to Problem 2 is negative (which is probably the case), what happens when the region is not required to be convex? Let's use the term *Jordan region* for a bounded subset of the plane whose boundary is a simple closed curve. Then the following is also open.

Problem 2.1

Can a Jordan region have two equichordal points?

It's easy to construct noncircular Jordan regions that have one equichordal point. Imagine a fixed point p in the plane, and a segment of fixed length λ that must pass through p but is free to slide back and forth as well as to rotate about p. The segment may be regarded as a sort of drawing arm, with markers attached at its ends x and y. Let x_0 and y_0 be the initial positions of the ends, and then move the end x along a curve J from x_0 to y_0 such that each point of J is at distance less than λ from p, and (except for x_0 and y_0) no two points of J are collinear with p. Then J lies on one side of the line through x_0 and y_0, and as x traces the curve J, y traces a curve K on the other side of the line (see Figure 2.1). The union $J\cup K$ bounds a Jordan region R that has p as an equichordal point, and with extra care in choosing J it may be assured that R is convex (Exercise 6 in Part Two). By permitting the arm

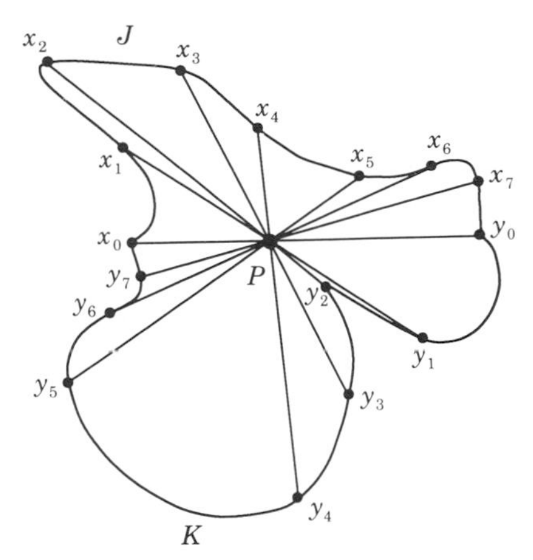

FIGURE 2.1
Use of a "drawing arm" to construct a noncircular region that has an equichordal point.

to vary in length, the same technique can be used to construct noncircular regions that have an equireciprocal point or equiproduct point (Exercise 1).

Now suppose that a Jordan region R has two equichordal points p and q, and imagine two drawing arms—an arm xy through p and an arm $x'y'$ through q, both of the same fixed length—with their ends x and x' joined by a hinge. Let both arms start with x and x' at a point x_0 collinear with p and q, so that the ends y and y' also start at the same initial position y_0 . As the hinge traces out half of R's boundary—a curve J from x_0 to y_0, each of the ends y and y' must trace out the same other half—a curve K from y_0 to x_0, even though the two arms are sliding and rotating about the distinct points p and q. That is hard to imagine, especially when R is required to be convex. It suggests that the answer to Problem 2 (and perhaps 2.1) is negative.

Of the principal problems treated in this book, Problem 2 may in a sense be the least important. Whatever the answer turns out to be, a solution probably won't have much influence on the rest of mathematics. However, because the problem is so old and so easy to state, it's irritating that no one knows the answer. It may have been partly a sense of irritation that has led people to work on the problem. Of the mathematicians attracted to it, some have thought they had established an affirmative answer, others a negative answer, but as yet, none of the proposed solutions in either direction has withstood close examination. A successful treatment of the problem will probably require a delicate interplay between geometry and analysis, and it is all too easy to gloss over crucial analytical details because they are "geometrically obvious."

Many attacks on the infamous equichordal problem have started by assuming that C is a plane convex body or Jordan region with two equichordal points and then deducing other properties of C. If C does exist, some of the properties might make it clear how to construct C. If C doesn't exist, perhaps that can be proved by establishing two mutually contradictory properties. The most important properties known for C (if it exists) are stated without proof in Theorems 2.1 and 2.3–2.6 below.

Theorem 2.1. *If a Jordan region has two equichordal points p and q, it is symmetric about the line pq and about the midpoint of the segment pq.*

Corollary 2.2. *No Jordan region has more than two equichordal points.*

Proof of 2.2. In terms of the usual vector operations, symmetry of a set about a point m means that for each point x of the set, the point $x + 2(m - x)$ is also in the set. If a Jordan region had more than two equichordal points, it would, by Theorem 2.1, have two different centers of symmetry, say m and m', and would then include each point of the unbounded sequence

$$z_1 = 2m' - m, \quad z_2 = 2m - z_1 = 3m - 2m', \ldots,$$

$$z_{2k-1} = 2km' - (2k-1)m, \quad z_{2k} = (2k+1)m - 2km', \ldots.$$

That's impossible, because Jordan regions are bounded. □

Theorem 2.3. *If a Jordan region C has two equichordal points, its boundary curve B is very smooth; in fact, B can be given by means of functions that are infinitely differentiable.*

In 1956 one mathematician claimed to have proved that when C is convex, B cannot be given by functions that are six times differentiable. A 1957 claim by another mathematician replaced "six" by "two." In 1958, a third mathematician proved that if C has two equichordal points, then B *can* be given by functions that are analytic and hence infinitely differentiable. Taken together, these results would show that no convex body has two equichordal points. However, there are errors in the "proofs" of the first two results. This shows how easily one can go astray in dealing with mathematical problems that, like this one, are easy to state and to "understand" but for which little rigorous machinery has been developed.

For a region with two equichordal points p and q (if there is such a region!), the *eccentricity* e is defined as the ratio of the distance pq to the length of the chord through p and q.

Theorem 2.4. *If e is the eccentricity of a Jordan region with two equichordal points then $e < 0.235$.*

Theorem 2.5. *For given p, q, and e, at most one Jordan region has equichordal points p and q and eccentricity e.*

Theorem 2.6. *There is a real-valued function h, defined by a power-series convergent in the open interval $(0, \sqrt{3}/2)$, such that $h(e) = 0$ whenever e is the eccentricity of a Jordan region with two equichordal points.*

Theorem 2.6 implies that the number of possibilities for c is at most countable and hence, in view of Theorem 2.5, there are (up to similarity) at most countably many possibilities for a Jordan region with two equichordal points.

Much of the above information may be viewed as evidence against the existence of a convex body (or even a Jordan region) with two equichordal points. On the other hand, Part Two contains some information about an almost successful attempt to construct a Jordan region with two equichordal points. Our own guess is that the answers to Problems 2 and 2.1 are both negative. However, at the time of writing, neither of the following possibilities is excluded by anything that has been published and subjected to careful and extensive examination:

a. there exists a nonconvex Jordan region with two equichordal points, but no such convex region exists;
b. there exists a convex Jordan region with two equichordal points, but no such nonconvex region exists.

Exercises

1. Suppose that a plane curve is given in polar coordinates by the equation $r = f(\theta)$ for $0 \leq \theta \leq \pi$, where f is a positive continuous function. Let $\mu = \max\{f(\theta) : 0 \leq \theta \leq \pi\}$. Under what conditions on μ can f be extended to a positive continuous function on $[0, 2\pi]$ so that the origin is an equichordal (respectively, equireciprocal, equiproduct) point of the region

$$\{(r, \theta) : 0 \leq r \leq f(\theta), 0 \leq \theta < 2\pi\}?$$

2. A subset S of the plane is said to be *symmetric* about the x-axis (respectively, y-axis, origin) if for each point (x, y) of S, the point $(x, -y)$ (respectively, $(-x, y)$, $(-x, -y)$) also belongs to S. Show that each of these conditions is implied by the conjunction of the other two.

3. Show that the foci of an ellipse are equireciprocal points.

4. Show that each interior point of a disk is an equiproduct point.

3. PUSHING DISKS TOGETHER

Problem 3

When congruent disks are pushed closer together, can the area of their union increase?

The answer to Problem 3 is certainly negative when there are only two disks. If two disks overlap, pushing them closer together increases the amount of overlapping, hence decreases the area of their union and increases the area of their intersection. It seems that this should be true for any number of disks, and perhaps it is. However, for more than two disks the effect of overlapping is more complicated than might at first be supposed, and must be examined closely. The following problem is also open.

Problem 3.1

When congruent disks are pushed closer together, can the area of their intersection decrease?

Even the notion of pushing the disks closer together must be examined carefully. Do we mean merely that in the new positions of the disks (after the pushing), the distance between the centers of any two disks is less than or equal to their distance in the original positions? Or do we mean that there is a *continuous shrinking*—a way of moving the centers continuously from their original to their new positions in such a way that at no time during the motion does any distance between centers increase? Both interpretations are of interest, but they may yield different answers. We'll take the first interpretation except where the second is specified.

Denoting the area of a set X by $\mu(X)$ (μ for measure), let us consider the area $\mu(\bigcup_i D_i) = \mu(\bigcup_{i=1}^n D_i)$, where $D_1, \ldots, D_n$ are disks in the plane. A first approximation A_1 to $\mu(\bigcup_i D_i)$ is the number $S_1 = \sum_i \mu(D_i)$, the sum of the individual areas. It is unchanged by moving the disks, but is greater than $\mu(\bigcup_i D_i)$ if any two of the disks overlap. Correcting for this, we obtain a second approximation A_2 by subtracting the sum of the areas of the two-at-a-time intersections of the disks; that is,

$$A_2 = A_1 - S_2 = S_1 - S_2, \qquad \text{where} \quad S_2 = \sum_{i<j} \mu(D_i \cap D_j)\,.$$

When the disks are pushed closer together, S_1 is unchanged and S_2 does not decrease, so A_2 does not increase. However, $A_2 < \mu(\bigcup_i D_i)$ if there is a point common to three interiors, and that suggests a third approximation

$$A_3 = A_2 + S_3 = S_1 - S_2 + S_3, \qquad \text{where} \quad S_3 = \sum_{i<j<k} \mu\left(D_i \cap D_j \cap D_k\right).$$

Continuing in this way, we see that an exact expression for the area of the union of n disks is

$$\mu\left(\bigcup_{i=1}^{n} D_i\right) = S_1 - S_2 + \cdots + (-1)^{n-2} S_{n-1} + (-1)^{n-1} S_n,$$

where

$$S_k = \sum_{1 \leq i_1 < i_2 < \cdots < i_k \leq n} \mu\left(D_{i_1} \cap \cdots \cap D_{i_k}\right).$$

For example,

$$\begin{aligned}\mu\left(D_1 \cup D_2 \cup D_3\right) = \mu(D_1) + \mu(D_2) + \mu(D_3) - \mu\left(D_1 \cap D_2\right) \\ - \mu\left(D_1 \cap D_3\right) - \mu\left(D_2 \cap D_3\right) + \mu\left(D_1 \cap D_2 \cap D_3\right),\end{aligned}$$

and for n disks there are $2^n - 1$ terms in the expansion of $\mu(\bigcup_i D_i)$. Problem 3 concerns the effect on this alternating sum of pushing the disks closer together, while Problem 3.1 involves only the effect on S_n. Certainly S_1 is unchanged, S_2 does not decrease, and it is known also that S_3 does not decrease. However, not much is known beyond that.

Problems 3 and 3.1 are of interest not only for the plane, but for a Euclidean space of any dimension, with the disks replaced by spherical balls or, in the one-dimensional case, by intervals; and they are of interest when the balls are not all of the same size. Thus the general problem is as follows, where $B(c, \rho)$ denotes the ball of radius ρ centered at the point c and μ denotes the appropriate measure for the space in question (length on the line, area in the plane, volume in 3-space, etc.).

Problem 3.2

Suppose that n radii $\rho_1, \ldots, \rho_n$, n centers $p_1, \ldots, p_n$, and n other centers $q_1, \ldots, q_n$ are given, where the q_is are at least as close together as the p_is (i.e., $\text{dist}(p_i, p_j) \geq \text{dist}(q_i, q_j)$). Under what additional conditions (if any are needed) is it true that

(i)
$$\mu\left(\bigcup_i B(q_i, \rho_i)\right) \leq \mu\left(\bigcup_i B(p_i, \rho_i)\right)?$$

(ii)
$$\mu\left(\bigcap_i B(q_i, \rho_i)\right) \geq \mu\left(\bigcap_i B(p_i, \rho_i)\right)?$$

Here are a few simple remarks related to Problem 3.2.

1. If $[a_1, b_1], \ldots, [a_n, b_n]$ are intervals in $\mathbb{R}$, with $a_i \leq b_i$ for each i, then the intersection $\bigcap_i [a_i, b_i]$ is equal to the interval $[a, b]$, where $a = \max_i a_i$ and $b = \min_i b_i$ ($[a, b]$ is empty when $a > b$). In particular, there are two intervals $[a_j, b_j]$ and $[a_k, b_k]$ that have the same intersection as all n of them.
2. The union $\bigcup_i B(p_i, \rho_i)$ is connected if and only if the balls can be arranged in a sequence such that each one intersects at least one of its predecessors.

Using (2) and the fact that two balls $B(p_j, \rho_j)$ and $B(p_k, \rho_k)$ intersect if and only if $\mathrm{dist}(p_j, p_k) \leq \rho_j + \rho_k$, we see that:

3. If the union $\bigcup_i B(p_i, \rho_i)$ is connected then so is the union $\bigcup_i B(q_i, \rho_i)$.

The various components of $\bigcup_i B(p_i, \rho_i)$ are pairwise disjoint, and hence the measure of their union is the sum of their measures. From this it follows that:

4. If (for balls of a given dimension) the inequality (i) holds whenever the union $\bigcup_i B(p_i, \rho_i)$ is connected, then it always holds.

In dealing with the one-dimensional case of (i), we write $I(p, \rho)$ instead of $B(p, \rho)$, to emphasize the fact that the "balls" are intervals. Thus $I(p, \rho)$ is the interval $[p - \rho, p + \rho]$.

5. If $\bigcup_i I(p_i, \rho_i)$ is a connected interval $[a, b]$, and j and k are such that $a \in I(p_j, \rho_j)$ and $b \in I(p_k, \rho_k)$, then
$$\mu\left(\bigcup_i I(p_i, \rho_i)\right) = \rho_j + |p_j - p_k| + \rho_k.$$

Theorem 3.1. *The inequalities* (i) *and* (ii) *always hold in the one-dimensional case.*

Proof. For (ii), this is immediate from the last statement in (1). In dealing with (i), we may by (4) assume that $\bigcup_i I(p_i, \rho_i)$ is connected. Then $\bigcup_i I(q_i, \rho_i)$ is connected by (3), and it follows from (5) that

$$\mu\left(\bigcup_i I(q_i, \rho_i)\right) = \rho_j + |q_j - q_k| + \rho_k$$

for some choice of j and k. But

$$\mu\left(\bigcup_i I(p_i, \rho_i)\right) \geq \rho_j + |p_j - p_k| + \rho_k \geq \rho_j + |q_j - q_k| + \rho_k,$$

which completes the proof. □

Here are the sharpest results currently known (related to the inequalities (i) and (ii)) in Euclidean spaces of dimension more than 1.

Theorem 3.2. *For arbitrary radii, both* (i) *and* (ii) *hold in* $\mathbb{R}^d$ *when the number of balls is at most* $d + 1$.

Theorem 3.3. *Suppose that* $d = 2$ *and the radii are all equal. Then under the "continuous shrinking" interpretation of "pushing closer together,"* (i) *holds and it is also true that the perimeter of* $\bigcup_i B(q_i, \rho_i)$ *is less than or equal to that of* $\bigcup_i B(p_i, \rho_i)$.

When the number of balls exceeds $d + 1$ and the less restrictive notion of togetherness is involved, then even when the radii are equal it is known in $\mathbb{R}^2$ only that $\mu(\bigcup_i B(q_i, 1)) \leq 9\mu(\bigcup_i B(p_i, 1))$. The corresponding inequality holds in $\mathbb{R}^3$ with the multiplier 27 on the right side, and in $\mathbb{R}^d$ with the multiplier 3^d. A proof of this is outlined in Part Two.

Exercises

1. Suppose that $p_1p_2p_3$ and $q_1q_2q_3$ are triangles in the plane, with $\text{dist}(p_i, p_j) \geq \text{dist}(q_i, q_j)$ for all i and j. Show that if in each case the vertices are listed in clockwise order, then there is a continuous shrinking that moves the points p_1, p_2, and p_3 into the positions q_1, q_2, and q_3, respectively. Show by example that the conclusion may fail when the sequences (p_1, p_2, p_3) and (q_1, q_2, q_3) do not have the same orientation.

2. Give an example of two sets $\{p_1, p_2, p_3, p_4\}$ and $\{q_1, q_2, q_3, q_4\}$ in the plane such that all the interpoint distances in the first set are greater than the correspond-

ing interpoint distances in the second set, and yet neither the first set nor its reflection in a line admits a continuous shrinking onto the second set.

3. (a) Show that if $n = 3$ and the plane sets $\{p_1, \ldots, p_n\}$ and $\{q_1, ..., q_n\}$ are such that $\text{dist}(p_i, p_j) \geq \text{dist}(q_i, q_j)$ for all i and j, then the perimeter of the set $U_p = \bigcup_i B(p_i, 1)$ is greater than or equal to that of the set $U_q = \bigcup_i B(q_i, 1)$.

(b) Show that (a) does not hold for all values of n. (It would be interesting to know the largest value of n for which (a) does hold.)

4. [FW] Show that if $[a_1, b_1], \ldots, [a_n, b_n]$ is a sequence of intervals in the line, with $a_1 \leq a_2 \leq \cdots \leq a_n$, then the measure of the union $\bigcup_i [a_i, b_i]$ can be computed by the following algorithm.

```
begin
  measure ⟵ 0;
  leftspan ⟵ a1 ; rightspan ⟵ b1 ;
  for i ⟵ 1 until n do
    begin
      if ai ≤ rightspan then rightspan ⟵ max{rightspan, bi}
      else begin
              measure ⟵ measure + rightspan − leftspan;
              leftspan ⟵ ai; rightspan ⟵ bi
           end
    end;
  measure ⟵ measure + rightspan − leftspan;
  output (measure)
end.
```

4. UNIVERSAL COVERS

Problem 4

If a convex body C contains a translate of each plane set of unit diameter, how small can C's area be?

The *diameter* of a bounded closed set is the largest number that is attained as the distance between two points of the set. For example, each of the following sets has diameter 1: a (circular) disk D of radius $1/2$; a square of edge-length $1/\sqrt{2}$; an equilateral triangle T of unit edge-length. A less familiar set with unit diameter is the so-called *Reuleaux triangle* R based on T; it is defined as the intersection of the three disks of unit radius centered at T's vertices (Figure 4.1). Reuleaux triangles have been used in designing linkages for converting reciprocating motion to rotating motion, and as shapes for manhole covers. A very similar shape is used for the rotor of the Wankel engine. We'll soon encounter some of the other interesting properties of Reuleaux triangles.

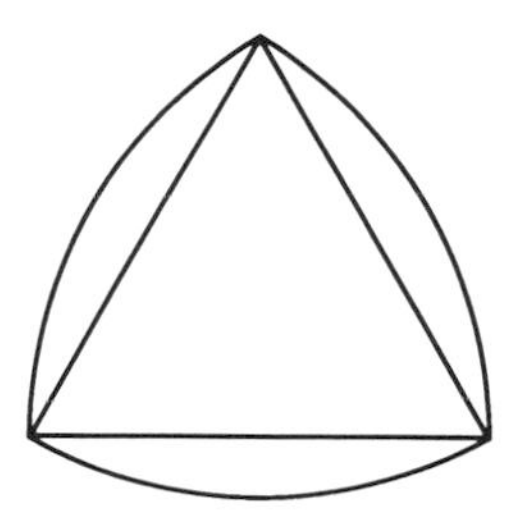

FIGURE 4.1
A Reuleaux triangle.

When $\mathcal{S}$ is a collection of sets in the plane, a *translation cover* for $\mathcal{S}$ is a set C that contains a translate of each member of $\mathcal{S}$; equivalently, each member of $\mathcal{S}$ lies in some translate of C. For example, the disk D is a translation cover for the collection $\mathcal{L}$ of all unit line-segments. Problem 4 asks for the minimum area of a convex translation cover for the collection $\mathcal{U}$ of all plane sets of unit diameter. This problem will serve as an entrée to some striking geometric facts and to some other, more famous, problems.

In order to gain experience with translation covers, let's consider the collection $\mathcal{L}$ briefly before passing on to the more complicated collection $\mathcal{U}$. Note that a set X is a translation cover for $\mathcal{L}$ if and only if X contains a unit segment in each direction. Hence the Reuleaux triangle R is also a translation cover for $\mathcal{L}$. Note that R's area of $(\pi - \sqrt{3})/2$ (≈ 0.705) is less than D's area of $\pi/4$ (≈ 0.785). It's not hard to verify that an equilateral triangle of edge-length $2/\sqrt{3}$ is also a translation cover for $\mathcal{L}$ (Exercise 1), and it is known to have the smallest area (of $1/\sqrt{3} \approx 0.577$) of any *convex* translation cover for $\mathcal{L}$. However, the picture changes dramatically when the condition of convexity is dropped. In fact, there are bounded plane sets of arbitrarily small area that are translation covers for $\mathcal{L}$. Part Two contains some information about these remarkable sets.

A plane convex body X is of *constant width* 1 if 1 is the distance between any two parallel lines that bound a minimal strip containing X. It might at first be thought that disks of unit diameter are the only convex bodies of constant width 1, but in fact the Reuleaux triangle R also has this property and there are many other noncircular bodies of constant width 1. It is known that each plane convex body of constant width 1 is of diameter 1 and perimeter π, but the areas of such bodies vary between $(\pi - \sqrt{3})/2$ (for R) and $\pi/4$ (for D) (see Exercises 2–3 for some examples). It is also known that each plane set of diameter 1 is contained in a convex body of constant width 1.

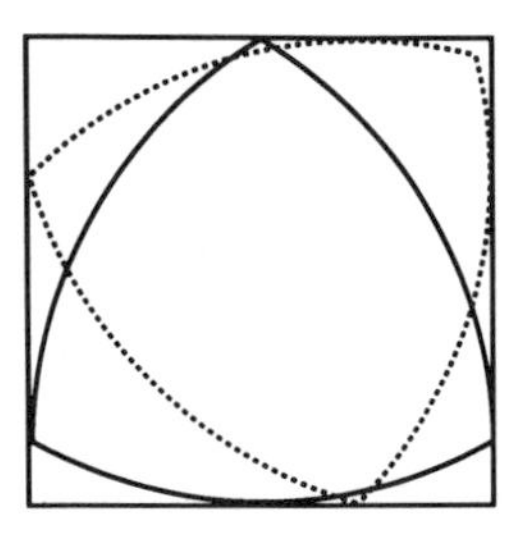

FIGURE 4.2
A Reuleaux triangle turning in a square hole.

Any convex body K of constant width 1 can be placed in a square S of unit edge-length, and can then be turned freely within S while maintaining contact with all four edges of S (Figure 4.2). That's obvious in the case of D, and for any choice of K it follows almost at once from the definition of constant width. The disk D merely turns around in its original position, but the Reuleaux triangle R covers most of S as it turns. That fact is used in the design of a bit that drills nearly square holes, and is also the basis of a possible answer to Problem 4. At each corner of the square is a small part that is not covered by R in its turning. The area β of each part turns out to be $1 - \sqrt{3}/2 - \pi/24$, and hence the covered part C of S has area

$$1 - 4\beta = 2\sqrt{3} + \pi/6 - 3 = 0.9877700392\ldots.$$

The convex set C is known to be a translation cover for $\mathcal{U}$, and to be *minimal* in the sense that no convex proper subset of C is a translation cover for $\mathcal{U}$. However, that doesn't rule out the possiblity that a smaller translation cover could be constructed in an entirely different way.

Here is a more famous relative of Problem 4. It was first proposed in 1914.

Problem 4.1

If a convex body C contains a rigid image of each plane set of unit diameter, how small can C's area be?

In Problem 4.1, we are not restricted to translations. All rigid motions of the plane—rotations, reflections, translations, and combinations of these—are permitted, and that makes it possible to reduce the area of the covering body C. A *rigid cover* for a collection $\mathcal{S}$ of plane sets is a plane set C such that for each member X of $\mathcal{S}$ there is a rigid motion of the plane that carries X onto a subset of X. For example, each unit segment is a rigid cover for the collection $\mathcal{L}$. Problem 4.1 asks for the minimum area of a convex rigid cover for the collection $\mathcal{U}$.

We have already seen that the square S is a rigid cover for $\mathcal{U}$. However, it doesn't provide the minimum area, for a regular hexagon H whose opposite sides are at unit distance is also a rigid cover for $\mathcal{U}$, and its area is only $\sqrt{3}/2 = 0.86602540\ldots$. In fact, it turns out that a rigid cover remains even after two corners of the hexagon have been shallowly truncated, and that reduces the area to $2 - 2/\sqrt{3} = 0.84529945\ldots$. This was known in 1920, along with a lower bound of $\pi/8 + \sqrt{3}/4 = 0.825711786\ldots$ on the answer to Problem 4.1. There have been three advances since then, the latest in 1980, but the problem remains open.

Among the many unsolved problems involving relationships between areas and diameters of convex bodies, Problem 4.1 is probably the best known because it is the oldest. Here is another old one.

Problem 4.2

What is the maximum possible area of a convex n-gon of unit diameter?

It has been known since 1922 that for odd n, the unique maximizing n-gon is the regular n-gon of unit diameter. For even n, the problem has been solved only for $n = 4$.

Exercises

1. Show that, in an equilateral triangle of edge-length $2/\sqrt{3}$, a unit segment can be moved continuously in such a way that its ends are interchanged.

2. Suppose that P is a regular n-gon of diameter δ, where n is odd and $0 < \delta \leq 1$. Let $\rho = (1-\delta)/2$. For each vertex v of P, let x_v and y_v denote the two adjacent vertices of P that are at distance δ from v. Let A_v and B_v be minor circular arcs that are centered at v and have radii $\delta + \rho$ and ρ, respectively, where the ends of A_v are on the rays from v through x_v and y_v and the ends of B_v are on the rays from x_v and y_v through v. (See Figure 4.3.) (When $\delta = 1$, B_v degenerates to the point v.) Let $R(n, \delta)$ denote the region that is bounded by the $2n$ arcs A_v and B_v. (Thus $R(3, 1)$ is the Reuleaux triangle.)

(a) Compute the diameter, perimeter, and area of $R(n, \delta)$.

(b) Show that a unit segment can be turned continuously in $R(n, \delta)$ in such a way that its ends are interchanged.

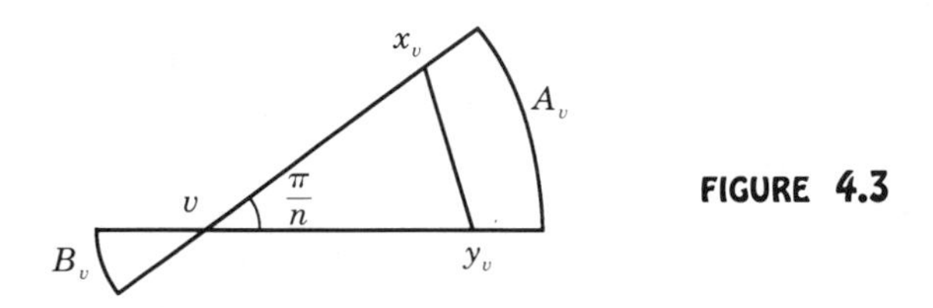

FIGURE 4.3

3. Suppose that C is a plane convex body of constant width 1.

(a) Prove that the diameter of C is 1.

(b) Suppose that L and M are parallel lines that touch C and bound a strip containing C. Prove that each of the lines intersects C in a single point, and the segment joining these points is perpendicular to both of the lines.

4. If C is a rigid cover for $\mathcal{U}$, then C contains both a circular disk D of diameter 1 and a Reuleaux triangle R of diameter 1. Compute the area of the union $D \cup R$ when D and R are such that a diameter of D joins two vertices of R. (Such a union is known to be a rigid cover that is *minimal* in the sense that no proper subset of it is a rigid cover. However, it is not *minimum* in the sense of having smallest area among the rigid covers.)

5. FORMING CONVEX POLYGONS

Problem 5

How many points are needed to guarantee a convex n-gon?

A convex n-gon has n vertices and n edges, and it is completely determined as soon as its vertices are specified. On the other hand, most ways of choosing n points in the plane will not result in the formation of a convex n-gon (see Exercise 1). If the chosen points are collinear, they won't even contain all the vertices of a triangle. A set of points in the plane is said to be *in general position* if no three of the points are collinear. The precise statement of Problem 5 is as follows.

Problem 5

What is the smallest number $f(n)$ such that whenever W is a set of more than $f(n)$ points in general position in the plane, then W contains all the vertices of some convex n-gon?

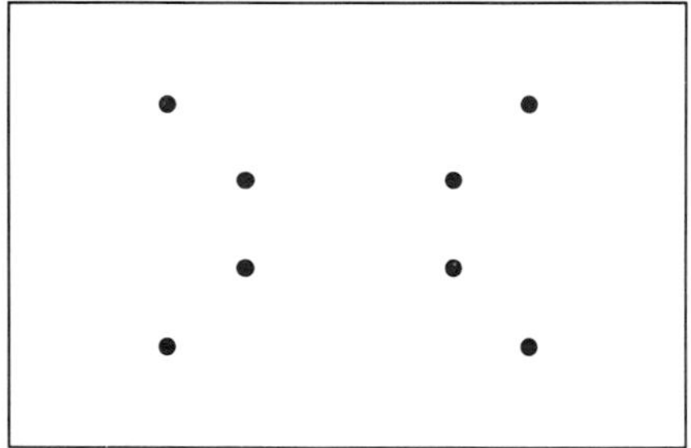

FIGURE 5.1
This set of eight points does not contain all the vertices of any convex 5-gon.

Clearly $f(3) = 2$, for if three points in the plane are in general position, they are the vertices of a triangle. To see that $f(4) = 4$, consider an arbitrary set W of five points in general position. If the convex hull of W is not a 4-gon or a 5-gon, it is a triangle T whose interior contains two points p and q of W. The line pq misses one of T's edges, and the ends of that edge together with p and q are points of W that determine a convex 4-gon. The set of eight points in Figure 5.1 is in general position but it does not contain the vertex-set of any convex 5-gon. On the other hand, it is known that if W is any set of nine points in general position in the plane, then W

does contain all the vertices of some convex 5-gon. Thus $f(5) = 8$. However, the proof of this is complicated, involving consideration of several cases.

Although $f(n) = 2^{n-2}$ when n is 3, 4, or 5, it does not seem intuitively obvious even that $f(n)$ is finite for large values of n. However, an elegant proof of this can be based on a famous theorem of F. Ramsey, which asserts that for each triple (r, s, t) of positive integers with $r \leq s$, there exists a positive integer $N(r, s, t)$ that has the following property:

> Whenever W is a set of cardinality at least $N(r, s, t)$ whose r-sets are partitioned into t classes, there is an s-set S in W such that all r-sets from S belong to the same class.

To gain some feeling for Ramsey's theorem, note that if we set $r = 1$ and $N(1, s, t) = st$, the following form of the pigeonhole principle is obtained:

> Whenever W is a set of cardinality at least $N(1, s, t) = st$ and the points (1-sets) of W are partitioned into t classes, there is an s-set in W all of whose points belong to the same class.

Fairly easy proofs of Ramsey's theorem appear in many books on combinatorics. The Ramsey numbers $N(r, s, t)$ are in most cases very difficult to determine precisely, but we require only the *existence* of $N(3, s, 2)$ to show that $f(n)$ is finite for all n.

Suppose that W is a plane set of at least $N(3, s, 2)$ points in general position. For each 3-set $\{a, b, c\}$ in W, let $|abc|_W$ denote the number of points of W that are interior to the triangle *abc*. By the definition of $N(3, s, 2)$, there is an s-set S in W such that either $|abc|_W$ is odd for all $\{a, b, c\} \subset S$ or $|abc|_W$ is even for all $\{a, b, c\} \subset S$. We claim that the points of S are the vertices of a convex s-gon. Suppose the contrary. Then there are four points a, b, c, and d of S such that d is interior to the triangle whose vertices are a, b, and c. No three points of W are collinear, and hence

$$|abc|_W = 1 + |abd|_W + |bcd|_W + |cad|_W.$$

But that is impossible when all four of the numbers $|\cdots|_W$ are of the same parity. □

In addition to the fact that $f(n) = 2^{n-2}$ when n is 3, 4, or 5, it has been proved that

$$2^{n-2} \leq f(n) \leq \binom{2n-4}{n-2}$$

and conjectured that $f(n) = 2^{n-2}$ for all n. However, these lower and upper bounds are far apart. For $n = 6$, they yield only $16 \leq f(6) \leq 70$. And the ratio of the upper bound to the lower bound increases exponentially with n (Exercise 2).

Part Two contains a proof of the upper bound. It is based on a close relative of the following fact, which is included here because of its elegant proof.

Theorem 5.1. *If m, n, and s are positive integers with $s > mn$, and $a_1, a_2, \ldots, a_s$ is a sequence of s distinct real numbers, then $\{a_i\}$ has a decreasing subsequence of more than m terms or an increasing subsequence of more than n terms.*

Proof. For each j, let m_j (respectively, n_j) denote the length (i.e., number of terms) of the longest decreasing (respectively, increasing) subsequence of $\{a_i\}$ that begins with a_j. When $j < k$, the pairs (m_j, n_j) and (m_k, n_k) are distinct, for if $a_j < a_k$ then $n_j > n_k$, and if $a_j > a_k$ then $m_j > m_k$. If, for all j, $m_j \leq m$ and $n_j \leq n$, then the total number of pairs (m_j, n_j) is at most mn, contradicting the fact that $s > mn$. □

If a set V consists of the n vertices of a convex n-gon, then no point of V is the midpoint of a segment joining two other points of V. In other words, the set V is *midpoint-free.* Hence it is natural, in connection with Problem 5, to ask how many points are needed to guarantee an n-set that is midpoint-free. An equivalent formulation is the following.

Problem 5.1

What is the largest integer $k(n)$ such that each set of n points in the plane contains a midpoint-free subset of size $k(n)$?

It has been proved that $k(2) = k(3) = 2$, $k(4) = k(5) = 3$, $k(6) = k(7) = k(8) = 4$, $k(9) = k(10) = k(11) = 5$, and

$$\left\lceil \frac{-1 + \sqrt{8n+1}}{2} \right\rceil \leq k(n) \leq \left\lceil \frac{5n}{11} \right\rceil \qquad \text{when } n \geq 11.$$

(Here $\lceil x \rceil$ denotes the smallest integer that is not less than x.) As in the case of the function f, there remains the problem of narrowing the gap between lower and upper bounds.

Exercises

1. Show (for $n \geq 3$) that a set $p_1, \ldots, p_n$ of n points in the plane is the vertex set of a convex n-gon if and only if for each j there is a line that has p_j on one side and all the rest of the p_is on the other side.

2. The De Moivre–Stirling approximation to $n!$ asserts that

$$\frac{n!}{\sqrt{2\pi} n^{n+(1/2)} e^{-n}} \to 1.$$

as $n \to \infty$. Using this, and letting

$$r_n = \binom{2n-4}{n-2} / 2^{n-2}$$

(the ratio of the upper and lower bounds on $f(n)$), prove that

$$r_n / \left(\frac{2^{n-2}}{\sqrt{\pi n}} \right) \to 1.$$

3. Show that for each pair of positive integers m and n, there is a sequence of mn distinct real numbers in which each decreasing subsequence is of length at most m and each increasing subsequence is of length at most n. (Hence Theorem 5.1 is the best possible result in a certain direction.)

6. POINTS ON LINES

Problem 6

If n points in the plane are not collinear, must one of them lie on at least $\frac{1}{3}n$ connecting lines?

Problem 6 is the most elementary of all the geometric problems presented in this book. It doesn't involve direction, distance, angle, or area, only the incidence of points and lines. A point and a line are *incident* if each is *on* the other; that is, the point belongs to the line and the line contains the point. Relative to a given set of points, a *connecting line* is one that's on *at least* two of the points and an *ordinary line* is one that's on *exactly* two of the points.

When n points are in general position (no three collinear), the situation is simple: the number of connecting lines is $\frac{1}{2}n(n-1)$ and each point is on $n-1$ of these lines; also, each connecting line is ordinary. The situation is even simpler when the points are all collinear: then there's only one connecting line, and when $n \geq 3$ it's not ordinary. Questions about the intermediate situation, when the points are neither collinear nor in general position, have been of interest for many years. As we'll see, some of them have been answered in striking ways.

For Problem 6, the stronger conclusion "at least $\frac{1}{2}n$ connecting lines" is suggested by the situation in which n is even and the n points are evenly divided between two lines (Figure 6.1).

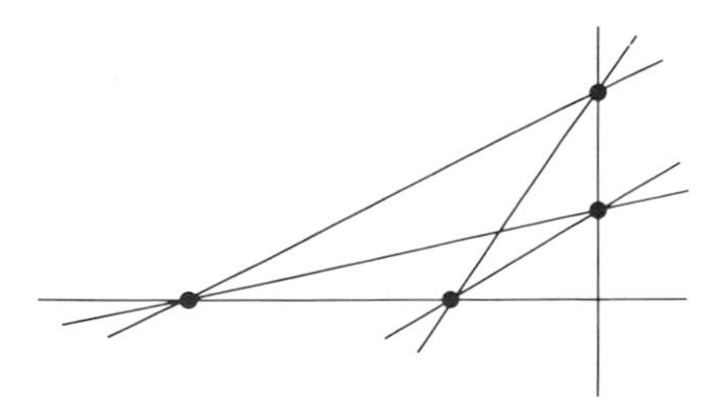

FIGURE 6.1
n points with $\frac{1}{4}n^2+2$ connecting lines in all; each point is on $\frac{1}{2}n+1$ of the lines (in the figure, $n=4$).

The situation is similar when n is odd (Exercise 1). Consideration of these and other examples led to a 1951 conjecture that the answer to Problem 6 is always affirmative, even with $\frac{1}{2}n$ in place of $\frac{1}{3}n$. Counterexamples to the $\frac{1}{2}n$ conjecture were later discovered for a few small values of n (9, 15, 19, 25, 31, 37). The known examples are not easy to draw convincingly on a small piece of paper, because apparently in each case the ratio of the greatest to the smallest of the interpoint distances must be very large. However, the idea of the nine-point example is conveyed by Figure 6.2. In this figure there are four triples of parallel lines: (L_1, L_2, L_3), (L_4, L_5, L_6), (L_7, L_8, L_9), and (L_{10}, L_{11}, L_{12}). For each triple it is possible to move some of the points $p_5, \ldots, p_9$ slightly so that the lines in the triple come together at a point. Now imagine this to be done in such a way that, simultaneously, *each* of the triples

$$(L_{3i-2}, L_{3i-1}, L_{3i}) \qquad (i = 1, 2, 3, 4)$$

has a common point p_i, and such that, further, the points $p_1, \ldots, p_4$ all lie together on a line L_{13} (not shown). Then the result is a configuration consisting of nine points and thirteen connecting lines, with each point on only four of the lines. It is not obvious that such simultaneous displacement of the points $p_5, \ldots, p_9$ is possible, but that is explained in Part Two with the aid of projective geometry.

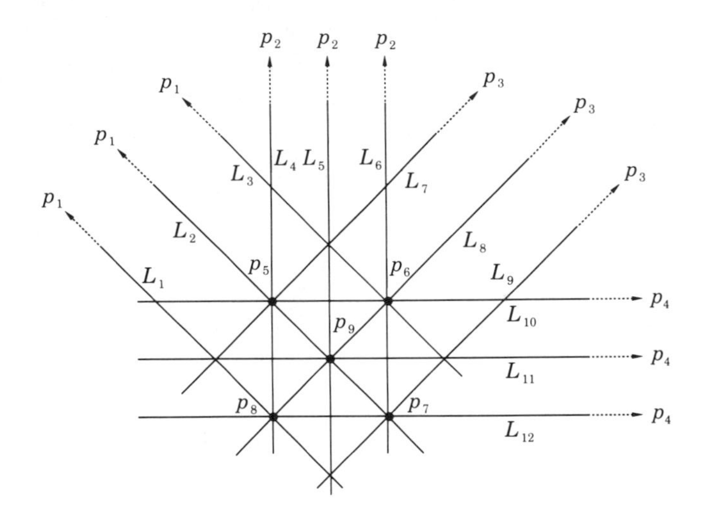

FIGURE 6.2
Schematic representation of a plane configuration of nine points and their thirteen connecting lines, with each point on precisely four of the lines.

Even though the $\frac{1}{2}n$ conjecture is false for some small values of n, it may be true for all sufficiently large n. Thus we state Problem 6.1.

Problem 6.1

Does there exist an integer n_0 such that whenever S is a noncollinear set of n points in the plane with $n \geq n_0$ then some point is on at least $\frac{1}{2}n$ connecting lines? If so, what is the smallest value for n_0?

The 37-point example shows that n_0 (if it exists) is at least 38. All six of the mentioned examples show that in Problem 6 the multiplier $\frac{1}{2}$ does not work for all choices of n. What about smaller multipliers? Does *any* positive multiplier work for all n? The answer is affirmative, but proving that seems to require sophisticated arguments that lead only to very weak estimates (such as 10^{-32}) for c. Thus the following close relative of Problem 6 is open.

Problem 6.2

What is the largest multiplier c such that whenever S is a noncollinear set of n points in the plane, then some point of S is on at least cn connecting lines? Is c at least $\frac{1}{3}$?

The choice of $\frac{1}{3}$ seems reasonable, because it works in all the known examples. In the 31-point example, each point is on only 12 connecting lines, so no value of c greater than $12/31$ would work for all n. We have chosen to devote a section to Problem 6 and its relatives because proving even the existence of a positive multiplier c is difficult, and because the gap between $\frac{1}{3}$ and the best proved estimate for c is so great. There aren't many similarly elementary problems for which the discrepancy between what is known and what is suspected is as great as for Problem 6.2.

In connection with Problem 6, the proof of the following theorem is easy and its basic idea has been used in the proofs of stronger results. However, since (for each fixed $c > 0$) $cn/\sqrt{n} \to \infty$ as $n \to \infty$, Theorem 6.1 doesn't come close to proving the existence of a positive multiplier that works for Problem 6.2.

Theorem 6.1. *For any noncollinear set of n points in the plane, some point is on at least $\sqrt{n}$ connecting lines.*

Proof. Let S denote the set of points in question, m the maximum number of connecting lines on any point of S, and k the maximum number of points of S on any connecting line. Then $m \geq k$, since for an arbitrary connecting line L and point $q \in S \backslash L$, the lines connecting q to the various points of $S \cap L$ are distinct.

Now let p be a point of S that is on m connecting lines. Since $k \leq m$, each line incident to p is incident to at most $m - 1$ additional points of S. But every point of S lies on some such line, and hence $n \leq m(m-1) + 1$. From this it follows that $m \geq \sqrt{n}$. □

We'll return in Part Two to Problem 6 and some of its relatives, but we want first to discuss another famous problem about incidence of points and lines. The following result was conjectured in 1893, but wasn't settled in print until almost 50 years later.

Theorem 6.2. *For each finite noncollinear set in the plane, there is at least one ordinary line.*

Proof. For the given set S of points, consider all pairs (p, L) consisting of a point p of S and a connecting line L that misses p. Since S is not collinear, there are such pairs, and since S is finite there are only finitely many of them. Hence among the pairs (p, L), there is one for which the distance from the point p to the line L is a minimum. We claim that in any such minimizing pair, the line L is ordinary. For suppose the contrary, and consider the foot q of the perpendicular from p to L. If L is not ordinary, there are at least two points of S on the same side of q; call them p' and s, with p' closer to q, and let L' denote the line on s and p (Figure 6.3). Then the pair (p', L') contradicts the minimizing property of (p, L), because the distance from p' to L' is less than the distance from p to L. □

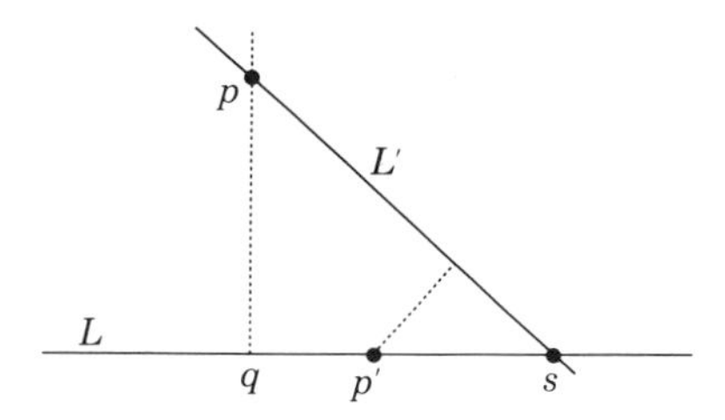

FIGURE 6.3
Proving the existence of an ordinary line.

Note that not only the noncollinearity, but also the finiteness of S is essential in Theorem 6.2. The elegant proof makes the existence of ordinary lines seem almost obvious. However, when there are many points in S, should there not be many ordinary lines? How many? This question has perhaps been settled by the following result, but see the comment in Part Two.

Theorem 6.3. *For each noncollinear set of n points in the plane, there are at least $\lfloor n/2 \rfloor$ ordinary lines.*

Remember that $\lfloor x \rfloor$ denotes the largest integer not exceeding x. The example requested in Exercise 2 shows that if we want the conclusion of Theorem 6.3 to hold for all values of $n \geq 3$, then the multiplier $\frac{1}{2}$ can't be pushed up to $\frac{4}{7}$; and if we want the conclusion to read "cn ordinary lines" (rather than merely $\lfloor cn \rfloor$), then the constant $c = \frac{3}{7}$ is the best possible. In problems concerning the incidence of points and lines, as in many other parts of mathematics, it may happen that small increases in the value of a constant can be attained only with great effort. For example, improving the constant in Theorem 6.3 from $\frac{3}{7}$ to $\frac{1}{2}$ required an argument of almost a hundred pages. Compare this with the much greater improvement in the value of the constant multiplier that may be attainable in connection with Problem 6!

As the reader by now suspects, obtaining best possible results concerning the incidence of points and lines can be much more difficult than is suggested by the simple formulation of the problems. We'll end this section by discussing an exception —an easy argument that does yield a best possible result concerning incidences in the plane. A set S of points is said to be a *near-pencil* if S is not collinear but there is a line that contains all but one point of S.

Theorem 6.4. *For n noncollinear points in the plane, the number of connecting lines is always at least n, and it is equal to n when the points form a near-pencil.*

Proof. The first statement is obvious for $n = 3$, and the second statement is obvious for all n. Supposing that the first statement fails for some $n \geq 4$, let us consider the least such n and the sorresponding set S. By Theorem 6.2, there is a line L that contains exactly two points, p and q, of S. If $S \setminus \{p\}$ is not collinear there are at least $n - 1$ connecting lines for $S \setminus \{p\}$, and since L is ordinary for S none of those lines is the same as L. □

Exercise 6 shows that the principal content of Theorem 6.4 (the fact that n noncollinear points determine at least n connecting lines) is purely combinatorial and does not really depend on the geometric result 6.2 that is used in the above proof of 6.4.

Theorem 6.4 is sharp in the sense that it precisely determines the minimum number of connecting lines for a set of n noncollinear points. There is also an easy proof (Exercises 7–8) that this minimum number n is attained only when the points form a near-pencil. The following theorem (not so easy to prove) goes considerably beyond that fact.

Theorem 6.5. *For each $n \geq 4$, let $\lambda(n)$ denote the minimum number of connecting lines for n points in the plane that do not form a near-pencil. Then*

$$\lambda(4) = \lambda(5) = 6 \quad \text{and} \quad \lambda(n) = \begin{cases} 2n-5 & \text{when } 6 \leq n \leq 9; \\ 2n-4 & \text{when } 10 \leq n. \end{cases}$$

Exercises

1. Suppose that J and K are two lines in the plane that intersect at a point q. Let the set S consist of j points on $J \setminus \{q\}$ and k points on $K \setminus \{q\}$. In terms of j and k, determine the number of connecting lines determined by S and by $S \cup \{q\}$, and the number of ordinary lines determined by each of these sets. Also, for each of these sets decide which integers occur as
   (a) the number of ordinary lines through some point of the set;
   (b) the number of connecting lines through some point of the set.

2. Construct a set of seven points in the plane for which there are exactly nine connecting lines and exactly three ordinary lines.

3. Assume the following, which is weaker than results stated above: for each finite noncollinear set X in the plane, there are at least $|X|/3$ ordinary lines. Use this assumption to prove that for each finite noncollinear set S and each point p that does not belong to S, S admits an ordinary line that misses p.

4. When p is the center of a circle C in the plane $\mathbb{R}^2$, *inversion in C* is the mapping ϕ that carries $\mathbb{R}^2 \setminus \{p\}$ into itself according to the following formula: for each point $q \in \mathbb{R}^2 \setminus \{p\}$,

$$\phi(q) = p + \frac{1}{\|q-p\|^2}(q-p).$$

Show that ϕ has the following properties:
   (a) for each circle K that misses p, $\phi(K)$ is a circle that misses p;
   (b) for each circle K that passes through p, $\phi(K \setminus \{p\}) \cup \{p\}$ is a line through p;
   (c) for each line L that misses p, $\phi(L) \cup \{p\}$ is a circle through p;
   (d) for each line L that passes through p, $\phi(L \setminus \{p\}) \cup \{p\}$ is a circle through p.

5. A set is said to be *cyclic* or *concyclic* if it lies on a circle. Prove that if a finite plane set is not collinear and not concyclic, then it admits an *ordinary circle* (i.e., a circle that contains exactly three points of the set). (Hint: Use the results of Exercises 3 and 4.)

6. Suppose that X is a set of n points, with $n \geq 3$, and $A_1, \ldots, A_m$ are proper subsets of X such that each pair of points of X is contained in a unique A_1. Prove that $m \geq n$. (Note that this is just a statement about abstract points and sets, where the points need not have anything to do with the plane.)

7. (a) Construct a set of five points in the plane for which there are precisely six connecting lines.

(b) Prove that if a set S of five noncollinear points in the plane is not a near-pencil, then it determines at least six connecting lines.

8. Prove that $\lambda(n) > n$ for all $n \geq 4$, where $\lambda(n)$ is the minimum number of connecting lines for a noncollinear plane set of n points that is not a near-pencil.

7. TILING THE PLANE

Problem 7

Is there a polygon that tiles the plane but cannot do so periodically?

For brevity, the term *polygon* is used to mean a simple polygonal region—a subset of the plane consisting of a polygonal simple closed curve and all the points inside it. A *tiling* of the plane is a collection $\mathcal{T}$ of regions that cover the plane in such a way that no point is interior to more than one of the regions. The individual regions (the members of $\mathcal{T}$) are then called *tiles*. There is special interest in tilings whose tiles are all alike in some sense. In the most familiar tilings, all the tiles are of the same size and shape—that is, there is a region T to which all the tiles are congruent; $\mathcal{T}$ is then said to be *monohedral*, and T to be a *prototile* for $\mathcal{T}$. We also say that T *tiles* the plane, and that the plane *can be tiled by* T. For example, a regular n-gon tiles the plane when n is 3, 4, or 6 (see Figure 7.1), but not otherwise (Exercise 1).

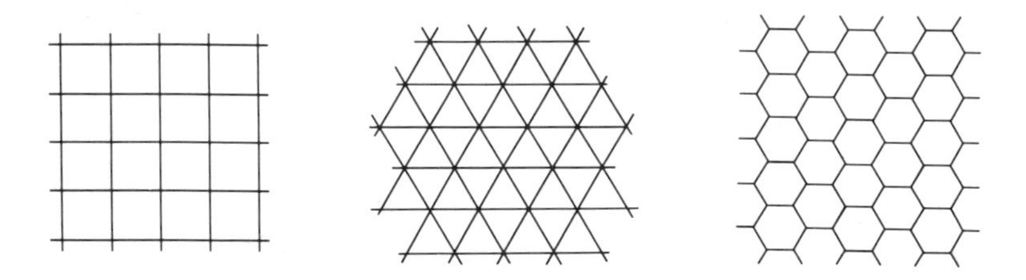

FIGURE 7.1
For $n = 3$, 4, or 6, there is a tiling by congruent regular n-gons.

The tilings in Figure 7.1 are highly symmetric, where a *symmetry* of a tiling $\mathcal{T}$ is defined as an isometry (a distance-preserving transformation) of the plane that maps $\mathcal{T}$ onto itself—in other words, the image of each tile is a tile (see Exercise 2). (Each isometry of the plane is produced by a rigid motion, a reflection, or a combination of the two.) A tiling $\mathcal{T}$ is *periodic* if its symmetries include translations in at least two nonparallel directions. When this happens, the entire tiling $\mathcal{T}$ can be generated by repetitive translation of $\mathcal{T}$'s intersection with a suitable parallelogram Q, or of the finite set of tiles intersecting Q. If u and v are two nonparallel translation vectors that describe symmetries of $\mathcal{T}$, then the vectors of the form $ju + kv$, where j and k are integers, form a plane *lattice*. For each parallelogram Q such that Q's vertices belong to this lattice and Q includes no other lattice point, *the entire tiling $\mathcal{T}$ can be generated from the part of $\mathcal{T}$ that appears in Q*. Simply use Q as the prototile of an appropriate periodic tiling, and, in translating Q, carry along the configuration formed by Q's intersections with the members of $\mathcal{T}$. Because each periodic tiling is closely related in this way to a periodic tiling by congruent parallelograms, the periodic tilings are especially easy to construct and analyze.

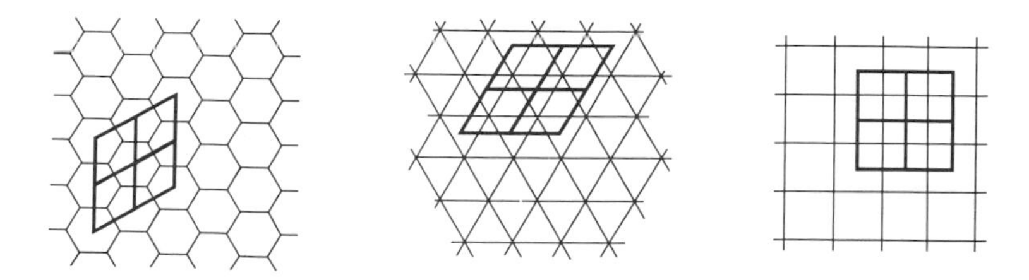

FIGURE 7.2
The tilings in Figure 7.1 are periodic.

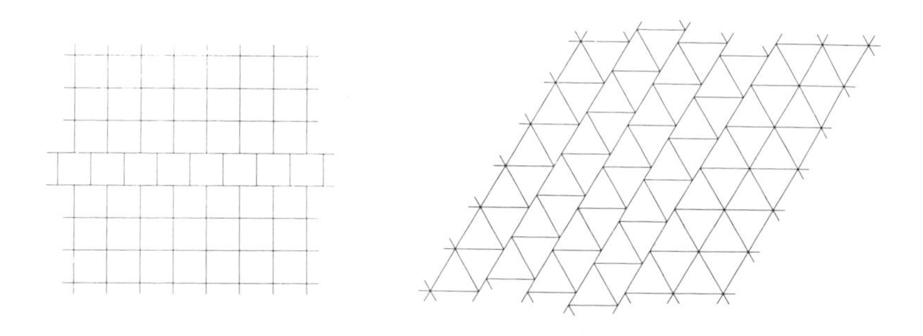

FIGURE 7.3
Some nonperiodic tilings by squares or triangles.

For each of the tilings in Figure 7.1, Figure 7.2 shows a lattice of translation vectors and an associated *period parallelogram.* In the tiling by congruent regular hexagons, the size and position of any one tile completely determines the tiling (Exercise 3); hence a regular hexagon can tile *only periodically.* However, in Figure 7.1 the tilings by squares and by equilateral triangles are formed from parallel strips in a natural way, and the tiles within each strip can be translated to produce tilings whose symmetry group consists only of translations in a single direction (Figure 7.3). Hence a tiling by either of those figures may be either periodic or nonperiodic.

A tiling by polygons is *edge-to-edge* if the intersection of any two of the tiles is either a vertex of both or an edge of both. It's easy to see that each monohedral edge-to-edge tiling by squares or by equilateral triangles is periodic (Exercise 3), but for less symmetric tiles there may be monohedral edge-to-edge tilings that exhibit a variety of symmetry groups (Exercise 4).

Problem 7 asks whether there is a polygon P that can tile *only nonperiodically,* so that *every* monohedral tiling with P as prototile is nonperiodic. If the answer is negative, then for each monohedral tiling whose prototile is a polygon, there is a periodic tiling that has the same prototile. To explain why this problem and its relatives are so interesting, let us extend the notion of a prototile. When each member of a tiling $\mathcal{T}$ is congruent to one of a minimal finite set $\mathcal{P} = \{P_1, \ldots, P_n\}$ of tiles, the P_is are called *prototiles* of $\mathcal{T}$. A striking discovery was that there are sets $\mathcal{P}$ of polygons (and even of convex polygons) such that $\mathcal{P}$ is the set of prototiles for a tiling but neither $\mathcal{P}$ nor any subset of $\mathcal{P}$ is the set of prototiles for any periodic tiling. Such sets $\mathcal{P}$ are said to be *aperiodic.* The existence of aperiodic sets is surprising, for the most natural ways of constructing tilings rely heavily on periodicity. Also striking is the success in finding ever simpler aperiodic sets. An early proof implied the existence of an aperiodic set consisting of more than 20,000 polygons, but now examples are known that consist of just two polygons or three *convex* polygons (two pentagons and a hexagon). The latter example is described in Part Two. Problem 7, motivated by the desire to discover just how simple an aperiodic set of prototiles can be, amounts to asking whether 1 or 2 is the minimum number of polygons in an aperiodic set. The problem is also open for convex polygons.

Problem 7.1

What is the minimum number of convex polygons that can be used to form an aperiodic set? Is it 1, 2, or 3?

The extensive theory of plane tilings is not restricted to convex tiles, but we focus on the convex case in the present Part One in order to keep matters simple and to gain rapid access to the results that are most relevant to Problem 7.1. In particular, we want to explain the close relationship between Problem 7.1 and the following.

Problem 7.2

Which convex pentagons tile the plane?

Let us, keeping Problem 7.1 in mind, assemble some relevant facts about tiling by convex polygons. A tiling by convex polygons is *normal* if there exist positive numbers ρ and δ such that each tile contains some disk of radius ρ and is contained in some disk of radius δ. We omit the proof of the following result.

Theorem 7.1. *For each n, the plane can be tiled by convex n-gons. However, in any normal tiling by convex polygons there must be infinitely many tiles that have six or fewer sides. In particular, for $n > 6$ there is no monohedral tiling by convex n-gons.*

Exercises 5 and 6 are related to Theorem 7.1, and Exercise 6 also states an open problem.

A tiling $\mathcal{T}$ is *isohedral* if for any two tiles T_1 and T_2 of $\mathcal{T}$ there is a symmetry of $\mathcal{T}$ that carries T_1 onto T_2. Monohedrality, periodicity, and isohedrality are all symmetry properties of tilings. It is obvious that isohedrality implies monohedrality, and we explain in Part Two why isohedrality also implies periodicity. However, even for monohedral tilings, periodicity does not imply isohedrality (Exercise 4).

Most convex pentagons and hexagons do not tile the plane. However, many of the convex polygons that do tile may be described as *p-hexagons,* where this means any of the following: a convex hexagon in which there are two opposite sides that are parallel and of equal length (*opposite* means separated, in each direction of traversing the boundary, by two other sides); a convex pentagon that has two parallel sides; a convex quadrilateral; a triangle. We shall use the term *nice p-hexagon* to refer to a p-hexagon that is not a pentagon, or, if it is a pentagon, it has two sides that are not only parallel but also of equal length.

Theorem 7.2. *Each p-hexagon is the prototile for an isohedral tiling of the plane, and each nice p-hexagon is the prototile for an edge-to-edge isohedral tiling.*

We omit the proof of Theorem 7.2, but Figure 7.4 is almost self-explanatory.

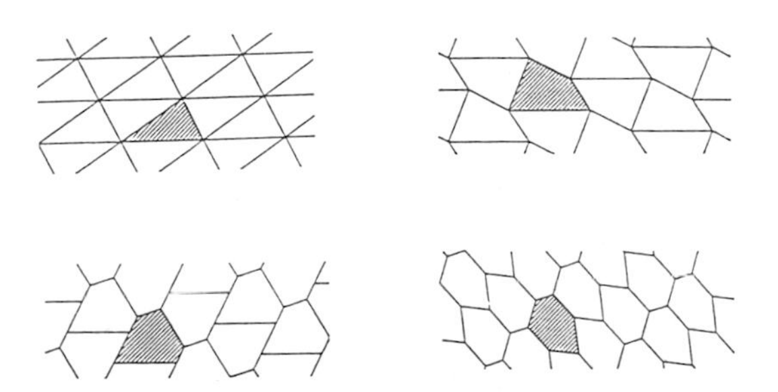

FIGURE 7.4
Some p-hexagons in isohedral tilings.

All convex hexagons that tile the plane have been known since 1918 (they are not all p-hexagons), and the following is a consequence of that knowledge.

Theorem 7.3. *If a convex hexagon tiles the plane, it can do so isohedrally and hence periodically.*

It follows from Theorems 7.1–7.3 that Problem 7.1 is equivalent to the same problem phrased for convex *pentagons* that do not have two parallel sides. The convex pentagons that tile isohedrally have all been classified, but the situation is unclear for the convex pentagons that are *anisohedral*, meaning that they tile but cannot do so isohedrally. All of the *known* anisohedral convex pentagons have been shown to tile periodically, and hence if the list of such pentagons is complete the answer to Problem 7.1 is negative. However, even for the anisohedral convex pentagons that are equilateral, the possibilities have been determined only with the aid of a computer search that has not been independently verified.

The assembled information suggests that no convex pentagon (and hence no convex polygon) is aperiodic, but it is far from proving that. In any case, anyone looking for aperiodic sets of polygonal prototiles should be aware of the following fact.

Theorem 7.4. *Whenever a finite set $\mathcal{S}$ of polygonal prototiles admits an edge-to edge tiling that has a translational symmetry, then $\mathcal{S}$ also admits a periodic tiling.*

If an aperiodic convex pentagon should be discovered, the next step might be to consider an even more outrageous possibility.

Problem 7.3

Is there a convex pentagon P that tiles the plane but is such that in every tiling $\mathcal{T}$ with P as prototile, the only symmetry of $\mathcal{T}$ is the identity mapping?

It is also conceivable (at least in the sense of not being excluded by any published proof of which we are aware) that some tiling $\mathcal{T}$ by an aperiodic convex pentgagon does admit translative symmetries in one direction. (By Theorem 7.4, such a tiling could not be edge-to-edge.) We mention this possibility and the one in Problem 7.3, not because we think they are likely, but to emphasize that even though tilings by convex pentagons have been studied for more than seventy years, some of the most basic questions concerning them are still open.

Problems 7 and 7.1–7.3 are among the most important unsolved problems concerning tilings of the plane. However, the study of tilings has led to many other unanswered but elementary questions, and some of them are mentioned in Part Two. There is also a variety of appealing open problems concerning tilings of bounded portions of the plane. We shall end the present section by discussing two of these.

In a monohedral tiling of the plane, each tile is surrounded by congruent nonoverlapping copies of itself. It might be thought that the property of being surroundable in this way guarantees that a polygon tiles the plane. However, that is not the case, as is shown by the pentagon Q in Figure 7.5. Even though Q is surrounded by a "first zone" formed from copies of itself, there is no way of surrounding Q by such copies so that the first zone can in turn be surrounded by a second zone formed from copies of Q.

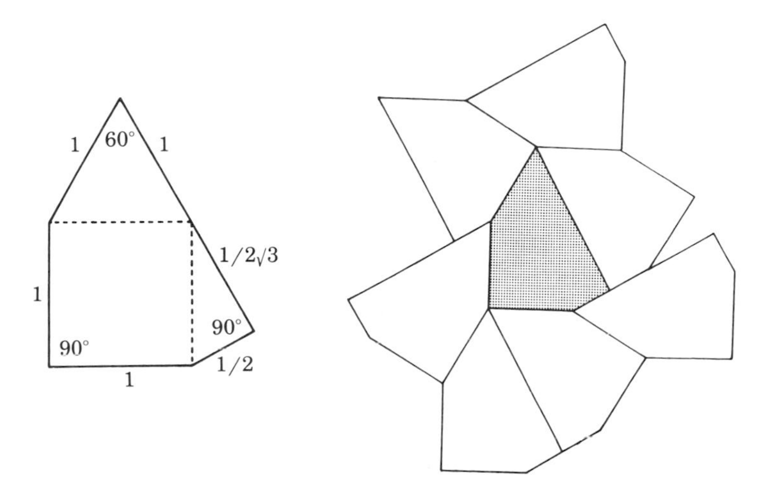

FIGURE 7.5
A pentagon surrounded by a first zone of congruent copies. It cannot in any way be surrounded by two zones.

Problem 7.4

For which positive integers r does there exist a convex polygon that can be surrounded by r successive zones formed from (nonoverlapping) copies of itself but cannot be surrounded by $r + 1$ such zones? In particular, is $r = 1$ the only possibility?

It is known that if $r = 1$ is indeed the only possibility, then the entire plane can be tiled by each convex polygon that can be surrounded by two successive zones formed from nonoverlapping congruent copies of itself. The first part of Problem 7.4 is open for nonconvex polygons as well, but in that case there are examples with $r > 1$.

Tilings of a given bounded portion of the plane by polygons are often called *dissections*. Dissection of polygons into triangles appeared in the proof of Theorem 1.4, and Theorem 9.1 is a striking result about the existence of such dissections. In those results, the triangles may vary widely in shape and in area. But what can be said about dissecting polygons into congruent triangles, or similar triangles, or triangles that all have the same area? Concentrating here on area, let us say that a polygon is *k-equidissectable* if it can be dissected into k triangles of equal area, and *equidissectable* if it is k-equidissectable for some k. The 1-equidissectable polygons are just the triangles, and the 2-equidissectable polygons are also easily characterized (Exercise 8). Each triangle is n-equidissectable for all n, and hence each k-equidissectable polygon is also nk-equidissectable. Beyond that, the situation is much more complicated, with a few striking results and many unsolved problems.

A set X is *centrally* symmetric if there is a point c of X such that for each point $x \in X$ there is a point $x' \in X$ such that c is the midpoint of the segment xx' (of course, $x' = 2c - x$). It is clear that if a convex n-gon is centrally symmetric, then n is even.

Theorem 7.5. *For the k-equidissectability of a centrally symmetric convex $2m$-gon, it is necessary that k be even.*

For the k-equidissectability of a *regular* $2m$-gon P, it is obviously *sufficient* that k should be a multiple of $2m$; first dissect P into $2m$ congruent triangles, and then dissect each of those into $k/(2m)$ triangles of equal area. For the k-equidissectability of a parallelogram P, it is obviously sufficient that k should be even; first dissect P into $k/2$ congruent parallelograms. and then dissect each of those into two congruent triangles. However, in each case the *necessity* of evenness is surprising.

The following is easily proved.

Theorem 7.6. *If all the vertices of a polygon P have rational coordinates, then P is equidissectable.*

Proof. When the vertices of a triangle have coordinates (ξ_1, η_1), (ξ_2, η_2), and (ξ_3, η_3), the triangle's area is half the absolute value of the determinant of the 3×3 matrix whose ith row is $(\xi_i, \eta_i, 1)$. Hence the area is rational if the vertices have rational coordinates. To prove Theorem 7.6, begin by dissecting P into triangles without adding any new vertices (see the proof of Theorem 1.4). Let the ith triangle have area a_i/b_i, where a_i and b_i are integers, and with d denoting the least common denominator of these fractions, write $a_i/b_i = c_i/d$. Then dissect the ith triangle into c_i triangles of area $1/d$, thereby obtaining a $(\sum_i^m c_i)$-equidissection of P. □

It might seem natural to try, by means of a limiting argument, to extend Theorem 7.6 to irrational polygons. However, that doesn't work because the smallest k for which a given rational polygon is k-equidissectable may depend on the denominators of the coordinates of its vertices. In fact, it *can't* work, as the following result makes clear.

Theorem 7.7. *When $n \geq 4$ it is true for almost all n-gons P that P is not equidissectable.*

We shall not make the notion of "almost all" precise here, but merely remark that the k-equidissectability of a polygon P implies some delicate algebraic relationships involving P's angles and edge-lengths; most polygons do not satisfy those relationships for any value of k.

Exercises

1. Show that if α is an interior angle of a regular n-gon, and the quotient $2\pi/\alpha$ is an integer, then n is 3, 4, or 6. Use this fact to show that for $n = 5$, and also for $n \geq 7$, there is no monohedral tiling of the plane by regular n-gons.

2. Show that if f and g are symmetries of a tiling $\mathcal{T}$, then the transformations $f \circ g$ and f^{-1} are also symmetries of $\mathcal{T}$. Here f and g are isometries of the plane, and for each point p of the plane, $(f \circ g)(p)$ and $f^{-1}(p)$ are defined as follows:

$$(f \circ g)(p) = f(g(p));$$

$$f^{-1}(p) \text{ is the unique point } q \text{ such that } f(q) = p.$$

3. Show that when the plane is tiled by congruent regular hexagons, the position of any one tile completely determines the positions of the others. Do the same for edge-to-edge tilings by congruent squares and for edge-to-edge tilings by congruent equilateral triangles.

4. With $0 < \alpha < \beta$, let $\mathcal{R}$ denote the tiling of the plane by the α-by-β rectangles R_{ij} whose vertices are $(i\alpha, j\beta)$, $((i+1)\alpha, j\beta)$, $(i\alpha, (j+1)\beta)$, and $((i+1)\alpha, (j+1)\beta)$ (i and j ranging independently over the integers). When each R_{ij} is split by a diagonal into two triangles, the result is a monohedral edge-to-edge tiling $\mathcal{T}$ of

the plane by α-by-β right triangles. Show that, depending on the placement of the diagonals that do the splitting, any of the following may be true of the tiling $\mathcal{T}$ and its symmetry group G: $\mathcal{T}$ is isohedral; $\mathcal{T}$ is periodic but not isohedral $(*)$; $\mathcal{T}$ is not periodic but G does include translations $(*)$; G includes a reflection but no translation; G includes more than one reflection but no translation; G consists of merely the identity transformation. (In the two cases indicated by $(*)$, it may be required, in addition, that among $\mathcal{T}$'s symmetries there are reflections—or that there are no reflections.)

5. Starting with a regular heptagon centered at the origin, produce two tilings $\mathcal{T}$ and $\mathcal{T}'$ of the plane that have the following properties:

(i) each tile is a convex heptagon;
(ii) each tiling has 7-fold rotational symmetry (i.e., is carried into itself by rotation through the angle $2\pi/7$ about the origin);
(iii) there is a positive ρ such that each member of $\mathcal{T}$ contains a disk of radius ρ;
(iv) there is a finite δ such that each member of $\mathcal{T}'$ is contained in a disk of radius δ.

(Note that by Theorem 7.1, $\mathcal{T}$ must include members of arbitrarily large diameter and $\mathcal{T}'$ must contain members that are arbitrarily thin.)

6. Produce a convex pentagon and a convex heptagon that can be used as the prototiles for a tiling composed of infinitely many congruent copies of each. (Experimental evidence favors the following conjecture from [GS4]: Whenever a normal tiling of the plane by convex polygons includes infinitely many tiles with seven or more sides, it also includes tiles that have five or fewer sides.)

7. (a) Show that each degenerate p-hexagon (one with less than six sides) is a limit of nondegenerate p-hexagons.

(b) Produce a convex hexagon that cannot tile the plane.

8. Provide easily computable characterizations of the quadrilaterals that can be dissected into

(a) two congruent triangles;
(b) two similar triangles;
(c) two triangles of the same area.

Which of these dissection properties is preserved by all invertible linear transformations of the plane?

9. Show that no convex n-gon can be dissected into exclusively nonconvex quadrilaterals.

8. PAINTING THE PLANE

Problem 8

What is the minimum number of colors for painting the plane so that no two points at unit distance receive the same color?

By a *painting* of a set X we mean a way of assigning one or more colors to each point of X. Whenever $\delta > 0$ and the distance between points of X is defined, we'll say that the painting is *δ-chromatic* if no two points of X at distance δ receive the same color. The *δ-chromatic number* $\chi(X, \delta)$ of X is the smallest number of colors that can be used in a δ-chromatic painting of X. When $\chi(X, \delta)$ has the same value for all $\delta > 0$ (as is clearly the case when X is the Euclidean plane), we speak simply of the *chromatic number* of X and denote it by $\chi(X)$.

Problem 8 asks for the chromatic number of the plane, denoted here by t_2; it is known only that $4 \leq t_2 \leq 7$. Among mathematicians who have studied the problem, several have stated their belief that $t_2 = 7$, and each of 4, 5, and 6 has also had its adherents.

Problem 8 is reminiscent of another plane coloring problem which for many years was one of the three most famous unsolved problems in mathematics (the others—Fermat's last theorem and the Riemann hypothesis—are discussed in other sections of this book). In the four-color map problem, a "map" is formed from a finite number of nonoverlapping plane regions ("countries"), each of which is to be painted with a single color in such that a way that different colors are used for any two countries that have a common boundary arc. It was conjectured in 1852 that four colors suffice to paint every planar map in this way. A supposed proof was published in 1879 and widely accepted until, in 1890, an error was pointed out and it was rigorously proved that five colors suffice. Over the years, several other erroneous "proofs" of the four-color conjecture found their way into print, but the conjecture was not finally established until 1976. If Problem 8 takes that long to settle, we should know the answer by the year 2084, for Problem 8 was first posed in 1960.

In discussing δ-chromatic paintings, it is convenient to say that two points are *friends* if the distance between them is δ, and that a set is *friendless* if it does not contain any pair of friends. The δ-chromatic number of X is the smallest number of friendless sets that can be used to cover X. To show that $t_2 \geq 4$, we shall derive a contradiction from the assumption that the plane is painted 1-chromatically in three colors—say red, white, and blue. Note first that all three colors must be used for the vertices of any equilateral triangle of edge-length 1. Let r, w, and b be the red, white, and blue vertices of such a triangle, and let $\overline{r}$ be the third vertex of the other equilateral triangle having the segment wb as an edge. Then $\overline{r}$ is red, for it has a white friend and a blue friend. By rotating the set $\{r, w, b, \overline{r}\}$ about r, we can obtain a similar set $\{r, w', b', \overline{r}'\}$ for which the the red points $\overline{r}$ and $\overline{r}'$ are friends (Figure 8.1). Hence the painting is not 1-chromatic, and in fact the 1-chromatic number of the 7-element set $\{r, w, b, \overline{r}, w', b', \overline{r}'\}$ is at least 4.

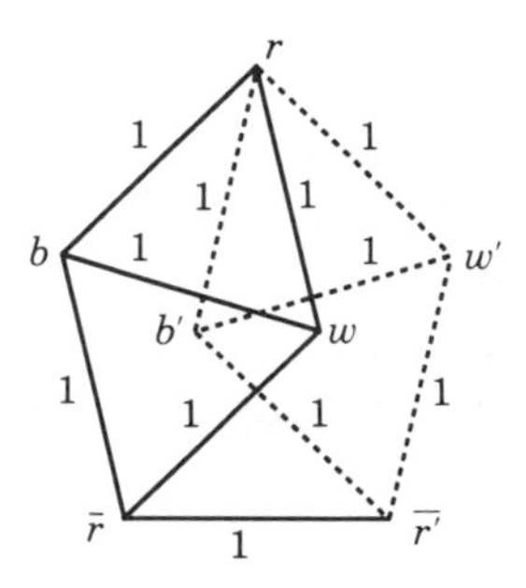

FIGURE 8.1
For each way of painting this set of seven points with three colors, there are two points at unit distance that have the same color.

That suggests another problem.

Problem 8.1

When k is 5, 6, or 7, what is the smallest number n_k of points in a plane set whose 1-chromatic number is k?

Obviously $n_2 = 2$, $n_3 = 3$, and the above example shows that $n_4 \leq 7$. It is known that, in fact, $n_4 = 7$; that is, each plane set of fewer than seven points can be painted 1-chromatically with only three colors (Exercises 1 and 2). Of course, the numbers n_5, n_6, and n_7 are meaningless if $t_2 = 4$. However, if $t_2 = 7$, then perhaps a good way to study t_2 is to focus initially on n_5 and n_6.

To show that $t_2 \leq 7$, we describe a 1-chromatic painting of the plane with 7 colors. The painting is based on a tiling of the plane by regular hexagons of side-length $2/5$ (Figure 8.2). Paint one of the hexagons with color #1, and then use colors #2–#7 for the 6 hexagons that surround it. Now consider the 18-sided colored "tile" T that is the union of these 7 hexagons, and note that the entire plane can be tiled by translates of T. The resultant 7-color painting of the plane turns out to be 1-chromatic. There's no need to worry about a little sloppiness (assigning more than one color to some points) in painting the edges of the hexagons, for it is known that when the painting is done precisely as described, no number in the interval $(4/5, \sqrt{28}/5)$ is realized as the distance between two points of the same color.

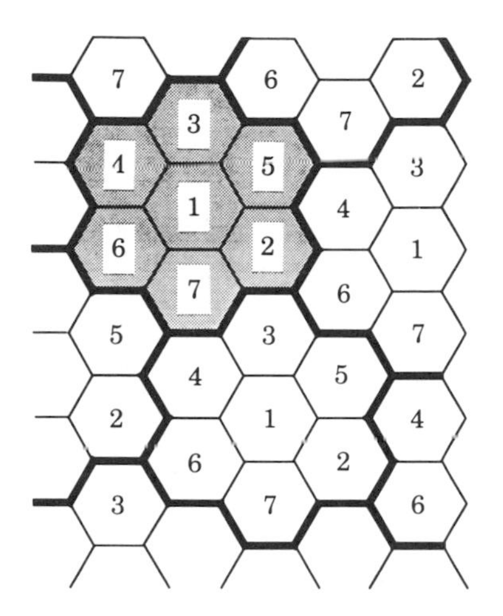

FIGURE 8.2
The plane can be painted in seven colors so that no two points at unit distance have the same color.

The preceding discussion concerns $\chi(X, \delta)$, the δ-chromatic number of X, which is the smallest number k such that X can be covered by k subsets none of which includes two points at distance δ. The *closed δ-chromatic number* of X, $\overline{\chi}(X, \delta)$, is similarly defined, taking into account only coverings by closed sets. For brevity, we'll use u_2 for the closed chromatic number of the Euclidean plane $\mathbb{R}^2$, and v_2 for the smallest number of closed *congruent* sets that can be used to cover the plane so that no set includes two points at unit distance. (The numbers t_d, u_d, and v_d are similarly defined for d-dimensional Euclidean space $\mathbb{R}^d$.) It is clear from the preceding results that $4 \leq t_2 \leq u_2 \leq v_2 \leq 7$, and it is known also that $u_2 \geq 6$ (see Part Two). However, even the following appears to be open.

Problem 8.2

Is $t_2 = u_2$? Is $u_2 = v_2$?

If $t_2 = 7$, then of course $u_2 = v_2 = 7$. However, it is also conceivable that $t_2 < u_2 < v_2$ (which could happen only if t_2 is 4 or 5, u_2 is 6, and v_2 is 7). Thus we see how easy it is, even for the much-studied Euclidean plane, to formulate intuitively appealing questions that seem to be beyond the scope of present methods.

To appreciate the significance of closedness in the definition of u_2, let's consider the real line $\mathbb{R}$. It's clear that two colors suffice for a 1-chromatic painting of $\mathbb{R}$. Simply use red for each half-open interval $[2k-1, 2k)$ and blue for each half-open interval $[2k, 2k+1)$ (k an integer). In particular, paint $[-1, 0)$ red and $[0, 1)$ blue. However, with real paint we couldn't use red for all the negative points without getting some red paint on the point 0. And in fact, $2 = t_1 < u_1 = v_1 = 3$ (Exercise 3). We might think of χ and $\bar{\chi}$ as applying, respectively, to ideally neat painters and to real-life painters. For each point p of a closed friendless subset of $\mathbb{R}$, the set S remains friendless even when a suitably small neighborhood of p is added to it. The size of the neighborhood determines how sloppy the painter can be in painting the point p.

Here is another variant of the problem of determining t_2.

Problem 8.3

What is the minimum number s_2 of sets into which the plane can be decomposed so that in each set, at least one distance is omitted?

A set that does not omit any distance is known as a Δ-*set*. For Problem 8.3, we want to paint the plane in a small number of colors so that for each color c there is *some* positive number δ_c that is not realized as the distance between any two points with color c. The number δ_c may vary with c. In Problem 8, by contrast, the *same* distance $\delta_c = 1$ has to be associated with each of the colors. Obviously, $s_2 \leq t_2$, and perhaps the two numbers are equal. However, it has been proved that $s_2 \leq 6$ (Exercise 4), so the sharpest *known* upper bound for s_2 is smaller than that for t_2. It is also known (Part Two) that $s_2 \geq 4$.

Exercises

1. [MM] Prove that for each set $X = x_1, \ldots, x_5$ of five points in the plane, the 1-chromatic number is at most 3. (Be prepared to consider several cases.)

2. [MM] Prove that for each set $\{x_1, \ldots, x_6\}$ of six points in the plane, the 1-chromatic number is at most 3.

3. Prove that $u_1 = v_1 = 3$.

4. [Rai] Consider the tiling of the plane in which the basic "prototile" is, as shown in Figure 8.3, a parallelogram that is the union of four regular hexagons and eight equilateral triangles, all of side-length 1. Let these twelve figures be painted in colors numbered 1–6, as shown, and extend by translation to the entire plane. Be

careful at the boundaries; each triangle is to omit its boundary B (i.e., colors #5 and #6 are not used on B), and each hexagon omits its two lowest vertices and its rightmost vertex. Show that the distance 1 is not realized within any of the sets #1, #2, #3, #4, and the distance 2 is not realized within either of the sets #5, #6.

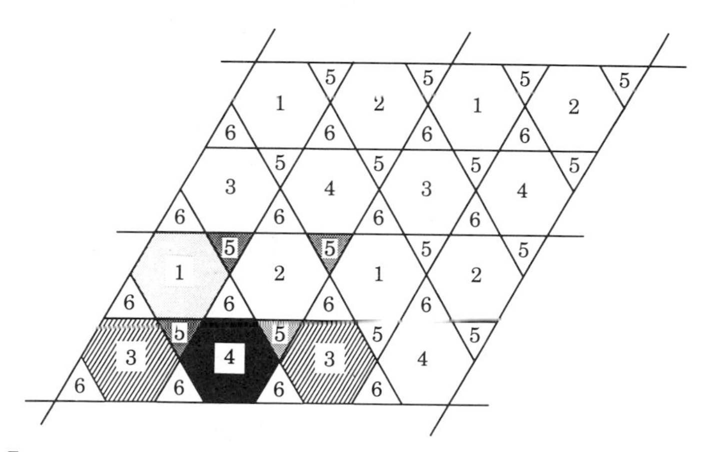

FIGURE 8.3
The tiling that is used to prove $s_2 \leq 6$.

9. SQUARING THE CIRCLE

Problem 9

Can a circle be decomposed into finitely many sets that can be rearranged to form a square? [1]

Only one mathematical problem is so famous that it has become part of the vernacular of English. "One might as well try to square the circle" is sometimes used to describe an impossible task. This usage stems from the ancient Greek problem of constructing a square equal in area to a given circle using only a straight-edge and compass. The Greeks knew of the existence of a number, which we now call π, such that the area of a circle of radius r is πr^2. Thus their problem, for the unit circle, say, was to construct a square of side-length $\sqrt{\pi}$. This they and generations of mathematicians after them could not do, but a proof of the impossibility of the task was not discovered until 1882, when F. Lindemann proved that π was a transcendental number (see Sections 21 and 22). The relationship between transcendence and geometric constructions stems from the fact that the coordinates of points in the plane obtainable as intersections of lines and lines, circles and lines, or circles and circles always satisfy quadratic equations with integer coefficients. But π, being transcendental, satisfies no such equation, and neither does $\sqrt{\pi}$.

A different circle-squaring problem arises if we replace the straight-edge and compass by a tool of modern mathematics, arbitrary sets, but retain the geometrical flavor by using rigid motions of the plane. Suppose A, a subset of $\mathbb{R}^2$, is partitioned into finitely many pieces A_i that can be rearranged using rigid motions of the plane to form another plane set B. Then A is said to be *equidecomposable* with B. Loosely speaking, you can imagine a jig-saw puzzle consisting of the pieces. If A is equidecomposable with B then the pieces can be put together one way to form A and another to form B. In the formal definition, note that every last point of A must be

---

[1] As this book was going to press, Problem 9 was solved by Miklos Laczkovich of Budapest, Hungary [Lac1]. Laczkovich's methods, which are quite complex (using ideas from the theory of uniform distribution of sequences), have many ramifications. First, he solves Problem 9 affirmatively. Second, he is able to accomplish the decomposition using only translations, which is quite surprising. Third, his work applies to polygons: Thus he improves Tarski's set-theoretic version of the Bolyai–Gerwien theorem discussed in Part Two by showing that any two polygons having the same area are equidecomposable using translations alone. And fourth, he can also solve, affirmatively, the one-dimensional version of the problem given in Part Two (Problem 9.2). Now perhaps the most important open question in this area is whether a polyhedron (or a ball) in $\mathbb{R}^3$ is translation-equidecomposable to a cube. An informal discussion of Laczkovich's breakthrough can be found in [GW].

accounted for, and the pieces in the decomposition of A and B can have no points in common. Now we can state the problem Tarski raised in 1925.

> Is a circle in the plane (including its interior) equidecomposable with a square?

In short, can the circle be squared set-theoretically?

To put this problem in a more familiar context, consider only polygons, and the weaker notion of equidecomposability where the borders of the pieces are ignored. More precisely, call two polygons *congruent by dissection* if the first can be cut up into finitely many polygons that can be rearranged using rigid motions to form the second, where we ignore the boundary lines of the pieces. This idea as a basis for computing areas goes back to classical geometry; for example, it is easy to see that any triangle is congruent by dissection to a rectangle whose length equals the base of the triangle and whose width is half the triangle's height (Exercise 1). More generally, it is not hard to prove the following theorem, which shows that the area of any polygon can, at least theoretically, be computed in this way.

Theorem 9.1. (Wallace–Bolyai–Gerwien Theorem). *Two polygons are congruent by dissection if and only if they have the same area. In particular, any polygon is congruent by dissection to a square of the same area.*

Proof. The easy direction of this theorem is the one asserting that two equidecomposable polygons have the same area. This follows immediately from the observation that the area of each is the sum of the areas of the pieces. For the harder direction we show that each polygon is congruent by dissection to a square. The result will then follow because, given two polygons of equal area, we get two dissections of a square which, by superimposing the decompositions on each other, yield a jig-saw puzzle that can be rearranged into either polygon.

In Exercise 1 it is shown that any triangle is congruent by dissection with a rectangle. If a rectangle has its length no greater than four times its width, then it can be turned into a square using three pieces as in Figure 9.1(a). But any rectangle can be turned into one of the desired sort by bisecting the rectangle and stacking the pieces, repeating as often as necessary (Figure 9.1(b)). We have thus shown that any triangle can be transformed to a square.

Now, any polygon can be decomposed into triangles, each of which can be squared by the preceding technique. Hence it remains only to show that two squares are congruent by dissection to a square, since this can be done repeatedly to transform finitely many squares into a single large square. To do this, we can use the relation $a^2 + b^2 = c^2$ together with the dissection proof of the Pythagorean theorem illustrated in Figure 9.1(c). All the steps of this proof can be combined to yield a single jig-saw puzzle decomposition of the final square that proves it is congruent by dissection to the original polygon. □

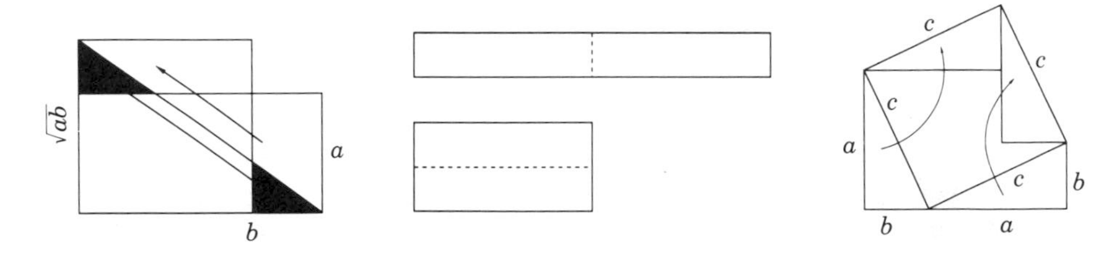

FIGURE 9.1
The main steps in the proof that any rectangle is equidecomposable to a square.

Theorem 9.1 is also true for the set-theoretic version of equidecomposability—two polygons are equidecomposable if and only if they have the same area—but the proof, sketched in Part Two, is much more difficult. This deeper theorem brings up a similarity of the modern circle-squaring problem to the classical Greek version. Like the Greeks, we completely understand the situation for polygons: by the generalization of Theorem 9.1 just mentioned, any polygon is equidecomposable to a square of the same area. But, again like the Greeks, we have made little progress for the simplest of curved regions, the circle; the answer to Problem 9 could easily be *yes,* or *no,* although the results that have been obtained suggest the latter (but see footnote at the beginning of this section).

Exercises

1. Prove that any triangle is congruent by dissection to a rectangle.

2. Take an equilateral triangle and follow through the steps of the proof of Theorem 9.1 to divide it into pieces that can be rearranged to form a square. You should end up with a five-piece decomposition. In fact, a four-piece decomposition exists (see [Eve, Chap. 5]).

3. (A. Engel [Eng, p. 385; AKW]) Let T be a triangle and T' its reflection. Show that, using only two cuts, T can be cut into pieces that can be rearranged,

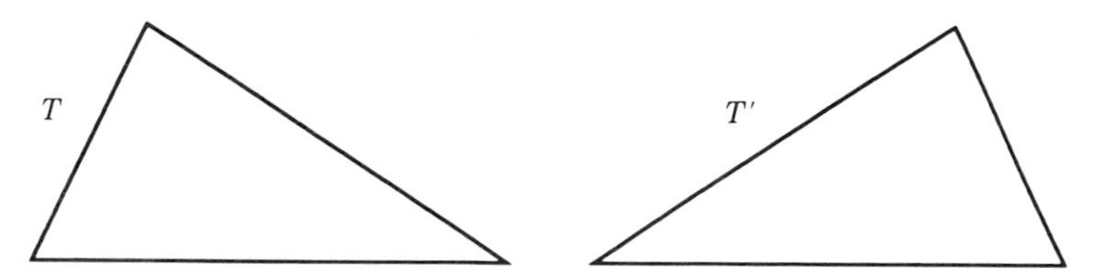

FIGURE 9.2
Transform T into its mirror image, T', using only two cuts.

using only rotations and translations (no reflections) into T'. (Hint: This is tricky; you might first try to find a dissection using three cuts.)

10. APPROXIMATION BY RATIONAL SETS

Problem 10

Does the plane contain a dense rational set?

For a set X, the associated *distance-set* $\Delta(X)$ consists of all positive numbers that are realized as distances between pairs of points of X. For example, if X is the set of integers $\{1, \ldots, n\}$, then $\Delta(X) = \{1, \ldots, n-1\}$, while $\Delta(X) = X$ when X is the set of all positive integers or the set of all positive rational numbers. A set X is called *rational* if each number in $\Delta(X)$ is rational. A set P is *rationally approximable* if it can be approximated to arbitrary precision by a rational set—that is, for each positive ϵ there is a rational set X such that each point of P is within ϵ of some point of X. It is obvious that if a set P is rationally approximable, then so is each subset of P.

A *dense* subset of the plane is one that intersects every disk. If the plane contains a dense rational set, then of course every set in the plane is rationally approximable. However, it seems likely that the plane does *not* contain a dense rational set, and thus the following problem is of interest.

Problem 10.1

Which subsets of the plane are rationally approximable?

The history of this problem is surveyed in Part Two. The survey is brief, because so little is known about the problem. Among problems that have been considered by first-rate mathematicians, there are few for which the range of ignorance is as great as it is for Problem 10.1 and its relatives. For example, it may turn out that the answer to Problem 10 is negative while the following has an affirmative answer.

Problem 10.2

Is the entire plane rationally approximable?

An affirmative answer to Problem 10.2 requires only that for each positive ϵ there is a rational subset X_ϵ of the plane that intersects each disk of radius ϵ. An affirmative answer to Problem 10 requires the existence of a single rational set X that can serve as X_ϵ for all positive ϵ simultaneously.

It is conceivable that the answer to Problem 10.2 is negative while the following has an affirmative answer.

Problem 10.3

Is each finite subset of the plane rationally approximable?

An equivalent formulation of Problem 10.3 is the following: When $D_1, \ldots, D_n$ is a finite sequence of disks in the plane, is it always possible to form a rational set $X = \{x_1, \ldots, x_n\}$ by choosing a point x_i from each D_i? If the answer to this question is negative, then the plane has finite subsets that are not rationally approximable and it would be interesting to know what is the minimum number of points that such a set could have.

Problem 10.4

What is the largest integer n (if it exists) such that each set of n points in the plane is rationally approximable?

In particular we pose the following.

Problem 10.5

Is each set of five points in the plane rationally approximable?

A very simple argument (Exercise 1) shows that each set of three points in the plane is rationally approximable, but it's not easy to prove this for sets of four points (see Part Two). Obviously each straight line is rationally approximable, and hence so is each set that is *collinear* (contained in a straight line). It is proved in Part Two that each circle is rationally approximable, and hence so is each set that is *concylic* (contained in a circle). A few other special cases are easily handled (Exercises 4–5). However, for most subsets of the plane, finite or infinite, the matter of rational approximability is still unsettled. At one extreme of what *might* be true, there is the possibility of a rational set X that is dense in the plane, in which case Problems 10, 10.2, 10.3, and 10.5 would all be answered affirmatively. At the other extreme, there might exist an integer m of the sort described in the following problem. Then it would be easy, as we shall see, to construct a set of m points that is not rationally approximable.

Problem 10.6

Does there exist an integer m such that each rational set of m points in the plane contains three collinear or four concyclic points?

It is known that any such m (if it exists) is at least 7.

Problem 10.7

Does the integer $m = 7$ have the property mentioned in Problem 10.6?

For each plane set P, let $S(P)$ denote the union of P with all lines that contain two or more points of P and all circles that contain three or more points of P. Note that when P is finite the set $S(P)$ is not the entire plane (Exercise 2). Thus a set $Q = \{q_1, \ldots, q_m\}$ can be constructed by choosing q_1 and q_2 arbitrarily and then, having chosen q_i for all $i < k$, choosing q_k in the complement of $S(\{q_1, \ldots, q_{k-1}\})$. When Q is constructed in this way, no three of its points are collinear and no four of its points are concyclic. But if Q were rationally approximable, we could find m sequences of points,

$$x_1^j, x_2^j, \ldots \to q_j \qquad (1 \leq j \leq m)$$

such that (by the definition of m) each set

$$X_i = \{x_i^1, x_i^2, \ldots, x_i^m\} \qquad (1 \leq i < \infty)$$

includes three collinear or four concyclic points. By a routine limiting argument (Exercise 3), this would be true of the set Q as well, a contradiction showing that Q is not rationally approximable.

Exercises

1. Suppose that a positive number ϵ and distinct points p, q, and r in the plane are given. Show that on the line pq there is a point y within ϵ of q such that the distance py is rational. Show that when rational numbers γ and δ sufficiently close to the distances pr and yr are chosen, the circles centered at p and q with radii γ and δ intersect at a point within ϵ of r.

2. Show that if S is the union of a finite number of lines and circles, and T is a line or a circle different from any of those used to form S, then the intersection $S \cap T$ is finite and hence there are infinitely many points of T not in S.

3. Suppose (for points x_i^j and q_j in the plane) that the points $q_1, \ldots, q_m$ are distinct, that

$$x_1^j, x_2^j, \ldots \to q_j \qquad (1 \leq j \leq m)$$

and that for $1 \leq i \leq m$ the set $X_i = \{x_i^1, x_i^2, \ldots, x_i^m\}$ includes three collinear or four concyclic points. Show that the set $Q = \{q_1, \ldots, q_m\}$ includes three collinear or four concyclic points. Give an example in which each X_i includes four concylic points but the limit set Q does not have this property.

4. [Day2] (In this and the next exercise, a polygon is called *rational* if its vertices form a rational set; i.e., its edges and diagonals all have rational lengths.) Show that if $\cos\theta$ is rational, then the rational triangles having an angle θ are dense in the class of all triangles with an angle θ. That is, each triangle in this class can be approximated to arbitrary precision by a rational triangle in the class.

5. [Bes2, Day2] Show that the class of all rational parallelograms is dense in the class of all parallelograms.

11. INSCRIBED SQUARES

Problem 11

Does every simple closed curve in the plane contain all four vertices of some square?

Intuitively, a simple closed curve corresponds to a walk along a path that does not intersect itself and eventually returns to its starting point. (Such a curve may not appear to be very "simple" at all; see Figure 11.1.) Simple closed curves in the plane are often called *Jordan curves,* and we'll use that term for brevity. In mathematical terms, a set J is a Jordan curve if and only if there are two continuous real-valued functions x and y on the unit interval [0,1] such that

(i) $x(0) = x(1)$ and $y(0) = y(1)$;
(ii) for $0 \leq s < t < 1$, the point $(x(s), y(s))$ is different from the point $(x(t), y(t))$;
(iii) the set of all points $(x(s), y(s))$ is precisely J.

Another way of putting it is to say that J is topologically equivalent to a circle.

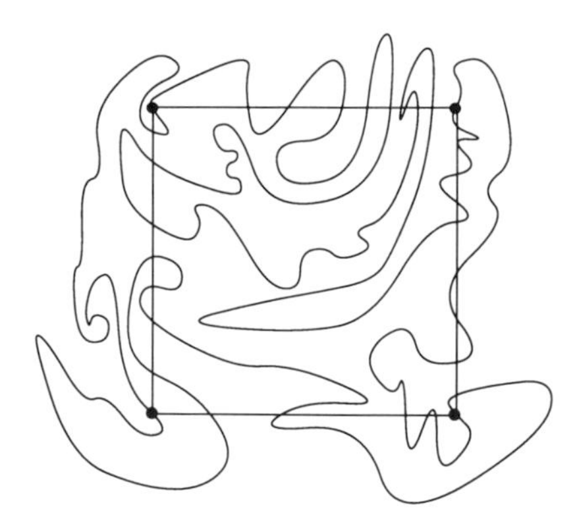

FIGURE 11.1
A Jordan curve (simple closed curve in the plane) that contains all four vertices of a square.

An important property of each Jordan curve J is expressed by the famous *Jordan curve theorem*, which asserts that the set $\mathbb{R} \setminus J$ (J's complement in the plane) has just two components and J is the boundary of each. One of the components is bounded and the other is unbounded. Each point of the bounded component is said to be *interior* to J. A subset of the plane that is contained in the union of J with the bounded (respectively, unbounded) component is said to be *inside* (respectively, *outside*) J.

As the term is used here, a polygon P is *inscribed* in a set Z if all of P's vertices belong to Z. It is not required that $P \subset Z$ or, when Z is a Jordan curve, that all of P is inside Z. Thus Problem 11 asks whether every Jordan curve has an inscribed square. The nicest Jordan curve—a circle—has many inscribed squares, all of the same size. There are also Jordan curves that have inscribed squares of many different sizes (Exercise 1). Some Jordan curves have only one inscribed square (Exercise 2), and perhaps (if the answer to Problem 11 is negative) some don't have any.

Since smoothness is not required in the definition of a Jordan curve, the functions x and y that appear there may even be nowhere differentiable. That certainly contributes to the difficulty of Problem 11, for it is known that each sufficiently smooth Jordan curve does have an inscribed square. An arbitrary Jordan curve J can be approximated by a sequence $J_1, J_2, \ldots$ of smooth Jordan curves, and it is tempting to argue as follows: For each i, let S_i be a square that is inscribed in J_i. Pass to a subsequence $S_{k(1)}, S_{k(2)}, \ldots$ that converges to a limit figure S inscribed in J. This almost works, because if the edge-length of the square $S_{k(i)}$ does not converge to 0 then S is indeed the desired square. However, if the sizes of the squares $S_{k(i)}$ can't be controlled then S may be merely a point of J, and that's neither interesting nor useful.

Although Problem 11 is still open, some striking results in the same spirit have been proved. We'll outline some of the simpler arguments to indicate the sort of methods that have been used. In our outlines of proofs, some details of compactness and continuity arguments are omitted because they are routine and a bit tedious. However, it should be mentioned that such omissions are dangerous in principle. Some of the erroneous "proofs" of similar results have failed precisely because of lack of attention to such details.

A convex polygon P is *circumscribed about* a set X if X is inside P and X intersects each edge of P. In this section we shall be concerned with finding inscribed or circumscribed polygons with specified shapes. Sometimes it is even possible to find an inscribed or circumscribed figure for which, in effect, a direction as well as a shape is specified. That involves the notion of homothety. A set W is a *homothet* of a set V if W can be obtained from V by a dilatation (magnification or contraction) followed by a translation. In terms of vector operations in the plane, this says that for some positive number λ and some point p, W is equal to the set

$$p + \lambda V = \{p + \lambda v : v \in V\}.$$

A set Z is *similar* to V if Z can be obtained by rotating some homothet W of V. (In standard usage, homothety permits reflection in a point and similarity permits also reflection in a line. That is not true of the notions used here, which are often called *positive* homothety and *direct* similarity. We omit the adjectives in the interest of brevity.)

The following result is almost obvious.

Theorem 11.1. *If T is a triangle in the plane and X is a compact subset of the plane consisting of more than one point, then some homothet of T is circumscribed about X.*

Proof. Since X is bounded, it is inside some homothet T' of T. Now slide the boundary lines of T' toward X, one at a time, parallel to themselves, until they just touch X. The resulting triangle is a homothet of T and is circumscribed about X. □

Here is a more substantial result.

Theorem 11.2. *If X is a compact subset of the plane consisting of more than one point, then X admits a circumscribed square.*

Proof. For easy visualization, we first restrict our attention to the case in which X is a convex body C. Let U denote the unit circle of $\mathbb{R}^2$. For each $u \in U$ let $\mathbb{R}u$ denote the line through u and the origin, and let v_u denote the point of U into which u is carried when the circle U is rotated counterclockwise through an angle of $\pi/2$. The line $\mathbb{R}v_u$ is orthogonal to $\mathbb{R}u$. We want to describe, in terms of certain continuous real-valued functions, the unique rectangle $F(u)$ that is circumscribed about the given convex body C and has two of its edges parallel to the line $\mathbb{R}u$. A circumscribed square is produced by "rotating" this rectangle.

The orthogonal projection of the plane onto $\mathbb{R}u$ carries the convex body C onto a closed line segment that joins two points of $\mathbb{R}u$—say $\phi(u)u$ and $\xi(u)u$, with $\phi(u) < \xi(u)$. For a given C, this defines two continuous real-valued functions ϕ and ξ on the circle U. Similarly, orthogonal projection of C onto the line $\mathbb{R}v_u$ produces a segment $[\eta(u), \zeta(u)]v_u$ with $\eta(u) < \zeta(u)$, and the functions η and ζ are also continuous. The rectangle $F(u)$ is bounded by two lines parallel to $\mathbb{R}v_u$, intersecting $\mathbb{R}u$ at the points $\phi(u)u$ and $\xi(u)u$, and two lines parallel to $\mathbb{R}u$, intersecting $\mathbb{R}v_u$ at $\eta(u)v_u$ and $\zeta(u)v_u$ (see Figure 11.2).

Now let $\delta(u) = \lambda(u) - \mu(u)$, where $\lambda(u)$ is the length of the edges of $F(u)$ parallel to the line $\mathbb{R}u$ and $\mu(u)$ is the length of the edges parallel to $\mathbb{R}v_u$. Then δ is a continuous function on the unit circle U, and the rectangle $F(u)$ is a square precisely when $\delta(u) = 0$. (If $\delta(u) = 0$ for all u, then *every* $F(u)$ is a circumscribed square. That would be the case if C were a disk, but it can happen in other ways as well (Exercise 3).) If the function δ is not identically zero, it must have both positive and negative values, because for all u, $\delta(v_u) = -\delta(u)$. If a continuous real function on the circle takes on both positive and negative values, then, of course, it takes on the value 0 as well, and the result is a square circumscribed about C.

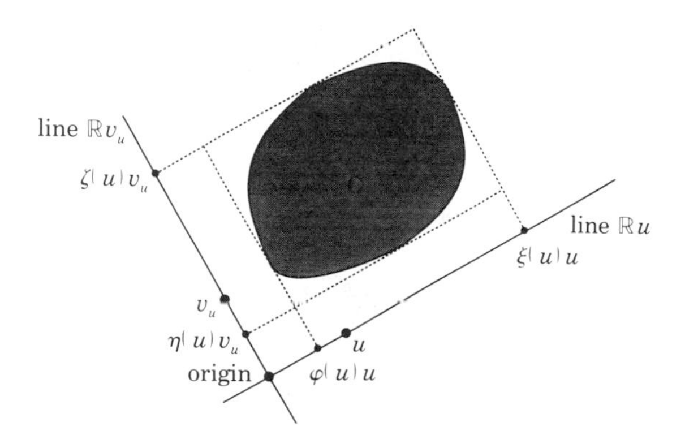

FIGURE 11.2
This figure illustrates the construction used in proving Theorem 11.2.

Now consider an arbitrary compact set X that consists of at least two points. If all of X is contained in a single line, then X is a subset of a segment pq joining two points of X. In this case the square with diagonal pq is circumscribed about X. If X is not collinear, the convex hull C of X is a convex body, and by the preceding paragraph C admits a circumscribed square S. That is, C is inside S and each edge of S intersects C. This implies that each edge of S intersects X, and hence S is circumscribed about X. □

As we have just seen, for circumscribed squares there is an easy way to pass from a seemingly very special case to the general case. However, there does not seem to be any way of making that transition for inscribed squares. The existing partial solutions of Problem 11 guarantee the existence of inscribed squares only under strong additional assumptions on the local behavior or global structure of J. These results are stated in Theorems 11.4, 11.5, 11.7, and 11.8, but we want first to deal with inscribed triangles.

Theorem 11.3. *For each triangle T and Jordan curve J, it is true that J admits an inscribed triangle similar to T.*

Proof. The construction is carried out in two stages. The first stage produces a triangle xyz that is similar to T, has yz as a longest edge, and is such that x and y both belong to J while z is inside J. In the second stage the vertices are moved continuously, keeping the triangle xyz similar to T and keeping x and y on J, in such a way that at the end of the motion, z is outside J. Since z's path starts inside J and ends outside J, it must at some time intersect J. At the time of intersection we have an inscribed triangle similar to T.

To start the first stage, let w be a point interior to J, x a point of J closest to W, and C the circle that is centered at w and passes through x. There is a homothet of T that is inscribed in C, and by suitably rotating this in C we obtain a triangle xyz that is similar to T and has yz as a longest edge. Denote these initial positions of the vertices by x_1, y_0, and z_0. Now keeping x fixed at x_1, move y and z away from x_1 on the rays from x_1 through y_0 and z_0, continuing this until at least one of y and z lies on J. With y_1 and z_1 denoting these new positions of y and z, we may assume without loss of generality that y_1 is on J and z_1 is inside J. No further motion is needed if z_1 is also on J.

Assume now that z_1 is interior to J, and let p and q be two maximally distant points of J—that is, $\text{dist}(u, v) \leq \text{dist}(p, q)$ for all $u, v \in J$. Move y along J from y_1 to p, causing z to move to a new position z_2. Then, keeping y fixed at p, move x along J from x_1 to q, causing z to move to a new position z_3. Now x and y are in positions p and q, while yz is a longest edge of the current triangle xyz. From the maximizing choice of p and q it follows that z_3 is outside J. Hence at some time during the motion, z must lie on J. □

A square may be characterized as a rectangle whose edges are of equal length, and also as a rhombus whose diagonals are of equal length or whose interior angles are equal. In the proof of Theorem 11.2, a circumscribed square was produced by continuous variation of a circumscribed rectangle. Similarly, in proving Theorem 11.4 we produce an inscribed square for a special sort of Jordan curve by continuous variation of an inscribed rhombus.

The following result deals only with a very special case of Problem 11, but it is the only one known to us that does not even implicitly require any sort of smoothness of J.

Theorem 11.4. *If the Jordan curve J is symmetric about a point c, and each ray issuing from c intersects J in a single point, then J admits an inscribed square.*

Proof. We employ polar coordinates in the plane, assuming without loss of generality that c is the origin. For each angle θ, let the positive numbers $\rho(\theta)$ and $\sigma(\theta)$ be defined by the condition that $(\rho(\theta), \theta)$ and $(\sigma(\theta), \theta + \pi/2)$ are points of J. The hypotheses imply that ρ and σ are continuous functions, with

$$\rho(\theta + \pi) = \rho(\theta) \quad \text{and} \quad \sigma(\theta + \pi) = \sigma(\theta)$$

for all θ. For each θ, the points

$$(\rho(\theta), \theta), (\sigma(\theta), \theta + \pi/2), (\rho(\theta), \theta + \pi), (\sigma(\theta), \theta + 3\pi/2)$$

are the successive vertices of a rhombus inscribed in J. For any value of θ such that $\rho(\theta) = \sigma(\theta)$, the rhombus has equal diagonals and hence is a square.

We may assume without loss of generality that $\sigma(0) \leq \rho(0)$. Then

$$\sigma(\pi/2) = \rho(\pi) = \rho(0) \geq \sigma(0) = \rho(\pi/2)$$

and it follows by continuity that $\rho(\theta) = \sigma(\theta)$ for some θ between 0 and $\pi/2$. □

The following result can also be proved by continuous variation of an inscribed rhombus. However, in the absence of a center of symmetry the technical details become considerably more complicated. (A Jordan curve is said to be *convex* if the set of all points inside it is convex.)

Theorem 11.5. *Each convex Jordan curve admits an inscribed square.*

Although smoothness is not assumed explicitly here, the convexity requirement implicitly imposes a sort of smoothness. A convex Jordan curve has only countably many "corners"—points at which there is no tangent—and even at a corner there are two *semitangents,* rays that act like tangents in one direction.

It is natural to wonder whether the triangles of Theorem 11.3 and the squares of Theorem 11.4 can be replaced by other polygons. Here is a relevant observation.

Theorem 11.6. *For a convex polygon P, the following two conditions are equivalent:*
(i) *each ellipse admits an inscribed polygon similar to P;*
(ii) *P is cyclic* (*can be inscribed in a circle*) *and has at most four vertices.*

Outline of Proof. Suppose first that condition (i) holds. Then P is obviously cyclic, and we want to show that P's vertex-set V consists of at most four points. If V consists of five or more points, there is a unique ellipse E containing V. Since two ellipses are similar if and only if their eccentricities are equal, no polygon similar to P can be inscribed in any ellipse whose eccentricity differs from E's. This contradicts (i), and completes the proof that (i) implies (ii).

Now suppose that (ii) holds and E is an ellipse with eccentricity ϵ. Let $p_1, \ldots, p_4$ be the vertices of P, and for each fifth point p_5 in the plane let $f(p_5)$ denote the eccentricity of the conic $C(p_5)$ that passes through all the p_is. Then $f(p_5) = 0$ when p_5 is concyclic with the points $p_1, \ldots, p_4$, and as p_5 is moved off to infinity, $f(p_5)$ varies continuously from 0 to values greater than ϵ. Hence there is a position of p_5 for which $f(p_5) = \epsilon$ and $C(p_5)$ is an ellipse similar to E. □

In view of Theorems 11.3, 11.5, and 11.6 and Exercise 5, the following problem is a natural complement to Problem 11.

Problem 11.1

If P is a convex polygon such that each Jordan curve (or each convex Jordan curve) admits an inscribed polygon similar to P, must P be a triangle or a square?

Exercises

1. (a) With $0 < a < b$, let K be the set formed by the two diagonals and one pair of opposite edges of a square of edge-length b. Tell how to modify K so as to obtain a Jordan curve that, for each s between a and b, has an inscribed square of edge-length s.

(b) With $1 \leq p < \infty$, let $J_p = \{(\xi, \eta) : |\xi|^p + |\eta|^p = 1\}$. Show that for each line L through the origin, J_p admits an inscribed square that is symmetric about the origin and has one of its diagonals parallel to L. Determine (in terms of p) the range in the lengths of the diagonals of such squares.

2. Suppose that E is the noncircular ellipse whose equation is

$$\frac{x^2}{a^2} + \frac{y^2}{b^2} = 1, \qquad \text{where } 0 < b < a.$$

Show that E admits a unique inscribed square, and the area of this square is

$$\frac{4a^2b^2}{a^2 + b^2}.$$

(Hint: Represent the ellipse parametrically by the equations $x = a \cos t, y = b \sin t$. Begin by showing that each parallelogram inscribed in E is concentric with E. Then use the fact that a parallelogram is a square if and only if its diagonals are perpendicular and of equal length.)

3. Let p, q, and r be the vertices of an equilateral triangle of side-length 1, and let the convex body C be the intersection of the disks of radius 1 centered at p, q, and r. Then C's boundary J is a Jordan curve formed from three circular arcs. Show that for each square S of side-length 1, there is a translate of S that is circumscribed about J.

4. Show that J, the Jordan curve of Exercise 3, admits an inscribed regular n-gon for $n = 3$, $n = 4$, and $n = 6$, but not for any other value of n.

5. Show that if Q is a quadrilateral whose successive angles are (for a small positive ϵ) $90°$, $(150 + \epsilon)°$, $90°$, and $(30 - \epsilon)°$, then Q cannot be inscribed in an equilateral triangle.

6. Show that the relations of homothety and of similarity are reflexive, symmetric, and transitive. That is:

(a) each set has that relation to itself;

(b) if V is related to W then W is related to V;

(c) if V is related to W and W is related to X, then V is related to X.

12. FIXED POINTS

Problem 12

Does each nonseparating plane continuum have the fixed-point property?

If each mathematician were asked to name twenty favorite theorems, the Brouwer fixed-point theorem would probably appear on most of the lists. Both for intuitive appeal and for its many applications, the theorem is outstanding. In its 2-dimensional form, it asserts that for each continuous mapping f of the unit disk

$$U = \{(x, y) : x^2 + y^2 \leq 1\}$$

into itself, there is at least one *fixed point*—a point p of U such that $f(p) = p$. Intuitively, the theorem says this: Suppose that you take a paper copy of U; fold, crumple, and wad it however you like, and finally crush it down into part of its original area. Then at least one point of U must be directly over its original position. (Tearing is excluded because it would interfere with continuity of the mapping. For example, you could tear out a piece P at U's center, rotate what's left, and push P away from the center. The resulting mapping would not have a fixed point, but it would be discontinuous at points of P's boundary.)

Part Two outlines a proof of the 2-dimensional form of Brouwer's theorem. It's easy to give a complete proof of the 1-dimensional form, which asserts that if I is a bounded closed interval in the real line $\mathbb{R}$ and the function $f : I \to I$ is continuous, then f has a fixed point. To prove this, just consider the continuous function $g : I \to \mathbb{R}$ defined by setting $g(x) = f(x) - x$ for each $x \in I$. If I is the interval $[a, b]$, it is clear that $g(a) \geq 0$ and $g(b) \leq 0$, so it follows from the intermediate-value theorem that there is a point p in I for which $g(p) = 0$. This implies $f(p) = p$, so p is a fixed point of f.

For any family of mappings of a set X into itself, it is natural to wonder (and sometimes essential to know) whether each member of the family has a fixed point. Because continuous mappings are so important, the set X is said to have the *fixed-point property* if each continuous mapping of X into X has a fixed point. The 1- and 2-dimensional versions of Brouwer's theorem assert that intervals and disks have the fixed-point property.

Two sets X and Y are said to be *homeomorphic* if there is a one-to-one mapping h of X onto Y such that both h and the inverse mapping h^{-1} are continuous. Such a mapping h is called a *homeomorphism*.

Theorem 12.1. *If X and Y are homeomorphic and X has the fixed-point property, then Y also has the fixed-point property.*

Proof. Consider an arbitrary continuous mapping g of Y into Y, and define the mapping $f = h^{-1}gh : X \to X$; that is, for each $x \in X$, $f(x) = h^{-1}(g(h(x))$. This definition makes sense because $h(x) \in Y$, whence $g(h(x))$ is defined as a point of Y and thus $h^{-1}(g(h(x))$ is defined as a point of X. The mapping f is continuous because h, g, and h^{-1} are all continuous. Since the set X has the fixed-point property, there is a point $x \in X$ such that $f(x) = x$; that is, $h^{-1}(g(h(x))) = x$. Applying the mapping h to both sides of this equation, we obtain $g(h(x)) = h(x)$ and hence the point $h(x)$ is a fixed point of the mapping g. □

An *arc* is a set that is homeomorphic to the interval $[0, 1]$, and a *topological disk* is a set that is homeomorphic to the disk U. By Theorem 12.1, all arcs and topological disks have the fixed-point property. Thus the following is of interest.

Theorem 12.2. *Each plane convex body is a topological disk.*

Outline of Proof. We suppose without loss of generality that the origin $(0, 0)$ is interior to the plane convex body C, and for simplicity we denote the origin simply by O. The natural radial mapping h of C onto the unit disk U turns out to be a homeomorphism. To define h, note that for each ray R that issues from the origin O, the intersection $R \cap C$ is a segment Oc and the intersection $R \cap U$ is a segment Ou, where c belongs to the boundary of C and u to the boundary of U. Then let h map the segment Oc linearly onto the segment Ou; that is $h(c) = u$, and $h(\lambda c) = \lambda u$ for $0 \leq \lambda \leq 1$. Obviously h is a one-to-one mapping of C onto U. We omit the verification that both h and h^{-1} are continuous. □

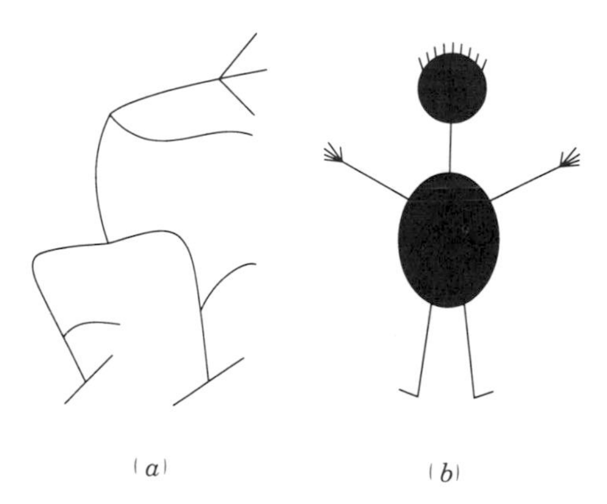

FIGURE 12.1
Some plane continua that have the fixed-point property.

A *continuum* is a set that is compact and connected. Not all plane continua have the fixed-point property. For example, a circle or an annulus can be rotated into itself to yield a continuous mapping with no fixed point. However, the fixed-point property is possessed by many subsets of the plane that are not arcs or topological disks. For example, if the union of a finite number of arcs is connected and does not contain any simple closed curve, then it has the fixed-point property (Figure 12.1a, Exercise 2). The same is true of the connected union K of a finite number of arcs and topological disks if each simple closed curve in K lies entirely in one of the disks (Figure 12.1b, Exercise 2). Other examples are provided by the "$\sin \frac{1}{x}$ arc" (Figure 12.2a)—part of the graph of $\sin \frac{1}{x}$ together with the limit segment—and by the "$\sin \frac{1}{x}$ circle" (Figure 12.2b). Additional examples can be formed by combining these in various ways (Exercise 1).

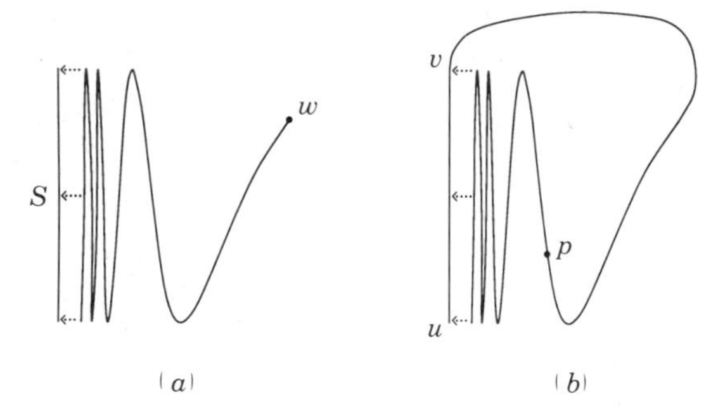

FIGURE 12.2
Some plane continua that have the fixed-point property.

The following theorem is especially useful in showing that various sets have the fixed-point property.

Theorem 12.3. *Suppose that X is a set with the fixed-point property, r is a continuous mapping of X onto a subset Y of X, and $r(y) = y$ for each point y of Y. Then Y has the fixed-point property.*

Proof. Consider an arbitrary continuous mapping $g: Y \to Y$, and define the mapping $f: X \to Y$ by setting $f = gr$; that is, for each $x \in X$, $f(x) = g(r(x))$. Since f is continuous and X has the fixed-point property, there exists $p \in X$ such that $f(p) = p$. Thus $p = g(r(p))$. But then $p \in Y$, since r maps X into Y and g maps Y into itself. From $p \in Y$ it follows that $r(p) = p$ and hence $p = g(p)$. □

When X and Y are related as in Theorem 12.3, the set Y is said to be a *retract* of X and the mapping r is a *retraction.* Continua like the ones in Figure 12.1 are retracts of topological disks (Exercise 2) and hence have the fixed-point property by Theorem 12.3. However, the continua of Figure 12.2 are not retracts of disks, so some other method must be used to show that they have the fixed-point property (Exercise 8 in Part Two).

Since circles and annuli lack the fixed-point property, it follows from Theorem 12.1 that any set homeomorphic to a circle or annulus also lacks the fixed-point property. This fact may be generalized as follows.

Theorem 12.4. *If K is a subset of the plane that contains a simple closed curve J but does not contain all the points inside J, then K lacks the fixed-point property.*

Proof. We discuss only the case in which J is the circle of unit radius centered at the origin O, and $O \notin K$. (For reduction of the general case to this, see Exercise 3.) For each point $k \in K$, let $r(k) = k/\|k\| \in J$. Then r is a retraction of K onto J. Since J lacks the fixed-point property, it follows from Theorem 12.3 that K also lacks the fixed-point property. □

Since the fixed-point property is so striking, many mathematicians have attempted to determine just which plane continua have it. Among continua K with the fixed-point property that have been mentioned so far, all except the "sin $\frac{1}{x}$ circle" are *nonseparating,* meaning that the complement $\mathbb{R}^2 \setminus K$ is connected. Problem 12, which asks whether all such plane continua have the fixed-point property, has been called "the most interesting outstanding problem in plane topology." Several erroneous proofs and counterexamples have been circulated informally.

Though Problem 12 is open, the intensive efforts to settle it have led to establishment of the fixed-point property for a variety of plane continua. Some of the best results apply only to continua K that are *arcwise connected,* meaning that any two points of K can be joined by an arc. Of the continua mentioned so far, all are arcwise connected except the "sin $\frac{1}{x}$ arc." In particular, the "sin $\frac{1}{x}$ circle" is arcwise connected, and it has the fixed-point property even though it separates the plane.

The following nice result settles Problem 12 for plane continua that are arcwise connected.

Theorem 12.5. *If K is an arcwise connected plane continuum that fails to separate the plane, then K has the fixed-point property.*

The next result is of special interest in connection with the "sin $\frac{1}{x}$ circle."

Theorem 12.6. *If K is an arcwise connected plane continuum that does not contain any simple closed curve, then K has the fixed-point property.*

References for Theorems 12.5 and 12.6 appear in Part Two, where some stronger related results are also stated.

Exercises

1. Suppose that X and Y are closed subsets of the plane, each of which has the fixed-point property. Prove that if the intersection $X \cap Y$ consists of a single point p, then the union $X \cup Y$ has the fixed-point property.

2. For each of the continua K of Figure 12.1, draw a topological disk D that contains K and provide an informal description of a retraction of D onto K. It follows from Theorem 12.3 that each of these continua has the fixed-point property. Produce another proof of this fact by using Exercise 1.

3. It is known that if J_1 and J_2 are simple closed curves in the plane $\mathbb{R}^2$, and the points p_1 and p_2 are inside J_1 and J_2 respectively, then there is a homeomorphism of $\mathbb{R}^2$ onto itself that carries J_1 onto J_2 and p_1 onto p_2. Using this fact, complete the proof of Theorem 12.4.

4. One of the many interesting relatives of the Brouwer fixed-point theorem is the fact that there is no retraction of the disk U onto its boundary B. Show that this is in fact equivalent to Brouwer's theorem.

TWO-DIMENSIONAL GEOMETRY: Part 2

1. ILLUMINATING A POLYGON

Problem 1

Is each reflecting polygonal region illuminable?

In using the term "illuminable," we are aware of the limitations of geometric optics as only a first approximation of physical (wave) optics. However, we couldn't resist this intuitively appealing statement of Problem 1. The problem's origin is uncertain, but apparently it was first published by Klee [Kle2]. The smooth region not illuminable from any point (Figure 1.2) first appeared in an article by Guy and Klee [GK] as a modification of an earlier example of Penrose and Penrose [PP]. The existence of smooth regions not illuminable from any finite set of points was noted in [GK], and Rauch [Rau] suggested the infinite extension of this idea. In contrast to these examples, Rauch and Taylor [RT] showed that in wave optics with a smooth boundary, "there are no perfect shadows."

A problem of Connett [Con] asks whether in $\mathbb{R}^3$ there is a container that "traps" all the light rays that enter it. The analogous problem may be formulated in the plane as follows. Suppose that a plane region R is bounded by a Jordan curve J that is formed from a line segment S and a finite number of smooth arcs. Consider all the light rays that enter R by crossing S and are thereafter reflected according to the usual rule. Can it happen that none of these rays again strikes a point of S?

In the analysis of billiard paths on a rational polygonal table, a crucial consequence of rationality is that for each direction of the initial segment of a path, there are only finitely many possible directions for the later segments (Exercise 10). The existence of dense billiard paths in rational polygonal tables (Theorem 1.1) was proved independently by Zemlyakov and Katok [ZK] and Boldrighini, Keane, and Marchetti [BKM]; Masur [Mas] proved the existence of periodic paths (Theorem 1.2). However, the existence of dense paths and the existence of periodic paths are open not only for general irrational polygons but even for irrational obtuse triangles (Problems 1.3 and 1.4). We saw in Part One that each acute triangular table admits a billiard path of period 3. It is easy to see that no such path is admitted by any right triangular table. Holt [Hol] proved that whenever the table is a right triangle it does not admit paths of period 5 or 7 but does admit paths of period 6 and (if not isosceles) of period 10. In fact, for each k there is a path of period $2+4k$ in all sufficiently thin right triangles.

The statements of Theorems 1.1 and 1.2 in Part One were chosen for their simplicity, but the known results for rational polygonal billiard tables go far beyond those simple statements. For a given interior point p of a rational table, and a given unit vector u, let $B(p, u)$ denote the billiard path that starts from the point p in the direction u. Then it is known [Mas] that the set of starting directions u for which some $B(p, u)$ is periodic is always dense in the unit circle. The set of all (p, u) forms a 3-dimensional manifold in $\mathbb{R}^4$, and it is known [ZK, BKM] that for all (p, u) except those in a set of 3-dimensional measure zero, the path $B(p, u)$ is dense in the table. There is a deep and important extension of this result due to Kerckhoff, Masur, and Smillie [KMS], but it is too technical to be appropriate for inclusion here.

Although the preceding results on rational tables cannot be discussed here in more detail, we should at least say a bit more about the easy case of a square table. Consider, first for an arbitrary convex polygonal table T_0, the following procedure that covers the plane with a sequence of congruent tables $T_0, T_1, T_2, \ldots$:

> for $i = 1, 2, \ldots$, find the smallest index j such that some edge of T_j is not an edge of any other T_k;
> choose some such edge E_i of T_j and form $T_i + 1$ by reflecting T_j across E_i.

The tables that arise from this construction may overlap in complicated ways, but when T_0 is a square they provide a simple covering of the plane. Then, when appropriate identifications are made between T_0's edges and those of the tables T_i, each billiard path may be represented by a single ray (half-line) in the plane. (For example, Figure 1.7 shows a billiard path of period four in a square table, and a portion of the corresponding ray.) With the aid of this identification, the following is not hard to prove (see Exercise 11).

Theorem 1.5. *On a square table, each billiard path is periodic or dense in the square. The periodic paths are exactly those for which the angle between the initial direction and an edge of the square has a rational tangent* τ.

The first proof of Theorem 1.5, provided by König and Szücs [KS] in 1913, used the tiling by squares and the following theorem of L. Kronecker to deal with the case in which τ is irrational: *For each irrational number* τ, *the numbers of the form* $n\tau - \lfloor n\tau \rfloor$ (n *a positive integer*) *are dense in the interval* $(0, 1)$. (See Hardy and Wright's *Introduction to Number Theory* [HW] for a concise version of the König and Szücs proof, and an article by Sudan [Sud] for a more directly geometric approach.) A similar tiling technique yields the same conclusion for a few other rational tables, and it was once conjectured that in an arbitrary convex polygonal table, denseness or periodicity were the only two possibilities for a billiard path that did not meet a corner. However, among the counterexamples of Galperin [Gal] there are rational n-gons for each $n \geq 4$, and there are also right triangles.

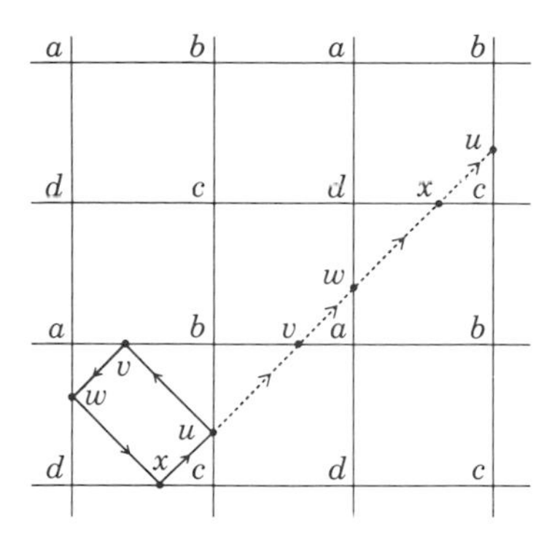

FIGURE 1.7
When the table is square, each billiard path may be represented by a half-line in the plane.

Now let us consider (still on a polygonal table) a new billiard game, similar to but more demanding than the game in Part One. In the new game, our opponent places on the table an invisible rectangle on which two opposite edges are marked `ENTER` and `EXIT`. We win the game only if there is a segment of our billiard path that enters the rectangle at the `ENTER` end and leaves at the `EXIT` end. We don't know the size, shape, or position of the rectangle. Hence to be sure of winning, our path must not merely come arbitrarily close to each point of the table (as in the earlier game), but must come arbitrarily close to each (point, direction) pair. In mathematical terms, the path should be dense in the *phase space.* When the table is rational, it follows from Exercise 10 that no such path exists. However, results of two papers mentioned earlier [ZK, KMS] show that for certain irrational tables—indeed, for "almost all" polygonal tables in a sense that will not be made precise here—there is a billiard path that is dense in the phase space. As was shown by Gruber [Gru], this holds also for "typical" smooth billiard tables.

One of the early results for smooth convex billiard tables was Theorem 1.3, asserting the existence of a billiard path of period k for each $k \geq 2$. It is due to G. D. Birkhoff [Bir]. It's not hard to produce a second path of period 2, for if L and M are parallel tangents of the table T for which the distance between L and M is a minimum, then there are points $p \in L \cap T$ and $q \in M \cap T$ such that the segment pq is perpendicular to both L and M. But how about paths of higher period? Croft and Swinnerton-Dyer [CS] showed that each smooth convex table admits a second billiard path of period 3, and they indicate an extension of their method to show that there are at least $\phi(k)$ essentially distinct billiard paths of period k, where ϕ is Euler's function. But is that the best possible?

Problem 1.5

For each positive integer $k \geq 3$, what is the smallest number of billiard ball paths of period k that a smooth convex table may have?

A *convex caustic* for a convex table T is a closed convex proper subset C of T such that whenever a billiard path in T is such that its initial segment lies in a supporting line of C, then every later segment also lies in a supporting line of C. (A line L *supports* C if L intersects C and C lies entirely in one of the two halfplanes bounded by C. When C is smooth, this just says that L is tangent to C.) We saw in Part One that when T is circular, each smaller concentric circle is a caustic for T. At least as early as 1745, it was known that for an elliptical table, each smaller confocal ellipse is a caustic (see Guillemin and Melrose [GM] for a reference and a simple proof). Interest in the existence of convex caustics for nonelliptical tables led eventually to the following deep result.

Theorem 1.6. *Suppose that the boundary curve of a plane convex body T is so smooth that (when parametrized in terms of arc length) its curvature is* 6 *times continuously differentiable. If, in addition, the curvature is everywhere positive, then the table T admits a convex caustic.*

A theorem of Mather [Mat] is said to imply that if the curvature is zero anywhere, there can be no caustic. For any theorem that involves a smoothness assumption, it is common mathematical practice to attempt to sharpen the theorem by assuming less smoothness. Great strides have been made in the case of Theorem 1.6, for the first published proof by Lazutkin [Laz] required that the curvature should be 553 times differentiable! The reduction to 6 is due to Douady [Dou]. However, the following problem is open.

Problem 1.6

How far can the 6 in Theorem 1.6 be reduced?

It is natural to suspect that the high smoothness requirement of Theorem 1.6 is an artifact of the proof rather than the underlying reality. However, Gruber [Gru] has observed that there is a sense in which, among the convex tables having merely a smooth boundary, almost all tables lack caustics. See Gruber for a survey of billiard properties, and Zamfirescu [Zam] for a survey of many other surprising properties of almost all convex bodies.

Let us now return briefly to periodic billiard paths. We remark first that to complete the proof of Theorem 1.3 in Part One, the argument indicated there should be accompanied by a proof of the *existence* of a convex k-gon of maximum perimeter inscribed in P. However, the existence of a longest inscribed convex polygon P with *at most* k sides follows from a routine compactness-and-continuity argument, and then, from the fact that T itself is not polygonal, it follows that P does in fact have k sides.

There are several results that actually characterize the shape of a convex table in terms of the behavior of periodic billiard paths on it. For example, the smooth tables of constant width are characterized by the property that each boundary point lies on a path of period 2 (see Exercise 12 and its solution), and also by the property that each path that is not of period 2 always turns right or always turns left (Sine and Kreĭnovič [SK]). Regular polygonal tables are characterized by admitting a billiard path similar to the table's boundary (DeTemple and Robertson [DR1]). Among sufficiently smooth tables, the circular ones are characterized by the property that for each $k \geq 2$, each boundary point lies on a regular k-gonal billiard path [DR2]. Probably the smoothness assumption in the latter paper can be weakened.

So far, all the problems discussed in this section have been set in the plane. However, most of the problems can also be formulated for higher-dimensional reflecting rooms (or billiard tables, or art galleries), and the range of ignorance in higher dimensions is far greater even than that in the plane. Croft and Swinnerton-Dyer [CS] conjectured that the answer to the following is affirmative. (With "period three" replaced by "period two," an affirmative answer was supplied by Kuiper [Kui].)

Problem 1.7

If T is a smooth convex d-dimensional table, must T admit at least d billiard paths of period three?

Gruber [Gru] observed that in any dimension, the confocal ellipsoids interior to a given ellipsoid T are caustics for T. He conjectured that these confocal ellipsoids (along with their intersection) are the only caustics for T, and that these are the only examples of higher-dimensional caustics. Thus he expects an affirmative answer to the following question.

Problem 1.8

For $d \geq 3$, are the d-dimensional ellipsoids the only smooth convex d-dimensional billiard tables that have caustics?

A partial affirmative answer has been obtained by Berger [Ber]. Compare this with Theorem 1.6, and with the fact that for each compact convex subset C of $\mathbb{R}^2$ it is easy to produce a convex table T for which C is a caustic. Turner [Tur1, 2] showed that such a T can always constructed as follows:

(a) place around C a loop of inelastic string whose length exceeds the perimeter of C;
(b) pull the string tight to produce a point p of the table's boundary;
(c) move the point p around C, keeping the string tight, to produce the rest of T's boundary (see Figure 1.8 and Exercise 13).

Note that if C is 2-dimensional and is not an ellipse, then T is not an ellipse. However, each T constructed in this way must be smooth and can have no segments in its boundary. (There is a 3-dimensional analogue of this string construction whereby, starting from an ellipsoid, one can produce a larger, confocal ellipsoid [HC]. However, that does not appear to be useful for attacking Problem 1.8.)

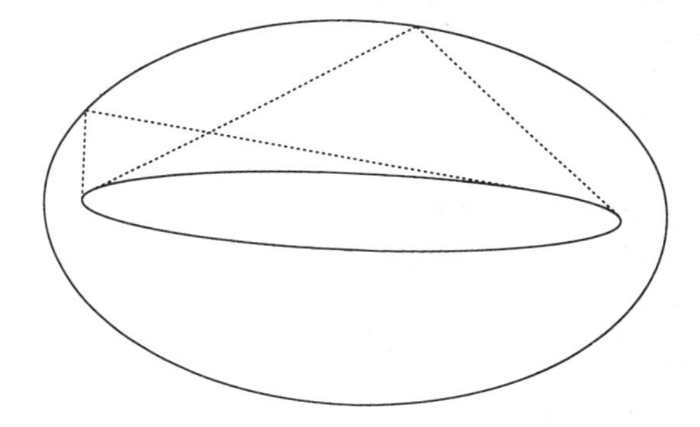

FIGURE 1.8
A string construction produces a billiard table for which a given convex curve is a caustic.

Most of the problems of Part One are also of interest when the walls of the room (or the billiard table or the art gallery) are made up of several Jordan curves, one forming the outer boundary of the room and the others creating holes in the room, forming inner boundaries as in Figure 1.9. For the problems that involve reflections, each of the holes redirects all the light rays that impinge upon it. Thus

FIGURE 1.9
The illumination problem and the art gallery problem are also of interest for polygonal rooms with polygonal holes.

it seems unlikely that the problem for h holes can be treated as a limiting case of the problem for $h+1$ holes by letting one hole shrink to a point or simply disappear.

Theorem 1.4 is due to Chvátal [Chv], and Fisk [Fis′] produced the elegant proof given in Part One. When holes are present in the art gallery problem, the number of necessary guards is affected by the number h of holes as well as the total number n of edges in the inner and outer boundaries. For example, any room with 24 edges and no holes requires at most 8 guards for supervision, but the room in Figure 1.9 requires 9 guards. The following problem was raised in O'Rourke's comprehensive study, *Art Gallery Theorems and Algorithms* [ORo], from which figures 1.6 and 1.9 were adapted.

Problem 1.9

If a polygonal art gallery has h holes, and n segments are used to form the inner and outer boundaries, are $\lfloor (n+h)/3 \rfloor$ guards always sufficient?

Extensions of Figure 1.9 show that $\lfloor (n+h)/3 \rfloor$ guards are sometimes necessary, but the sufficiency of this number has been proved only when $h \leq 1$. See Shermer [She′] for a survey of art gallery results that came after O'Rourke's book.

Let us end this section with the most important of the many open higher-dimensional problems in the spirit of the art gallery theorems. In effect, our "gallery" will be the part of d-space $\mathbb{R}^d$ that is left after the interior of a convex body C has been removed, and we'll be concerned with direct illumination of the entire boundary B of C by a small number of light sources that are "infinitely far" from C, so that all the rays from a given source are parallel to each other. More precisely, when $b \in B$ and R is an open half-line issuing from the origin, we'll say that b is *R-lighted* if there is a point p of $\mathbb{R}^d \setminus C$ such that the half-line $p + R$ hits b before it hits any other point of C (in other words, b can be seen by an observer looking in the direction R from some point outside C). And b is *strongly R-lighted* if, in addition, the half-line $p + R$ intersects the interior of C (i.e., if the point b were removed then an observer looking from p in the direction R could see C's interior).

The following is a consequence of a more general theorem of Ewald, Rogers, and Larman [ERL] (see Exercise 14 for the case $d = 2$).

Theorem 1.8. *For each convex body C in $\mathbb{R}^d$, there is a line M through the origin that is not parallel to any segment in the boundary of C.*

If R and S are the two open half-lines in M that issue from the origin, then each point of B is R-lighted or S-lighted. Hence two directions of illumination always suffice to light the entire boundary B. But how about strong lighting? For each convex body C in R^d, let $L(C)$ denote the minimum number k of half-lines $R_1, \ldots, R_k$ such that each point of C's boundary is strongly R_i-lighted for some i. Let $L(d)$ denote the maximum of $L(C)$ over all C in R^d. It is not hard to see that $L(C) = 2^d$ when C is a d-dimensional parallelotope, so $L(d) \geq 2^d$ for all d (Exercise 15).

Problem 1.10

Is it true that $L(d) = 2^d$ for all d? If so, is the maximum attained only by the d-parallelotopes?

Affirmative answers were established by Levi [Lev] for $d = 2$, and by Lassak [Las1] in the case of centrally symmetric bodies for $d = 3$. However, the best known general bound, applying in all dimensions and not assuming symmetry, appears to be the bound,

$$(d+1)d^{d-1} - (d-1)(d-2)^{d-1},$$

of Lassak [Las2]. Schramm [Sch] has a better bound for convex bodies of constant width. See Bezdek [Bez] for additional results related to Problem 1.10, and for a discussion of the problem's history.

Exercises

10. Suppose that T is an n-gonal billiard table in which the interior angles $\beta_1, \ldots, \beta_n$ are commensurate (i.e., each ratio β_i/β_j is rational). Prove that each β_i is a rational mutiple of π. With $\alpha_1, \ldots, \alpha_n$ denoting the angles of inclination of T's edges, and with θ denoting the initial travel direction of a billiard path in T, write down a finite set of angles that must contain the travel directions of all later segments of the path.

11. For billiard paths on a square table, establish the characterization of periodicity stated in Theorem 1.5.

12. Construct a noncircular smooth convex table in which each point of the boundary lies on a billiard ball path of period 2.

13. What sort of table T results from the string construction when the inside body (the caustic C) is an ellipse? an equilateral triangle?

14. Show that if C is a 2-dimensional convex body, then there is a line through the origin that is not parallel to any segment in C's boundary.

15. Consider the d-dimensional cube $C = [-1, 1]^d \subset \mathbb{R}$. With $k = 2^d$, produce k half-lines $R_1, \ldots, R_k$ issuing from the origin such that each point of C's boundary B is R_1-lighted or R_2-lighted, and such that each point of B is strongly R_i-lighted for some i. Prove that the strong lighting of B cannot be accomplished with fewer than 2^d half-lines (i.e., $L(C) = 2^d$).

2. EQUICHORDAL POINTS

Problem 2

Can a plane convex body have two equichordal points?

Problem 2 was raised in 1916 by Fujiwara [Fuj], and in 1917 by Blaschke, Rothe, and Weitzenböck [BRW]. Thus it is one of the oldest of the geometric problems discussed here. For convex bodies, Theorem 2.1 was proved by Süss [Süs] in 1925, Theorem 2.5 by Dirac [Dir] in 1952. (There are related later results that employ much weaker assumptions about the behavior of the "chord functions." See C. A. Rogers [Rog] and Larman and Tamvakis [LT] for central symmetry, R. Gardner [Gar′2] for uniqueness.) In 1958 Wirsing [Wir] abandoned the convexity assumption and considered the Problem 2.1 for an arbitrary Jordan region. In the forms stated in Part One, all of the results 2.1–2.6 except for 2.4 are due to him. He and Dirac showed first that the boundary curve is smooth at the ends of the chord through the two equichordal points, then used this fact to establish smoothness of the entire boundary. Wirsing then strengthened the smoothness result to establish analyticity of the boundary. Ehrhart [Ehr] showed $e < 0.5$ (assuming convexity) in 1967, and Theorem 2.4's sharper result on e was proved in the late 1980s by Michelacci and Volčič [MV]. The asymptotic behavior of the function h (from Theorem 2.6) was studied in detail by Schäfke and Volkmer [SV], and one of their results implies that there are (up to similarity) at most finitely many equichordal curves.

The cautionary words at the end of Part One were necessary because there have been so many claimed solutions of the equichordal problem that turned out to be erroneous. However, it seems that the delicate asymptotic analyses of J. Rogers [Rog′1, 2] and of Schäfke and Volkmer provide the right way of approaching the problem, and we expect that when they have been carried far enough, the infamous equichordal problem will have been laid to rest with a result asserting that no Jordan region can have two equichordal points.

For two points p and q of a line M, and for certain constants λ, Hayashi [Hay] in 1926, Butler [But] in 1968, and Hallstrom [Hal] in 1974 came close to constructing a Jordan region R with p and q as equichordal points and λ as the length of all chords through p or q. Rather than concentrating on the boundary behavior near M, they initially ignored this and were able to construct a closed region R such that for each line L that passes through p or q but not both, the intersection $L \cap R$ is a segment of length λ. Aside from its intersection with M, R's boundary is formed from curves J and K on opposite sides of M. However, in this construction the curves J and K don't actually reach the line M and it's not clear what happens as they approach M. If the construction is to produce a genuine Jordan region, M's intersection with the closure of $J \cup K$ must consist of just two points. However, it appears that as the constructed curves approach M they oscillate so wildly that this intersection is instead a union of two segments—though it has not been proved that this must occur.

Now we'll describe two constructions of convex regions that *do* have two equichordal points. However, the regions are not in the Euclidean plane. The first construction (Spaltenstein [Spa] 1984) is on the sphere; the second (Petty and Crotty [PC] 1970) is in the vector space $\mathbb{R}^d$ with $d \geq 2$, but this space is equipped with a non-Euclidean distance.

Let's use the term *spherical ellipse* for a region R on an open hemisphere such that under orthogonal projection onto the equatorial plane, R's boundary B is carried onto a noncircular ellipse (see Figure 2.2).

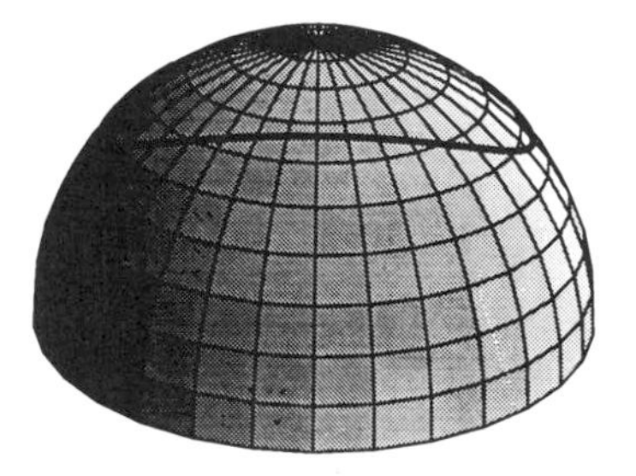

FIGURE 2.2
A spherical ellipse.

An interior point p of of a spherical region R is *equichordal* if there is a constant λ that is the length of each great-circle arc in R that passes through p and joins two points of B. We'll present the argument [Spa] showing that certain spherical ellipses have two equichordal points. Perhaps the following problem can be settled by an extension of the method.

Problem 2.2

Do all spherical ellipses have two equichordal points?

For $1 < \xi < \eta$, consider a spherical ellipsc whose boundary B consists of all points (x, y, z) for which $z > 0$,

$$x^2 + y^2 + z^2 = 1 \qquad \text{and} \qquad \xi^2 x^2 + \eta^2 y^2 = 1.$$

Then B consists of all points of the hemisphere that lie in the cone given by

$$z^2 = (\xi^2 - 1)x^2 + (\eta^2 - 1)y^2.$$

The minor axis of the equatorial ellipse ends at $(0 \pm 1/\eta, 0)$, and the length of the corresponding great-circle arc G in R is the inverse cosine of the inner product

$$\left(0, \frac{1}{\eta}, \frac{\sqrt{\eta^2 - 1}}{\eta}\right) \cdot \left(0, \frac{-1}{\eta}, \frac{\sqrt{\eta^2 - 1}}{\eta}\right) = \frac{(\eta^2 - 2)}{\eta^2}.$$

Now we show that for certain choices of ξ and η there is a point $p = (0, \gamma, \delta)$ of G that is an equichordal point of R. Of course, $\gamma \neq 0$ and hence the point $(0, -\gamma, \delta)$ is a second equichordal point.

Suppose that p is given, and for each point b of B let

$$b' = p - \mu(p \cdot b)b,$$

where μ is chosen so that

$$b \cdot b' = \|b'\|(\eta - 2)/\eta.$$

Since $\|b\| = 1$, this guarantees that the length of the minor great-circle arc joining the points b and $b'/\|b'\|$ is the inverse cosine of $(\eta^2 - 2)/\eta$. Thus to obtain the point

$$p = (0, \gamma, \delta) = (0, \sqrt{1 - \delta^2}, \delta)$$

as an equichordal point it suffices to choose δ in such a way that for each point $b \in B$, the point b' belongs to the cone mentioned earlier.

The computations may be complicated in the general case, but they're easy when $1 < \xi < \eta = \sqrt{2}$. Then μ is given the value 1 to assure that $b \cdot b' = 0$. The equation of the equatorial ellipse has the form

$$2(\cos^2 \phi)x^2 + 2y^2 = 1$$

with $0 < \phi < \pi/4$, and the equation of the cone mentioned earlier becomes

$$z^2 = (2\cos^2 \phi - 1)x^2 + y^2.$$

Now set $\gamma = \sin \phi$, to make $p = (0, \sin \phi, \cos \phi)$. If (x, y, z) belongs to B then with

$$b' = p - (p \cdot b)b$$

we obtain

$$b' = (0, \sin\phi, \cos\phi) - (y\sin\phi + z\cos\phi)(x, y, z).$$

Substitution in the equation of the cone yields (after some manipulation) an identity and completes the proof.

Now for the example of Petty and Crotty [PC]. With $\frac{1}{2} \leq \alpha < 1$, let p be a point of $\mathbb{R}^d$ for which $p \cdot p = 1 - \alpha$. Define the function ψ by setting $\psi(0) = 0$ and

$$\psi(x) = (x \cdot x)/\sqrt{\alpha(x \cdot x) + (x \cdot p)^2} \qquad \text{for each } x \in \mathbb{R}^d \setminus 0.$$

It is clear for each $x \in \mathbb{R}^d$ that $\psi(x) > 0$ if $x \neq 0$ and $\psi(x) = |\lambda|\psi(x)$ for all real λ. It can be verified that the set $\{x : \psi(x) \leq 1\}$ is convex, and from this it follows (Exercise 7) that

$$\psi(x + y) \leq \psi(x) + \psi(y) \qquad \text{for all } x, y \in \mathbb{R}^d.$$

Thus the function ψ can be used to introduce a new distance δ in $\mathbb{R}^d$, defining $\delta(x, y) = \psi(x - y)$ (Exercise 7). Let the length of a segment in $\mathbb{R}^d$ be defined as the δ-distance between its endpoints.

Now let C denote the set of all $x \in \mathbb{R}^d$ for which $x \cdot x \leq 1$. This is just the usual unit ball of $\mathbb{R}^d$, and is symmetric about the origin. With respect to the new distance function δ, the point p turns out to be an equichordal point of C and hence the point $-p$ is also equichordal.

In view of the example, the following theorem of Petty and Crotty [PC] is of interest.

Theorem 2.7. *In a normed linear space, no convex body has more than two equichordal points.*

Now let us return to the equireciprocal points mentioned in Part One. Suppose that a simple closed curve B bounds a region with two equireciprocal points p and q, and let p' and q' be the points where B intersects the line pq. Falconer [Fal2] showed that, as in the case of an ellipse, B is smooth at p' and q' and the tangents there are perpendicular to the line pq; and if B is twice differentiable at p' or q', C must in fact be an ellipse. However, he also showed how to construct nonelliptical plane convex bodies with two equireciprocal points, and even how to make their boundaries infinitely differentiable except at p' and q'. A tool in his investigation was a result from the field of differentiable dynamics that provides information about the iterates of a mapping. It seems probable that any rigorous solution of the equichordal problem will depend on similarly deep machinery, and this is one respect in which the equichordal problem may turn out to be important after all. Perhaps, like many other special problems, it will stimulate the development of mathematical tools that turn out to have applications far beyond their original purpose.

Convex bodies with equiproduct points have been studied by Yanagihara [Yan1, 2], Rosenbaum [Ros], J. Kelly [Kel], and Zuccheri [Zuc]. They show that the set of equiproduct points of a d-dimensional convex body is the intersection of a flat with the body's interior, and for $0 \leq k \leq d-2$ there are nonspherical d-dimensional convex bodies for which the set of equiproduct points is k-dimensional. However, a d-dimensional convex body B must be spherical if B's set of equiproduct points is of dimension $d-1$, and also if B has at least two equiproduct points and is either smooth or 2-dimensional.

For each exponent $k \neq 0$, R. Gardner [Gar'3] defines a *k-equipower point* of a region R as a point p such that the sum $(px)^k + (py)^k$ is constant for all chords xy through p. The equichordal and equireciprocal points correspond, respectively, to $k = 1$ and $k = -1$, and the equiproduct points are suggested to cover the missing value $k = 0$. Thus for $k = 0$ and $k = -1$, there exist plane convex bodies that have two k-equipower points. Our Problem 2 asks about the still open case $k = 1$. Gardner conjectures that "There is no convex body with two k-equipower points, for any $k \geq 1$. Possibly, there are none for any $k \neq 0, 1$."

Equichordal curves in space are studied by Craveiro and Robertson [CR], who raise some new problems. It seems safe to guess that, in one guise or another, the infamous equichordal problem will be with us for some time to come.

Exercises

5. For each positive continuous function f on $[0, 2\pi]$ with $f(2\pi) = f(0)$, let $R(f)$ denote the Jordan region whose boundary curve has the equation $r = f(\theta)$ in polar coordinates. Show that if f is twice continuously differentiable, then the region $R(\gamma + f)$ is convex for all sufficiently large constants γ. (Use the following facts, whose proofs can be found in many calculus books:

(a) The boundary's curvature is given by $k = (2r'^2 - rr'' + r^2)/(r'^2 + r^2)^{3/2}$.
(b) $R(f)$ is convex if k is everywhere nonnegative.)

6. Suppose that g is a positive continuous function on $[0, 2\pi]$ with $g(2\pi) = g(0)$ and such that $g(\theta) + g(\pi + \theta)$ has the same value for all θ. Show that the origin is an equichordal point of $R(g)$. Use this in conjunction with Exercise 5 to show that for each choice of constants $\alpha_1, \alpha_2, \ldots, \alpha_k$, it is true for all sufficiently large α that the curve whose equation is

$$r = \alpha + \alpha_1 \cos\theta + \alpha_2 \cos 3\theta + \cdots + \alpha_k \cos(2k-1)\theta$$

bounds a convex region that has the origin as an equichordal point. (See P. Kelly [Kel′] for more information concerning the construction of noncircular regions that have equichordal points.)

7. Suppose that a function $\phi: \mathbb{R}^d \to \mathbb{R}$ satisfies the following conditions:

(i) $\phi(x) > 0$ except when x is the origin;
(ii) $\phi(\lambda x) = |\lambda| x$ for all $\lambda \in \mathbb{R}$ and $x \in \mathbb{R}^d$;
(iii) the set $\{x: \phi(x) \leq 1\}$ is convex.

Show that $\phi(x+y) \leq \phi(x) + \phi(y)$ for all $x, y \in \mathbb{R}^d$. Show that if the *distance* $\delta(x, y)$ between any two points x and y is defined by setting $\delta(x, y) = \phi(x - y)$, then δ is a genuine distance function in the sense that it satisfies the following three conditions for all x, y, and z:

(symmetry) $\delta(x, y) = \delta(y, x)$;
(positivity) $\delta(x, y) > 0$ if and only if $x = y$;
(triangle inequality) $\delta(x, z) \leq \delta(x, y) + \delta(y, z)$.

3. PUSHING DISKS TOGETHER

Problem 3

When congruent disks are pushed closer together, can the area of their union increase?

Problem 3 was raised independently by Thue Poulsen [Thu] in 1954 and Kneser [Kne1] in 1955. Kneser proved that if the pairwise distances between points $q_1, \dots,$ q_n in $\mathbb{R}^d$ do not exceed those between the corresponding points p_i, then

$$\mu\left(\bigcup_i B(q_i, 1)\right) \leq 3^d \mu\left(\bigcup_i B(p_i, 1)\right).$$

Here is his proof. Let $U = \bigcup_{i=1}^n B(p_i, 1)$, $V = \bigcup_{i=1}^n B(q_i, 1)$, and suppose that $\|p_i - p_j\| \geq \|q_i - q_j\|$ for all i and j (where we are writing $\|x - y\|$ for $\operatorname{dist}(x, y)$). Let I be a subset of $\{1, \dots, n\}$ that is maximal with respect to the condition that $\|p_i - p_j\| \geq 2$ for all distinct $i, j \in I$. Then

$$\begin{aligned}
\mu\left(\bigcup_{i=1}^n B(q_i, 1)\right) &\leq \mu\left(\bigcup_{i \in I} B(q_i, 3)\right) \leq \sum_{i \in I} \mu(B(q_i, 3)) \\
&= 3^d \sum_{i \in I} \mu(B(q_i, 1)) = 3^d \sum_{i \in I} \mu(B(p_i, 1)) \\
&= 3^d \mu\left(\bigcup_{i \in I} B(p_i, 1)\right) \leq 3^d \mu\left(\bigcup_i B(p_i, 1)\right).
\end{aligned}$$

The first inequality follows from the fact that each q_k is at distance at most 2 from some q_i with $i \in I$. The third equality follows from the nonoverlapping of balls of radius 1 centered at points p_i with $i \in I$.

In the conjectured inequality (i) of Part One, the multiplier 3^d is replaced by 1. It would not surprise us greatly if this conjecture turns out to be false, even when $d = 2$, but we certainly expect the best multiplier for $d = 2$ to be much closer to 1 than to 9. In any case, the replacement has been rigorously justified only under the assumption that one of following strong additional conditions holds:

$d = 1$;

$d = 2$, and there is a continuous shrinking of the points $p_1, \dots, p_n$ into the positions $q_1, \dots, q_n$;

the ratios $\text{dist}(p_i, p_j)/\text{dist}(q_i, q_j)$ are the same for all $i \neq j$
the number n of balls is at most $d + 1$.

The sufficiency of each of these conditions was stated by Hadwiger [Had2]. A proof of the result for $d = 2$ (part of Theorem 3.3 in Part One) was published by Bollobás [Bol2], of the result for constant ratios by Bouligand [Bou] and by Avis, Bhattacharya, and Imai [ABI]). For the case in which $n \leq d+1$ (Theorem 3.2), Gromov [Gro2] outlined a proof of the inequality (ii) and Capoyleas and Pach [CP] proved both (i) and (ii). Kneser [Knc2] made the stronger conjecture (implying (i) and (ii)) that is stated in Exercise 5, and announced that he had a proof of this conjecture for $n \leq d + 1$ as well as for $d = 1$.

Other affirmative results in the spirit of the inequalities (i) and (ii) may be found in papers by Alexander [Ale], Capoyleas and Pach [CP], Gale [Gal], Lieb [Lie], Lieb and Simon [LS], Rehder [Reh], and Sudakov [Sud]. In particular, Sudakov proved the following, for which a much simpler proof was supplied by Capoyleas and Pach [CP]. The latter authors also noted that the result is implied by a negative answer to Problem 3 (see Exercise 6).

Theorem 3.4. *Suppose that the points $p_1, \ldots, p_n$ and $q_1, \ldots, q_n$ in the plane are such that $\|p_i - p_j\| \geq \|q_i - q_j\|$ for all i and j, and let P and Q denote the convex hulls of the p_i and the q_i, respectively. Then $\text{per}(P) \geq \text{per}(Q)$, where* per *denotes perimeter.*

When either of the convex hulls in Theorem 3.4 is a line segment S, $\text{per}(S)$ is to be interpreted as twice the length of S. To see that the hypotheses of Theorem 3.4 do not imply $\mu(P) \geq \mu(Q)$, consider the situation in which the points p_i are collinear but greatly spread out while the points q_i are bunched closely together but in general position.

For experimentation in connection with (i), (ii), and similar conjectured inequalities, it is useful to have fast algorithms that compute the measures of unions and intersections of important sorts of geometric figures. The algorithm of Exercise 4 (Part One) is due to Fredman and Weide [FW], whose paper also contains a precise determination of the computational complexity of finding measures of unions of intervals. See van Leeuwen and Wood [VW] for measures of unions of rectangles and their higher-dimensional analogues, Spirakis [Spi] and Avis, Bhattacharya, and Imai [ABI] for measures of unions of balls.

The notion of continuous shrinking leads to some interesting problems in its own right. The method of Gromov [Gro2] was used by Capoyleas and Pach [CP] to prove that if $p_1, \ldots, p_{d+1}, q_1, \ldots, q_{d+1}$ are points of $\mathbb{R}^d$ with $\text{dist}(p_i, p_j) \geq \text{dist}(q_i, q_j)$, and if the $(d + 1)$-tuples $(p_1, \ldots, p_{d+1})$ and $(q_1, \ldots, q_{d+1})$ are similarly oriented, then there exists a continuous shrinking of the p_is onto the q_is. However, the following old problem from Danzer, Grünbaum, and Klee [DGK] appears to be open even for $d = 2$ (see Exercise 8).

Problem 3.4

For each dimension $d \geq 2$, decide whether there exists an integer m_d with the following property: Whenever $p_1, \ldots, p_n, q_1, \ldots, q_n$ are points of $\mathbb{R}^d$ such that each set of m_d or fewer of the p_is admits a continuous shrinking onto the corresponding q_is, then there is a continuous shrinking of all the p_is onto all the q_is.

If m_d does exist, what is the best (smallest) value for m_d?

Of the known facts related to intersections of balls, the following result of Kirszbraun [Kir] is the most interesting and useful. It can be deduced from Theorem 3.3 (Exercises 10 and 11), but here we give the direct proof of Schoenberg [Sch′].

Theorem 3.5. *Suppose that n radii $\rho_1, \ldots, \rho_n$, n centers $p_1, \ldots, p_n$, and n other centers $q_1, \ldots, q_n$ are given, where the q_is are at least as close together as the p_is (i.e., $\|q_i - q_j\| \leq p_i - p_j\|$). If the intersection $\bigcap_i B(p_i, \rho_i)$ is nonempty then so is the intersection $\bigcap_i B(q_i, \rho_i)$.*

Proof. We use the following fact: *If C is a compact convex subset of $\mathbb{R}^d$ and w is a point of $\mathbb{R}^d \backslash C$, then arbitrarily close to w is a point z such that $\|z - c\| < \|w - c\|$ for all $c \in C$.* To prove this in $\mathbb{R}^2$, note that by C's convexity and closedness there is a unique point c_0 of C that is closest to w. Let H denote the line through c_0 orthogonal to the segment c_0w, and note that all of C is on the side of H opposite from w. From this it follows that each point z of the open segment c_0w is closer to each point of C than w is. (To extend this argument from $\mathbb{R}^2$ to $\mathbb{R}^d$, merely replace "line" by "hyperplane" in the description of H.)

To prove the theorem, we assume that a point $v \in \bigcap_i B(p_i, \rho_i)$ is given and we show how to produce a point $w \in \bigcap_i B(q_i, \rho_i)$. We may assume v is not among the p_is, for if $v = p_j$ we can set $w = q_j$.

For each $u \in \mathbb{R}^d$, let $\phi(u) = \max_i \|u - q_i\| / \|v - p_i\|$. The function ϕ is continuous and $\phi(u) \to \infty$ as $\|u\| \to \infty$, so ϕ attains its minimum λ at some point $w \in \mathbb{R}^d$. Our aim is to show that $\lambda \leq 1$, for then w is the desired point. Suppose, to the contrary, that $\lambda > 1$, and let the notation be chosen so that

$$\frac{\|w - q_i\|}{\|v - p_i\|} \begin{cases} = \lambda & \text{for } 1 \leq i \leq k; \\ < \lambda & \text{for } k < i \leq n. \end{cases}$$

With $P_i = p_i - v$, $Q_i = q_i - w$, and $\langle , \rangle$ denoting the inner product,

$$\langle Q_i, Q_i \rangle > \langle P_i, P_i \rangle \qquad \text{for } 1 \le i \le k.$$

Also, since the q_is are at least as close together as the p_is,

$$\langle Q_i - Q_j, Q_i - Q_j \rangle \le \langle P_i - P_j, P_i - P_j \rangle \qquad \text{for all } i \text{ and } j.$$

It follows from these two sets of inequalities that

$$\langle Q_i, Q_j \rangle > \langle P_i, P_j \rangle \qquad \text{for } 1 \le i \le j \le k. \tag{$*$}$$

By the observation in the first paragraph, w belongs to the convex hull C of the set $\{q_1, \ldots, q_k\}$, for otherwise the function ϕ attains a value smaller than the minimum $\phi(w)$. Since $w \in C$, there are nonnegative constants γ_i with sum 1 such that $w = \sum_1^k \gamma_i q_i$, whence, setting $Q = \sum_i^k \gamma_i Q_i$, we have $Q = \sum_i^k \gamma_i (q_i - w) = 0$ and thus $\langle Q, Q \rangle = 0$. However, since the γ_is are nonnegative with sum 1, it follows with the aid of $(*)$ that

$$\langle Q, Q \rangle > \left\langle \sum_i^k \gamma_i P_i, \sum_i^k \gamma_i P_i \right\rangle \ge 0.$$

The contradiction completes the proof. □

See Grünbaum [Grü1] for a similar proof of a significant generalization of Theorem 3.4. For applications of these results and their relatives, see Minty [Min] and Exercises 12–14.

Exercises

5. Suppose that $p_1, \ldots, p_n, q_1, \ldots, q_n$ are points of $\mathbb{R}^d$ such that $\text{dist}(p_i, p_j) \ge \text{dist}(q_i, q_j)$ for all i and j, and that $\rho_1, \ldots, \rho_n$ are positive numbers. For $1 \le k \le n$, let α_k (respectively, β_k) denote the volume (d-measure) of the set of all points that belong to the ball $B(p_i, \rho_i)$ (respectively, $B(q_i, \rho_i)$) for at least k distinct values of i. Kneser [Kne2] conjectured that for each monotonically decreasing sequence $\tau_1 \ge \tau_2 \ge \cdots \ge \tau_n$ of real numbers, it is true that $\sum_{k=1}^n \tau_k a_k \ge \sum_{k=1}^n \tau_k b_k$. Show that the conjectured inequalities (i) and (ii) are corollaries of this conjecture.

6. (J. Pach) Assume that the answer to Problem 3 is negative, and show how this assumption can be used to prove Theorem 3.4.

7. For each d-tuple $[a_1, b_1], \ldots, [a_d, b_d]$ of pairs of real numbers with $a_i < b_i$, let $R(a_1, \ldots, a_d; b_1, \ldots, b_d)$ denote the rectangular parallelepiped consisting of all

$x = (x_1, \dots, x_d) \in \mathbb{R}^d$ such that $a_i \leq x_i \leq b_i$ for all i. Write down a simple formula for the d-measure of the intersection of a finite number of sets of this form.

8. With m_d as in Problem 3.4, prove:
(a) $m_1 = 2$;
(b) for all $d \geq 2$, $m_d \geq 2d+1$ (a bound attributed to A. H. Cayford [DGK]).

9. Prove the following theorem of E. Helly [Hel]: If $n > d$ and $C_1, \dots, C_n$ are convex subsets of $\mathbb{R}^d$, each $d+1$ of which have nonempty intersection, then there is a point in common to all the sets.

10. Prove Theorem 3.5 by using Theorem 3.2 and Exercise 9.

11. (J. Pach) Without using Helly's theorem, prove Theorem 3.5 by embedding $\mathbb{R}^d$ in $\mathbb{R}^{n-1}$ and applying Theorem 3.2.

12. Suppose, as in Theorem 3.5, that radii $\rho_1, \dots, \rho_n$, centers $x_1, \dots, x_n$, and other centers $y_1, \dots, y_n$ are given, with $\|x_i - x_j\| \geq \|y_i - y_j\|$ for all i and j. Show that:

(i) If the intersection $\bigcap_i B(x_i, \rho_i)$ contains a ball of radius ρ, then the same is true of the intersection $\bigcap_i B(y_i, \rho_i)$.

(ii) If the set $\{x_1, \dots, x_n\}$ is contained in a ball of radius ρ, then the same is true of the set $\{y_1, \dots, y_n\}$.

13. When X and Y are sets equipped with distance functions, a *contraction* of X into Y is a mapping $f : X \to Y$ such that $\text{dist}(f(x_1), f(x_2)) \leq \text{dist}(x_1, x_2)$ for all $x_1, x_2 \in X$. Show that if X is a finite subset of $\mathbb{R}^d$, $f : X \to \mathbb{R}^d$ is a contraction, and p is a point of $\mathbb{R}^d \backslash X$, then f can be extended to a contraction of $X \cup \{p\}$ into $\mathbb{R}^d$.

14. Prove that if X is a subset of $\mathbb{R}^d$ and $f : X \to \mathbb{R}^d$ is a contraction, then f can be extended to a contraction of $\mathbb{R}^d$ into itself. (Hint: Show first, by means of a compactness argument, that the result of Exercise 13 holds even when X is infinite. Then establish the result of Exercise 14 for the case in which $m = d$. Finally, derive the general case from this.)

4. UNIVERSAL COVERS

Problem 4

If a convex body C contains a translate of each plane set of unit diameter, how small can C's area be?

It is not known whether the answer to Problem 4 is changed by omitting the condition that C is convex. However (as we saw in Part One), with or without the convexity of C, the answer is not changed by replacing "set of unit diameter " by "convex body of constant width 1." The same comments apply to Problem 4.1, which we state again for the reader's convenience.

Problem 4.1

If convex body C contains a rigid image of each plane set of unit diameter, how small can C's area be?

Without the convexity requirement, Problem 4.1 was posed in 1914 by Henri Lebesgue, the French mathematician who was responsible for the theory of measure and integration that bears his name. Over the years, the problem has become so well known that the term *universal cover* is often used for the sets that in Part One were called "rigid covers" for the collection $\mathcal{U}$ of sets of diameter 1. It was Pál [Pál1] who discovered in 1920 that the regular hexagon H of area $\sqrt{3}/2$ is a universal cover. He showed also that by shallowly truncating H at two vertices, a convex universal cover of area $2-2/\sqrt{3}$ could be obtained (Exercise 5). Still shallower further truncation by Sprague [Spr] in 1936 produced a convex universal cover of area $0.84413770\ldots$. In 1975 H. Hansen [Han] showed that two *very* shallow truncations of Sprague's example (removing an area of about $2\cdot 10^{-19}$) yield a convex universal cover that is minimal in the sense that no convex proper subset of it is a universal cover. A different modification of Pál's truncated hexagon was produced by Duff [Duf1] in 1980, to obtain a nonconvex universal cover of area $0.84413570\ldots$. Perhaps the examples of Hansen and Duff provide the answers to Problem 4.1 in the convex and nonconvex case, but there do not appear to be any strong reasons for believing this to be the case. However, it does seem safe to guess that progress on Problem 4.1, which has been painfully slow in the past, may be even more painfully slow in the future.

Although there are many *minimal* universal covers that are not *minimum,* in $\mathbb{R}^2$ it is at least true that each minimal universal cover is of diameter less than 3 [Grü3]. In connection with higher-dimensional analogues of Problem 4.1, a hint of new complexities is provided by Eggleston's result [Egg2] that for $d \geq 3$, $\mathbb{R}^d$ contains minimal universal covers of arbitrarily large finite diameter. The set of Exercise 4 plays a role in his construction.

Problem 4 is relatively new, having been proposed by B. C. Rennie in 1977. The example described in Part One, obtained by turning a Reuleaux triangle inside a square, was produced by Duff [Duf2].

Among the many other unsolved problems concerning the covering of plane sets by other plane sets, we want to mention the following.

Problem 4.3

If a plane convex body contains a rigid image of each plane closed curve of unit length, how small can its area be?

Although Problem 4.3 is open when the shape of the body C is unrestricted aside from the requirement of convexity, there are satisfactory results for several prescribed shapes. When the curves to be covered are not required to be closed, problems of this sort are sometimes called "worm problems," so perhaps Problem 4.3 should be known as the "ringworm problem." For surveys of results on this and a variety of other covering problems, see Chakerian and Klamkin [CK] and Bezdek and Connelly [BC].

Now let's return to the subject of translation covers for the collection $\mathcal{L}$ of unit segments in the plane. It's easy to see (Exercise 1) that an equilateral triangle of area $1/\sqrt{3}$ is such a cover, and Pál [Pál2] showed that it is the smallest one that is convex. When convexity is not required, two separate problems arise:

What is the smallest area of a plane set that contains unit segments in all directions?

What is the smallest area of a plane set in which a unit segment can be moved continuously in such a way that its ends are interchanged? (This is known as the *Kakeya problem,* and the sets in question are called *Kakeya sets.*)

Besicovitch [Bes1] solved both of these problems. He showed that the answer to the first problem is "zero," in the sense that there are plane sets of Lebesgue measure zero that contain not only unit segments, but in fact entire lines in all directions. Later Davies [Dav] produced sets of measure zero that are translation covers for the collection of all polygonal arcs in the plane.

Besicovitch showed that there is no "smallest area" for the second problem, since there are Kakeya sets of arbitrarily small positive area. The first Kakeya sets of small area were multiply connected (they had many "holes") and of large diameter. However, much nicer Kakeya sets are now known. In particular, Cunningham [Cun] showed that for each $\epsilon > 0$ there exists a simply connected Kakeya set of area less than ϵ contained in a disk of radius 1.

A condition weaker than convexity but stronger than being simply connected is that of being *starshaped,* meaning that the set is a union of segments issuing from some point. Starshaped Kakeya sets of areas approaching $(5-2\sqrt{2})\pi/24$ have been constructed by Cunningham and Schoenberg [CS‴], and it has been proved [Cun] that the area of any such set is at least $\pi/108$. The problem is to eliminate the gap.

Problem 4.4

What is the minimum area of a starshaped Kakeya set?

Since Section 4 is concerned mainly with covering problems, Problem 4.2 ("What is the maximum possible area of a convex n-gon of unit diameter?") is perhaps out of place. However, we could not resist including it as a further illustration of our general theme—that there remain many interesting unsolved problems concerning even the most familiar mathematical objects. The basic papers on Problem 4.2 are those of Reinhardt [Rei2] and Graham [Gra1], and, for a higher-dimensional analogue, Kind and Kleinschmidt [KK].

We end this section by mentioning what is probably the most famous of all unsolved problems that relate diameters to covering. The form dealing with arbitrary bounded sets was raised in the 1930s by Borsuk [Bor].

Problem 4.5

Can every bounded subset B of Euclidean d-space be covered by $d + 1$ sets of smaller diameter than B? If not, does such a covering exist at least when B is finite?

The range of ignorance on this problem is impressive. Let us denote by $f(d)$ the smallest integer k such that every bounded subset B of $\mathbb{R}^d$ can be covered by k sets of smaller diameter than B. It is easy to establish specific upper bounds for $f(d)$ (Exercise 6), but all the known bounds increase exponentially with d. On the other hand, an affirmative answer to Problem 4.5 would say that $f(d) = d + 1$. It is clear that $f(d) \geq d + 1$, because a theorem of Borsuk and Ulam [Bor] says that for each covering of the unit ball B in $\mathbb{R}^d$ by d closed sets, at least one of the sets includes two antipodal points and hence has the same diameter as B itself.

Again, convex bodies of constant width come into the picture, for this notion extends to $\mathbb{R}^d$ in a natural way (using distances between supporting hyperplanes) and it is again true that each set of diameter δ is contained in a convex body of constant width (and diameter) δ. Hence Problem 4.5 is equivalent to its special case in which the set B is assumed to be a convex body of constant width. (See Chakerian and Groemer [CG] for a survey of properties of these bodies.) The restriction to *finite* sets B may change the answer to Problem 4.5. However, a theorem of deBruijn and Erdős [DE2] implies that if the answer is affirmative for finite sets, then every bounded set of diameter δ in $\mathbb{R}^d$ set can be covered by $d + 1$ sets none of which includes two points at distance δ. (This does not imply that the covering sets are of diameter less than 1.)

For $d \leq 3$, Problem 4.5 has been answered affirmatively by finding a convex body C such that

(i) C is a rigid cover for the collection $\mathcal{U}_d$ of all subsets of $\mathbb{R}^d$ that are of diameter 1, and
(ii) C can be partitioned into $d + 1$ sets, each of diameter less than 1.

(See Exercise 6 for the case $d = 2$.) However, Problem 4.5 remains open for all $d \geq 4$, even when the set B is required to be finite. For these and many other facts related to Borsuk's problem, the reader may consult the excellent survey article of Grünbaum [Grü3].

Exercises

5. Let H denote a regular hexagon in which parallel edges are at unit distance, so that H is a universal cover. Let G be obtained by rotating H through an angle of $\pi/6$ about the center of H, so that G cuts off a portion of H around each of H's vertices. Show that when two of these six portions (corresponding to two vertices that are neither adjacent nor opposite) are taken away, what is left of H is still a universal cover. Find the area of this universal cover.

6. With H as in Exercise 4.5, show that H can be split into three congruent parts of diameter $\sqrt{3}/2$.

7. By using the fact that each subset of $\mathbb{R}^d$ of diameter 1 is contained in a d-cube of edge-length 1, show that every bounded subset B of $\mathbb{R}^d$ can be covered by $(\lfloor\sqrt{d}\rfloor + 1)^d$ sets, each of which is of smaller diameter than B. (A better method, due to Lassak [Las1], $d - 1$ yields the much smaller number of $2^{d-1} + 1$ covering sets. However, this is still far from the conjectured $d + 1$.)

5. FORMING CONVEX POLYGONS

Problem 5

How many points are needed to guarantee a convex n-gon?

What is the smallest integer $f(n)$ such that whenever W is a set of more than $f(n)$ points in general position in the plane, then W contains all the vertices of some convex n-gon? Investigation of this function f was begun in 1935 by Erdős and Szekeres [ES1]. They noted the relevance of Ramsey's theorem (from [Ram]), proved Theorem 5.1, and showed that $f(n) \leq \binom{2n-4}{n-2}$. (In Part One, the elegant use of Ramsey's theorem is due to S. Johnson [Joh′], the proof of Theorem 5.1 to Seidenberg [Sei].) Returning to the problem twenty-five years later, Erdős and Szekeres [ES2] showed that $f(n) \geq 2^{n-2}$ and conjectured that equality always holds. There are several published proofs that $f(5) = 8$ [KKS, Bon, Lov], and Lovász [Lov] also has the most readable proof that $f(n) \geq 2^{n-2}$.

Problem 5.1 (at the end of Part One) was proposed by Moser and Pach [MP]. The results on $k(n)$ that are stated there were obtained by a seminar group led by V. Bálint [BBB].

To prepare for the proof that $f(n) \leq \binom{2n-4}{n-2}$, note that for each finite plane set T there are many ways of introducing a rectangular coordinate system so that no two points of T agree in either coordinate. With $t = |T|$, the set T then has the form

$$\{(x_1, y_1), (x_2, y_2), \ldots, (x_t, y_t)\} \qquad \text{with } x_1 < x_2 < \cdots < x_t.$$

When $t > n^2$, Theorem 5.1 guarantees the existence of a subsequence $x_{k(1)} < x_{k(2)} < \cdots < x_{k(n)}$ for which the corresponding sequence $\{y_{k(i)}\}$ is also monotone. To find the vertices of a convex n-gon in T, it suffices to find an increasing subsequence $\{x_{k(i)}\}_{i=1}^{n}$ for which the sequence of slopes

$$\frac{y_{k(i)} - y_{k(i-1)}}{x_{k(i)} - x_{k(i-1)}} \qquad (i = 2, 3, \ldots, n)$$

is strictly monotone. When this sequence is increasing, the segments joining successive points $(x_{k(i)}, y_{k(i)})$ form the graph of a convex function and the lower boundary of a convex n-gon; when the sequence is decreasing, they form the graph of a concave function and the upper boundary of a convex n-gon. The segment joining the first point to the last forms the rest of the boundary.

Now suppose that X is a finite subset of $\mathbb{R}$ and ϕ is a real-valued function defined for all pairs (x, x') of points of X such that $x < x'$. By a *ϕ-convex* (respectively, *ϕ-concave*) sequence in X, we mean an increasing sequence $x_1 < x_2 < \cdots < x_m$ (with $m \geq 3$) such that

$$\phi(x_1, x_2) \leq \phi(x_2, x_3) \leq \cdots \leq \phi(x_{m-1}, x_m)$$

$$(\text{respectively, } \phi(x_1, x_2) \geq \phi(x_2, x_3) \geq \cdots \geq \phi(x_{m-1}, x_m).)$$

The proof that $f(n) \leq \binom{2n-4}{n-2}$ is based on the following fact.

Theorem 5.2. *For each pair of integers $m, n \geq 3$, let $h(m, n)$ denote the smallest integer s such that for each function ϕ on a set of more than s points of $\mathbb{R}$, there exists a ϕ-convex sequence of length m or a ϕ-concave sequence of length n. Then*

$$h(m, n) = \binom{m+n-4}{m-2}$$

for all $m \geq 3$ and $n \geq 3$.

Proof. It is apparent that for all m and n, $h(m, n) = h(n, m)$ and $h(3, n) = n - 1$. Now consider an arbitrary m and n, each greater than 3, and set $s = h(m - 1, n) + h(m, n - 1)$. To prove that $h(m, n) \leq s$, we show that if X is a set of $s + 1$ points in $\mathbb{R}$ and $\phi(x, x')$ is defined for all $x < x'$ in X, then X contains a ϕ-convex sequence of length m or a ϕ-concave sequence of length n. Suppose that X contains neither sort of sequence, and let X_0 consist of the first $h(m - 1, n) + 1$ points of X. Then X_0 contains a ϕ-convex sequence of length $m - 1$. Let v be the last member of this sequence, and form X_1 from X_0 by removing v_0 and adding the first point of X beyond X_0. Then X_1 contains a ϕ-convex sequence of length $m - 1$ whose last member v_1 is different from v_0. Form X_2 from X_1 by removing v_1 and adding the first point of X beyond X_0. Continue in this manner, ending with the formation of $X_{h(m,n-1)}$ from its predecessor by removing $v_{h(m,n-1)-1}$ and adding the final point of X. Then $X_{h(m,n-1)}$ also contains a ϕ-convex sequence of length $m - 1$ whose last member $v_{h(m,n-1)}$ is different from all the preceding v_is. Since the set of all v_is is of cardinality $h(m, n-1) + 1$, it contains a ϕ-concave sequence B of length $n - 1$. Note that the first member q of B is itself the last member of a ϕ-convex sequence A of length $m - 1$. Let a denote the predecessor of q in A, and b the successor of q in B. If $\phi(a, q) \leq \phi(q, b)$ then b can be added at the right end of A, extending A to a ϕ-convex sequence of length m. If $\phi(a, q) \geq \phi(q, b)$ then a can be added at the left end of B, extending B to a ϕ-concave sequence of length n.

The preceding paragraph shows that $h(m,n) \le s$, and the reverse inequality is left to the reader. From the recursion

$$h(m,n) = h(m,n-1) + h(m-1,n)$$

in conjunction with the fact that $h(m,3) = h(3,m) = m-1$, a straightforward induction shows that $h(m,n) = \binom{m+n-4}{m-2}$. □

With $t = \binom{2n-4}{n-2}$, we can now show that $f(n) \le t$. Consider an arbitrary plane set T of more than t points in general position, and choose a rectangular coordinate system so that no two members of T have the same x-coordinate. Then there are numbers $x_1 < x_2 < \cdots < x_t$ and numbers y_i such that

$$T = \{(x_1,y_1),(x_2,y_2),\ldots,(x_t,y_t)\}.$$

For each i and j such that $i < j$, define

$$\phi(x_i,x_j) = \frac{y_j - y_i}{x_j - x_i}.$$

Since $t = h(n,n)$, it follows from Theorem 5.2 that the sequence $x_1,\ldots,x_t$ has a subsequence of length n that is ϕ-convex or ϕ-concave. With respect to the successive pairs of members of this subsequence, the slopes defined by ϕ are not merely monotone but strictly monotone, because of general position. Hence, as mentioned earlier, a convex n-gon is formed by the points (x_i,y_i) corresponding to the subsequence.

Bisztriczky and Fejes Tóth [BF1, 2, 3] considered the analogue of Problem 5 in which individual points are replaced by disjoint ovals (compact convex subsets of the plane). They call three ovals *collinear* if the convex hull of the union of two of them contains the third, and they call a collection $\mathcal{F}$ of n ovals an *n-gon* if no member is contained in the convex hull of the union of the remaining members. They proved [BF1] that for each $n \ge 4$ there exists $b(n)$ such that whenever $\mathcal{F}$ is a collection of more than $b(n)$ disjoint ovals, no three of of which are collinear, then some n-gon is a subcollection of $\mathcal{F}$. They conjectured that the answer to the following question is affirmative, and proved the conjecture for $n = 4$ [BF1] and $n = 5$ [BF2].

Problem 5.1

Is $b(n) = f(n)$ for all n?

Motzkin and O'Neill [MO] extended some of the results of Erdős and Szekeres [ES1] to subsets of $\mathbb{R}^d$.

The following variant of Problem 5 was proposed by Erdős [Erd4]:

> How many points are needed to form an empty convex n-gon? Specifically, what is the smallest integer $g(n)$ (if it exists) such that whenever W is a set of more than $g(n)$ points in general position in the plane, then W contains all the vertices of a convex n-gon whose interior misses W?

It is obvious that $g(3) = 2$ and easy to see that $g(4) = 4$ (Exercise 4). The configuration in Figure 5.2 shows that $g(5) \geq 9$, and Harborth [Har3] proved that $g(5) = 9$.

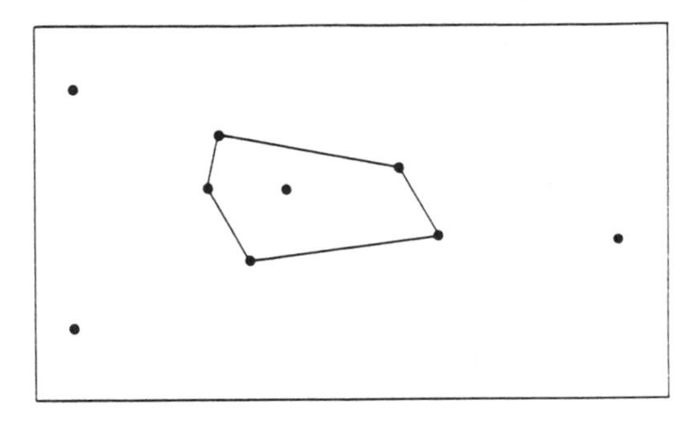

FIGURE 5.2
In this set of nine points, it is impossible to find the vertex-set of a convex pentagon whose interior misses the set.

Here is the principal open problem related to Erdős's question.

Problem 5.2

Is $g(6)$ finite? If so, what is its value?

Starting with the work of Avis and Rappaport [AR], successive computer searches have increased the lower bound on $g(6)$. The latest of which we are aware is that of Overmars, Scholten, and Vincent [OSV], based on a search-technique developed by Dobkin, Edelsbrunner, and Overmars [DEO]; it showed that $g(6) \geq 28$ [OSV]. Probably more extensive searching can increase this lower bound further, but of course cannot in itself decide whether $g(6)$ is finite. However, a surprising construction of Horton [Hor] shows that for $n \geq 7$, $g(n)$ is *not finite*; that is, there are arbitrarily large sets W, in general position in the plane, such that for each convex 7-gon P with vertices in W, the interior of P includes other points of W.

It is common in mathematics that the insight required to settle a conjecture may arise from formulating a stronger conjecture that includes the original one as a special case. In 1881–82, Perrin [Per] approached some problems concerning points and lines in the plane by studying certain periodic sequences of permutations. His approach has been revived and extended by Goodman and Pollack [GP1,2,3], and has led to a more general conjecture from whose proof it would follow that $f(n) = 2^{n-2}$.

Consider an indexed set $\{p_1, \ldots, p_n\}$ of n points in the plane, and a directed line L that is free to rotate counterclockwise about a fixed point. For each position of L that is not orthogonal to any line determined by two of the points p_i, a permutation of $(1, \ldots, n)$ is produced by projecting these points orthogonally onto L; i precedes j in this permutation if, in the ordering provided by L's direction, the projection of p_i precedes the projection of p_j. As the line L rotates, it occasionally assumes a position that is orthogonal to one or more of the lines determined by pairs of the p_is. As soon as L passes this position, it yields a new permutation in which the ordering of each substring corresponding to such a line is reversed. Eventually L returns to its starting position and the process repeats itself, thus producing a periodic sequence of permutations of the indices $1, \ldots, n$. Figure 5.3 shows a half-period of such a sequence of permutations, and indicates by underlining which substrings of indices are reversed.

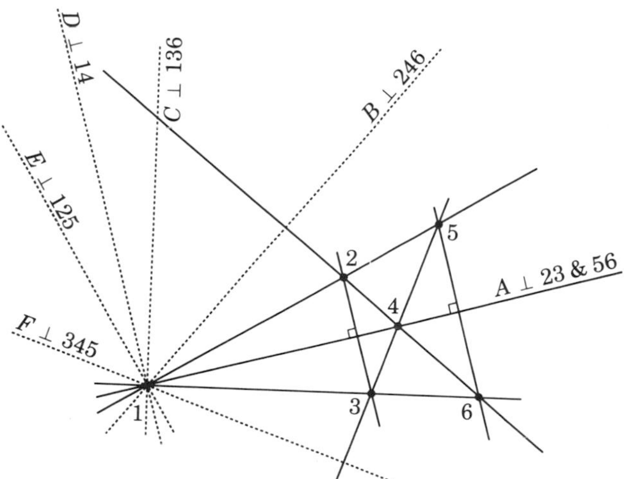

FIGURE 5.3
The first half-period of a sequence of permutations associated with the given set of indexed points is

$$123456 \xrightarrow{A} 1\underline{32}4\underline{65} \xrightarrow{B} 13\underline{642}5 \xrightarrow{C} \underline{631}425 \xrightarrow{D} 63\underline{41}25 \xrightarrow{E} 634\underline{521} \xrightarrow{F} 6\underline{543}21.$$

The following two conditions are satisfied by each sequence of permutations that arises in the above manner from a finite set of indexed points in the plane:

(i) The move from one permutation to the next consists of reversing one or more nonoverlapping substrings.
(ii) If a move results in reversal of a pair (i, j) then every pair of indices whose order is not reversed in the same move is reversed by some later move that precedes the next reversal of (i, j).

Any periodic sequence of permutations of $(1, \ldots, n)$ satisfying these conditions is called an *allowable n-sequence.* Condition (ii) guarantees that permutations a half-period apart in the sequence are the reverses of each other. Thus, for example, the rest of the allowable 6-sequence in Figure 5.3 can be constructed without further reference to the points themselves.

When a configuration is formed from a set $\{p_1, \ldots, p_n\}$ of n indexed points along with all the lines determined by pairs of those points, various properties of the configuration can be translated directly into properties of the corresponding allowable n-sequence. For example, general position of the point set is equivalent to the condition that only strings of length two are reversed in a move from one associated permutation to the next. A subset $\{p_{k(1)}, \ldots, p_{k(m)}\}$ is the vertex-set of a convex m-gon if and only if, in the allowable n-sequence associated with $\{p_1, \ldots, p_n\}$, there is for each $j \in \{1, \ldots, m\}$ a permutation in which $k(j)$ precedes all the other $k(i)$s (this follows readily from Exercise 9). For translations of two other geometric properties, see Exercise 10 and also Part Two of Section 6.

Goodman and Pollack [GP1] show that not all allowable sequences arise from sets of points in the manner described above. In particular the following 5-sequence is not geometrically realizable:

$$12534 \xrightarrow{12} 21534 \xrightarrow{53} 21354 \xrightarrow{54} 21345 \xrightarrow{13} 23145 \xrightarrow{23} 32145 \xrightarrow{14} 32415 \xrightarrow{15} 32451 \xrightarrow{24} 43251 \xrightarrow{25} 43521.$$

It follows that although many questions about points and lines in the plane can be translated into the language of allowable sequences, answers for the new purely combinatorial problems may be different from those for the geometric problems from which they stem. In particular, the function f_c of the following problem may be different from the function f associated with Problem 5. In any case, $f \leq f_c$ and thus any upper bound for f_c applies also to f.

Problem 5.3

What is the smallest integer $f_c(n)$ such that in each allowable $(f_c(n)+1)$-sequence in which only strings of length two are reversed, there are n

indices such that each occurs before all the others in some member of the sequence.

Goodman and Polack conjectured [GP3] that $f_c(n) = 2^{n-2}$, and noted that proving this would amount to extending a dual equivalent (formulated in [GP2]) of the Erdős–Szekeres conjecture from lines to "pseudolines."

Exercises

4. Prove that $g(4) = 4$.

5. [ES1] Suppose that $a_1, a_2, \ldots, a_{s+1}$ is a sequence of real numbers, with $s = \binom{2n-4}{n-2}$ and $n \geq 3$. Prove that there is an increasing sequence $k(1) < k(2) < \cdots < k(n)$ in $\{1, \ldots, s+1\}$ for which the sequence of differences $\{a_{k(i)} - a_{k(i-1)}\}_{i=2}^{n}$ is monotone. Show that the conclusion may fail when the sequence $\{a_i\}$ is of length s rather than $s + 1$.

6. [ES1] For n and s as in Exercise 5, construct an increasing sequence of real numbers $x_1 < x_2 < \cdots < x_s$ and a sequence $\{y_i\}_1^s$ of s distinct real numbers such that the sequence of slopes,

$$\frac{y_i - y_{i-1}}{x_i - x_{i-1}}, \qquad i = 2, \ldots, s$$

does not have any monotone subsequence of length n.

7. A subset of $\mathbb{R}^d$ is *in general position* if, for $1 \leq k < d$, no k-dimensional flat contains more than $k + 1$ points of the set. Show that for each d, and for each $n > d$, there is an integer $C(d, n)$ that has the following property: Whenever a set X of more than $C(d, n)$ points in $\mathbb{R}^d$ is in general position, it has a subset Y of cardinality n such that no point of Y is a convex combination of other points of Y. (It appears that for most d and n, no good estimates are known for the smallest value of $C(d, n)$ with the stated property.)

8. Prove that for each pair of positive integers m and n, there are positive integers p and q (depending on m and n) such that $n < p < q$ and the following is true: Whenever W is a plane set of more than q points in general position, W contains the vertex-set of a convex p-gon P such that the number of points of W interior to P is divisible by m. (For an independently discovered close relative of this result, see Bialostocki, Dierker, and Voxman [BDV].)

9. Prove that the following two conditions are equivalent for a sequence $(p_1, \ldots, p_n)$ of n points in the plane ($n \geq 3$):

(i) $\{p_1, \ldots, p_n\}$ is the vertex-set of a convex n-gon;

(ii) for each j, the allowable n-sequence corresponding to $(p_1, \ldots, p_n)$ includes a permutation that starts with j.

10. Formulate, in terms of allowable sequences of permutations, the strengthened form of the conjecture that $g(6)$ is finite.

6. POINTS ON LINES

Problem 6

If n points in the plane are not collinear, must one of them lie on at least $\frac{1}{3}n$ connecting lines?

With the multiplier $\frac{1}{3}$ replaced by $\frac{1}{2}$, Problem 6 was first posed by Dirac [Dir1], who also proved Theorem 6.1. Despite Grünbaum's examples [Grü4, 5] showing that $\frac{1}{2}$ doesn't work for all values of n, perhaps it does work for all sufficiently large n (see Problem 6.1). It was Erdős [Erd2] who first suggested the difficulty of establishing *any* lower bound of the form cn where c is a positive constant. The independent efforts of Beck [Bec] and Szemerédi and Trotter [ST] settled this difficult question, but their constants are extremely small compared to the constant $\frac{1}{3}$ that may work for all n. In any case, the constant c_3 of [ST] can be considerably improved by using a more recent result of Clarkson, Edelsbrunner, Guibas, Sharir, and Welzl [CEGSW]. The situation is reviewed in the following paragraph.

The constant c_3 is of the form $10^{-7}c_1^{-18}$, where the following property is required of the constant c_1: For any set of n points and t lines in the plane such that $\sqrt{n} \leq t \leq \binom{n}{2}$, the total number of point-line incidences is at most $c_1 n^{2/3}t^{2/3}$. It is shown in [ST] that this holds for $c_1 = 10^{60}$, and that yields $c_3 = 10^{-1087}$. However, it follows from a theorem of [CEGSW] that we may take $c_1 = 23.233$, and that yields a value less than 10^{-32} for c_3—a great improvement over 10^{-1087} but still far from the conjectured $\frac{1}{3}$. It is surely possible to refine this analysis to obtain a further improvement in c_3, but this approach doesn't seem to offer much hope of finding the best value for the constants in the main problems of this section.

Aside from the value of the constant, any upper bound (on point-line incidences) of the form $cn^{2/3}t^{2/3}$ is sharp, for a construction of Erdős based on the $n \times n$ square grid provides a lower bound of the same form. (See also Edelsbrunner [Ede] and Kranakis and Pocchiola [KP].)

Grünbaum's examples are most easily presented in the real projective plane Π rather than the Euclidean plane E. A *parallel pencil* in E consists of all lines parallel to a given line. To construct Π, start with the usual incidence relations in E and augment them as follows:

(i) For each parallel pencil $\mathcal{P}$, add a "point at infinity" $q(\mathcal{P})$ that is incident to all members of $\mathcal{P}$, with the understanding that $q(\mathcal{P}) \neq q(\mathcal{P}')$ when $\mathcal{P} \neq \mathcal{P}'$;
(ii) Add a "line at infinity" that is incident to all points at infinity.

Here are the properties of Π that are most relevant to Problem 6 and its relatives (for proofs, see any text on projective geometry):

(a) Π has not only the Euclidean property that for any two points, there is a unique line incident to both of them, but also the dual property that for any two lines there is a unique point incident to both of them. In fact, for any theorem about the incidence of points and lines in Π there is a dual theorem in which the words "point" and "line" are interchanged.

(b) Any two lines in Π can be interchanged by a one-to-one transformation of Π's points that preserves all incidences between points and lines.

Now let us reinterpret Figure 6.2 in terms of the projective plane Π. For $i = 1, \ldots, 4$, the parallel lines L_{i-2}, L_{i-1}, and L_i belong to a parallel pencil and hence determine a point p_i on the line L_{13} at infinity. By (b) there is a one-to-one incidence-preserving transformation of Π that carries this line onto a "finite" line. That yields, in the Euclidean plane, the example of nine points determining thirteen lines, with each point on precisely four of the lines. Since $4 < \frac{9}{2}$ this shows that the multiplier $\frac{1}{2}$ doesn't always work in Problem 6.

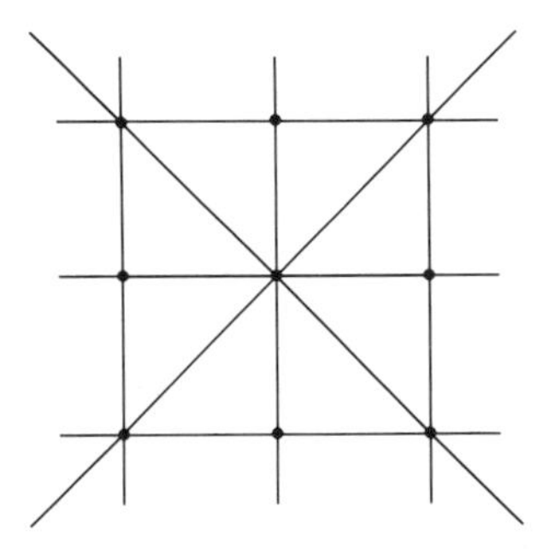

FIGURE 6.4
Nine lines and thirteen points (including the unshown line and four points at infinity), with each line on precisely four of the points.

Figure 6.4 is a dual illustration of the fact that the multiplier $\frac{1}{2}$ doesn't always work. The eight lines represent four parallel pencils $\mathcal{P}_1, \ldots, \mathcal{P}_4$. When the line at infinity and the four points $q(\mathcal{P}_i)$ are added to the configuration, we have in Π a system of nine lines determining thirteen points, with each of the lines on precisely four of the points.

The emphasis on the multiplier $\frac{1}{3}$ in Problem 6 and $\frac{1}{2}$ in Problem 6.1 may be a consequence of the fact that the supply of critical examples is so limited. The following is perhaps the most natural relative of Problem 6.

Problem 6.3

What is the least upper bound of constants c' such that whenever n is sufficiently large, S is a noncollinear set of n points in the plane, and $c < c'$, then some point of S is on at least cn connecting lines?

Theorem 6.2 of Part One has an interesting history, for it settles a problem first raised in 1893 by the famous mathematician, J. J. Sylvester. He may have had a solution, but the first published solutions to this specific problem appeared in response to Erdős's revival of the problem in 1943 (see [Erd1]). A solution also follows (using the duality of projective geometry) from a 1940 result of Melchior [Mel]. The elegant Euclidean proof given in Part One is due to L. M. Kelly (see [Cox1]). For an account of more recent research on relatives of Sylvester's problems, see the survey papers of Borwein and Moser [BM] and Erdős and Purdy [EP2]. In particular, Dirac [Dir1] and Motzkin [Mot] both conjectured in 1951 that for a noncollinear set of n points in the plane, there should be at least $\lfloor n/2 \rfloor$ ordinary lines. Kelly and Moser [KM] came close to proving this, with their $3n/7$ result mentioned in Part One. The final solution with $\lfloor n/2 \rfloor$ was perhaps obtained in the doctoral dissertation of S. Hansen [Han′3], with the result that except when $n = 7$ and $n = 13$, there must be at least $n/2$ ordinary lines. However, since his solution is so long and is not widely available, and since several mathematicians have had difficulty in following its details, it would be worthwhile to check it carefully and look for ways of shortening the argument. A more recent independent approach of Csima and Sawyer [CS′] provides a readable proof that there are at least $6n/13$ ordinary lines, and also provides verification of one of the main results stated by Hansen on the way to the $\lfloor n/2 \rfloor$ result.

Our proof of Theorem 6.4 is similar to that of de Bruijn and Erdős [DE1]. Theorem 6.5 resulted from the combined efforts of Kelly and Moser [KM] and Elliott [Ell]. Elliott was concerned primarily with the circles determined by the triples from a set S of n points in the plane, assuming of course that S is neither collinear nor concyclic. He stated that for each such S, the total number of ordinary circles (those containing exactly three points of S) is at least $2n(n-1)/63$, and each point of S lies on at least $2(n-1)/21$ such circles. (The existence of an ordinary circle was first noted by Motzkin [Mot].) W. Webber has remarked that while Elliott's arguments do indeed show the total number of ordinary circles, and the number through a given point of S, to be bounded below by, respectively, quadratic and linear functions of n, they do not establish the specific bound for the total number of circles given by Elliott. It would be of interest to find best-possible results on these numbers. Bálintova and Bálint [BB] showed that each point of S lies on at least $15(n-1)/133$ ordinary circles, and for $n \geq 6$ the total number of ordinary circles is at least $(15n^2 - 15n + 1678)/266$. J. Pach remarked that if the Kelly–Moser bound $(3n/7)$ on the total number of ordinary lines is replaced by the bound $(\lfloor n/2 \rfloor)$ of Hansen, then the argument of [BB] shows that each point lies on at least $3(n-1)/19$ ordinary circles.

In view of the existence, under reasonable assumptions, of ordinary lines and ordinary circles, it is natural to wonder about *ordinary parabolas* and *ordinary conics,* with the natural definitions involving four points and five points, respectively. The case of parabolas is open but that of conics has been settled, for Wiseman and Wilson [WW] proved that if a finite plane set S is not contained in any conic, then there is a conic C such that C contains precisely five points of S and is determined by these points. They conjectured that this conclusion extends to plane algebraic curves of higher order. That is, they expect an affirmative answer to the following question.

Problem 6.4

> Suppose that S is a finite plane set that is not contained in any algebraic curve of order n. Must there exist such a curve that contains exactly $n(n+3)/2$ points of S and is determined by those points?

We have already referred to two publications of Grünbaum [Grü4, 5] for specific examples related to the main problems of this section. However, it should be added that these papers contain many other results and problems that were new at the time (1971), and stimulated much of the later work on arrangements of lines in the plane. In fact, these two papers are still excellent sources of unsolved problems in the spirit of our present exposition. (As was noted in a review of [Grü5], it "is packed with interesting theorems and conjectures, enough to occupy a whole generation of future geometers." [Cox2])

There are higher-dimensional theorems and unsolved problems related to Problem 6 and to Sylvester's problem. The result on the existence of ordinary lines extends immediately to $\mathbb{R}^d$ (Exercise 8), but how about ordinary planes for a finite noncoplanar subset of $\mathbb{R}^d$? One might be tempted to call a plane "ordinary" if it contains only three points of the determining set. However, if the given points lie on two skew lines in $\mathbb{R}^3$ and there are at least three of the points on each line, then there are no ordinary planes in this sense. (See [EP2] for references to other examples.) Motzkin [Mot] therefore called a plane P *ordinary* with respect to a given set S in $\mathbb{R}^3$ if P is determined by some three points of S and there is a line L in P that contains all but one point of $S \cap P$. More generally, when $d \geq 2$, H is a hyperplane in $\mathbb{R}^d$, and S is a subset of $\mathbb{R}^d$ that does not lie in any hyperplane, H is *ordinary* with respect to S if H is the affine hull of some $d+1$ points of S and there is a $(d-2)$-flat in H that contains all but one point of S. Hansen showed for each d that there always exists an ordinary hyperplane, and that when S consists of n points in $\mathbb{R}^3$ there are at least $\frac{2}{5}n$ ordinary planes [Han1, 2]. Probably the constant $\frac{2}{5}$ can be improved.

Purdy [Pur2] proved the existence of positive constants c and k having the properties expressed in the next theorem.

Theorem 6.5. *For an arbitrary set of n points in $\mathbb{R}^3$, not all on a plane and not all contained in two skew lines, let l and p denote, respectively, the number of lines and the number of planes determined by these points. Then*

(a) *one of the points is incident to at least cl of the planes;*

(b) $l^2 \geq knp$.

Problem 6.4

What are the best values of the constants c and k in Theorem 6.5?

The range of indecision for the constant c is as great as that for Problem 6.3, because the proof of 6.5(a) is based on the theorem of Beck [Bec] and Szemerédi and Trotter [ST] mentioned earlier. However, as was mentioned by Purdy [Pur2], there is reason to believe that the best value for k is at least 1 and perhaps even 1.5. Another attractive 3-dimensional incidence problem is the following, due to Erdős and Purdy [EP1, Pur1] who established an affirmative answer when no three of the points are collinear.

Problem 6.5

With hypotheses and notation as in Theorem 6.5, is it always true that $n-l+p \geq 2$? If not, is it at least true that $n-l+p \geq 0$ for all sufficiently large n?

Concerning points and lines in the plane or in higher-dimensional spaces, there is a bewildering variety of problems, both solved and unsolved, that are similar in spirit of the ones considered here except that they involve such notions as direction, distance, angle, and area—all less purely combinatorial than the notion of incidence. See Erdős and Purdy [EP2] for an excellent survey of such problems. We end this section by discussing two of them. The following problem was raised by P. R. Scott [Sco] in 1970: What is the minimum number of directions that can be determined by a noncollinear set of n points in the plane? Here each direction is represented by an unoriented line through the origin.

It's easy to arrange n points in the plane so that no two connecting lines are parallel; then the number of directions is $\frac{1}{2}n(n-1)$. As is seen in Exercise 12, it's also easy to arrange n noncollinear points so that they determine only $2\lfloor n/2 \rfloor$ directions. Scott conjectured that this is the minimum, but was able to prove only a much weaker lower bound (see Exercise 14). Ungar [Ung] eventually proved Scott's conjecture by using the allowable sequences of Goodman and Pollack [GP2] that were discussed in Part Two of Section 5. Ungar's theorem inspired the following problem of Jamison [Jam].

Problem 6.6

If S is a set of n points in the plane, no three of which are collinear, is it necessarily possible to arrange the points in a sequence $s_1, \ldots, s_n$ so that the $n-1$ segments $[s_{i-1}, s_i]$ all have different slopes?

Jamison conjectured that the answer is affirmative, even under the weaker assumption that no connecting line contains more than $\frac{1}{3}n$ of the points. He also proved the following.

Theorem 6.6. *To any noncollinear set of n points in the plane it is possible to add $n-1$ line segments (joining points of the set) so as to form a "direction tree"—a graph that is a tree in which no two edges are parallel.*

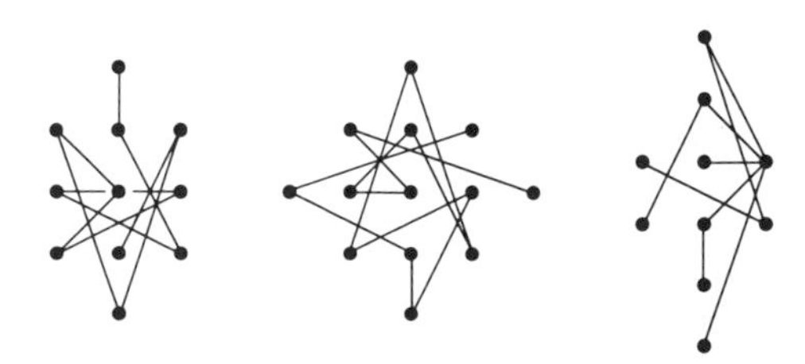

FIGURE 6.5
A direction tree and two direction paths.

Here the word "tree" is used in the usual graph-theoretic sense: a graph that is connected and whose number of edges is one less than its number of vertices. Of the direction trees (adapted from [Jam]) shown in Figure 6.5, the last two are *direction paths*—direction trees of the sort requested in Problem 6.6. However, no direction path exists for the set of ten points in the first part of Figure 3.5.

Each line L in the plane separates the plane into two open half-planes L^- and L^+. Let us say that a finite plane set X is *k-balanced* with respect to L if the two sets $X \cap L^-$ and $X \cap L^+$ differ in cardinality by at most k.

Problem 6.7

Does there exist an integer k such that every finite plane set X of two or more points admits a connecting line with respect to which X is k-balanced? What about $k = 2$?

It is easy to see that $k = 1$ works whenever X is in general position (Exercise 15). Y. Kupitz conjectured that $k = 1$ always works, but counterexamples were found by N. Alon and D. Wilbour. M. Perles proved that if $b(n)$ denotes the smallest value of k that works for all sets of n points, then $b(n) = O(\log n)$; that is, the quotient $b(n)/\log n$ is bounded.

Exercises

9. Prove that for n noncollinear points in $\mathbb{R}^d$, there is at least one ordinary line. (Hint: Project onto an appropriate plane.)

10. Suppose that J and K are two skew lines in $\mathbb{R}^3$, and that the finite set S consists of j points on J and k points on K. How many lines are determined by pairs of points of S? How many ordinary lines? How many planes are determined by triples of points of S.? How many ordinary planes?

11. Prove that if n points in $\mathbb{R}^d$ do not lie in any hyperplane, then they determine at least n ordinary hyperplanes. (Hint: Imitate the proof of Theorem 6.4, and use the fact that there must be at least one ordinary hyperplane.)

12. Show that if V is the vertex-set of a regular n-gon, then n is the number of directions of the connecting lines of V, and if V is the vertex-set of a regular n-gon plus the center point, then $2\lfloor n/2 \rfloor$ is the number.

13. With S as in Exercise 1, assume in addition that with respect to J and also with respect to K, q is not between any two points of S. Prove that the

number of directions of lines determined by two points, one of which belongs to $J \cap S$ and the other to $K \cap S$, is at least $j + k + 1$.

14. For an arbitrary noncollinear set S of n points in the plane, use the result of Exercise 13 to show that the number d of directions of connecting lines is at least $\sqrt{n}/2$. (Hint: Let q be a vertex of S's convex hull and let $L_1, \ldots, L_r$ denote the lines connecting q to the remaining points of the set. For $i = 1, \ldots, r$ let p_i denote the number of points of T on L_i. Apply the result of Exercise 13.)

15. Show that if X is a plane set that is in general position, then each point of X lies on a connecting line with respect to which X is balanced. Conclude that there are at least $|X|/2$ such lines.

7. TILING THE PLANE

Problem 7

Is there a polygon that tiles the plane but cannot do so periodically?

Because tilings are useful in construction and for decorative purposes, interest in tilings goes back to antiquity. However, the concept of an *aperiodic tiling* (one whose prototiles cannot be used to produce any periodic tiling) is a relative newcomer on the mathematical scene. For almost all aspects of the theory of tilings of the plane, the book of Grünbaum and Shephard [GS4] is unrivaled in its accuracy and completeness. With their kind permission, we have borrowed freely from their exposition. Theorem 7.4 is theirs, and Problems 7–7.4 all appear (sometimes in a different form) in their book. Theorems 7.1 and 7.2 are due to Reinhardt [Rei1] (see Niven [Niv] for a nice proof of 7.1). Problem 7.4 and the pentagonal example associated with it are due to Heesch [Hee], and Figure 7.5 is from [GS4, p. 156]. The term *p-hexagon* was initiated by Kuperberg [Kup], and Figure 7.4 is from a survey article by him and G. Fejes Tóth [FK] that describes the effective use of p-hexagons in the study of packings and coverings by plane convex bodies.

We concentrated in Part One on Problem 7.1 (which is essentially the restriction of Problem 7 to polygons that are convex), and we saw that it is closely related to the problem of finding all convex pentagons that tile the plane. That is because of Reinhardt's 1918 proof [Rei1] that for $n \neq 5$, each convex n-gon that can tile the plane monohedrally can also do so isohedrally and hence periodically (see also [HK′, Bol1]). Reinhardt's incomplete discussion of the case $n = 5$ was extended in 1968–69 by Kershner [Ker1, 2], who was the first to discover convex pentagons that admit monohedral periodic tilings but do not tile isohedrally. Grünbaum and Shephard [GS4] have a complete description not only of all the convex polygons that can tile isohedrally but also of the various types of isohedral tilings that they admit. According to their classification, there are 14 types of such isohedral tilings with triangles, 57 with quadrangles, 24 with pentagons, and 13 with hexagons. (They constructed 56 types for quadrangles, and an additional type was discovered by Shtogrin [Sht].)

Kershner claimed to describe all convex pentagons that tile monohedrally. His claim was correct so far as isohedral tilings are concerned, but renewed interest in the subject stimulated by M. Gardner [Gar2] led to the discovery of new anisohedral convex pentagons by R. James and M. Rice. An excellent survey of the pentagonal situation was provided by Schattschneider [Sch]. Although the regular pentagon does not tile, there are many equilateral convex pentagons that do tile. With the aid of a computer, Hirschhorn and Hunt [HH] found a presumably complete list of these, but their list has not been checked independently. In 1985, a new (not equilateral) convex pentagon that tiles monohedrally was found by R. Stein [Ste′]. For additional details, see [GS4].

As this is being written, all of the *known* convex pentagons that tile monohedrally also tile periodically. Hence if the list of such pentagons is now finally complete, it is true that Problem 7.2 has been solved, that the answer to Problem 7.3 is negative, and the answer to Problem 7.1 is 2 or 3. However, there is no published proof of the list's completeness, and because of the tricky nature of this subject, even a published proof should be viewed with suspicion until it has been checked by several experts in the field.

Since nonperiodic tilings are usually harder to construct than periodic ones, the following theorem could be useful in attacking unsolved problems on aperiodic tilings. It is a variant of the extension theorem of Grünbaum and Shephard [GS4, p. 161], and its proof is similar to theirs. (A *Jordan region* is a part of the plane that consists of all points on or inside a simple closed curve.)

Theorem 7.7. *Suppose that $R_1, \ldots, R_k$ are Jordan regions, and that for $1 \leq j \leq k$, $R_j^1, R_j^2, \ldots$ is a sequence of Jordan regions converging to R_j. Suppose that $\rho_1, \rho_2, \ldots$ is an unbounded sequence of positive numbers, and that for each i there is a collection $\mathcal{T}_i$ of nonoverlapping regions such that*

(i) *$\mathcal{T}_i$ covers the disk of radius ρ_i centered at the origin, and*

(ii) *each member of $\mathcal{T}_i$ is congruent to some R_j^i.*

Then some subset of $\{R_1, \ldots, R_k\}$ is the set of prototiles for a tiling $\mathcal{T}$ of the entire plane. If, moreover, there is a disk D such that for each i and j, D is intersected by a member of $\mathcal{T}_i$ congruent to R_j, then there exists $\mathcal{T}$ having all of $\{R_1, \ldots, R_k\}$ as its set of prototiles.

The condition that $R_j^1, R_j^2, \ldots$ converges to R_j may be defined in terms of the behavior of convergent sequences $p_1, p_2, \ldots$ such that $p_i \in R_j^i$ for all i. It amounts to saying that each such sequence converges to a point of R_j, and each point of R_j is the limit of such a sequence (see Exercise 10).

Now the time has come to describe some aperiodic sets of polygons. Grünbaum and Shephard [GS4] provide a comprehensive and readable account of the many sets now known, and some of the more popular sets are also discussed by M. Gardner [Gar3]. Hence we shall merely tell how to construct two aperiodic sets that are the simplest known in their respective categories. One consists of two nonconvex polygons, the other of two convex pentagons and a convex hexagon.

In Figure 7.6 (from [GS4, p. 542]) the angle θ is $\pi/5$ and the two rhombi have the same edge-length. Of course, there exist periodic tilings using one or both of these prototiles. However, the arrows on the edges and the black or white dots at the vertices are intended to restrict attention to edge-to-edge tilings in which, whenever an edge E is shared by two adjacent tiles T_1 and T_2, the following conditions are satisfied:

(i) whichever color (black or white) an end of E has as a vertex of T_1, it has the same color as a vertex of T_2;
(ii) if E carries an arrow as an edge of T_1, it carries a similarly directed arrow as an edge of T_2.

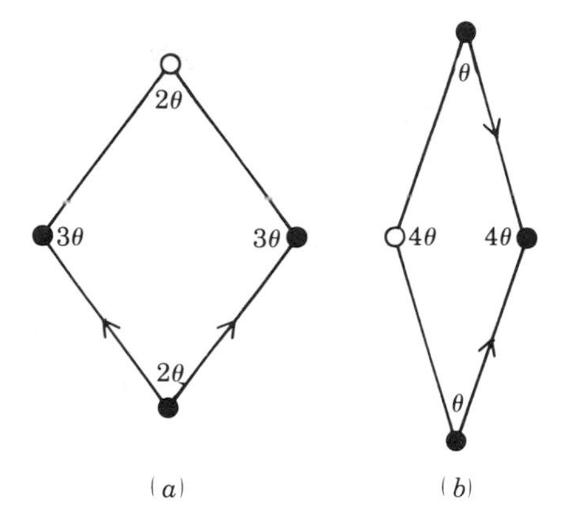

FIGURE 7.6
Penrose rhombs with identification rules indicated by arrows and colors ($\theta = \pi/5$).

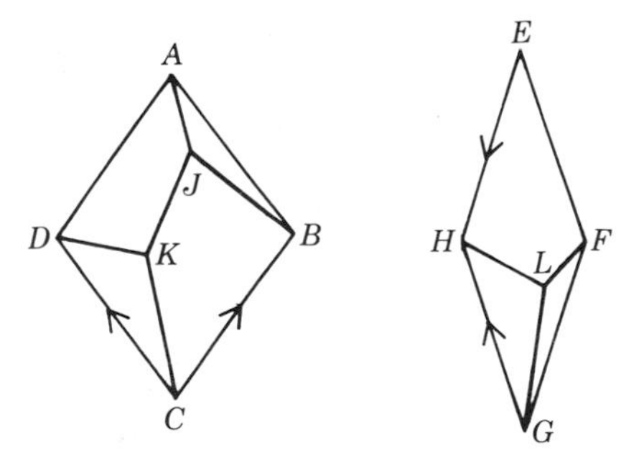

FIGURE 7.7
A dissection of the Penrose rhombs.

These tiles are called the *Penrose rhombs* for R. Penrose, who established that, subject to the restrictions (i) and (ii), they tile only nonperiodically. It is shown in Grünbaum and Shephard [GS4] how, by suitably indenting or "outdenting" the edges of the two rhombi, matching rules imposed by the colors and arrows can be enforced instead by shape alone. When suitable polygonal indents and outdents are used, the result is an aperiodic set consisting of two nonconvex polygons.

Figures 7.7 and 7.8 are from Grünbaum and Shephard [GS4, p. 548]. They show how, by dissecting the Penrose rhombs into convex polygons and then recomposing these in a different manner (due to R. Ammann), one can obtain an aperiodic set consisting of two convex pentagons and a convex hexagon. The aperiodicity of this set of three convex tiles is enforced by shape alone.

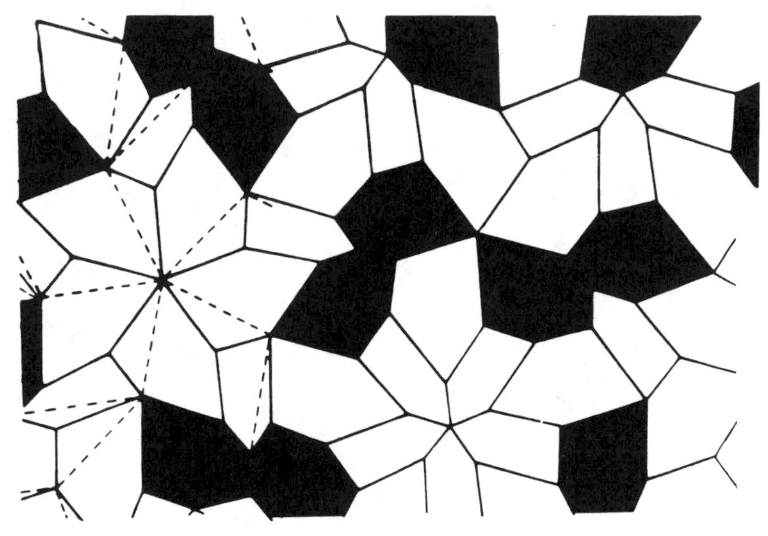

FIGURE 7.8
An aperiodic tiling due to Ammann. The three prototiles arise from the dissection of the Penrose rhombs shown in Figure 7.7, followed by recomposition into a convex hexagon and two convex pentagons. To obtain, from shape alone, an effect equivalent to that of the matching rules for Figure 7.6, the point L in Figure 7.7 must be chosen so that the segments GL, HL, and JK are of different lengths.

Problems 7 and 7.1 were motivated by measuring the complexity of an aperiodic set of polygons by the number of polygons in the set. Another reasonable measure would be the total number of edges used, which is 16 for the three prototiles of Figure 7.8.

Problem 7.6

For an aperiodic set of polygons, what is the minimum total number of edges? What is the minimum for aperiodic sets of convex polygons?

See Girault-Beauquier and Nivat [GN] for a different approach to the problem of finding simplest aperiodic sets.

We mentioned in Part One that if a polygon tiles isohedrally, then it does so periodically. A little reflection shows that the set G of symmetries of an isohedral tiling of the plane by polygons satisfies the following two conditions in addition to the two expressed in Exercise 7.2:

(i) there is a bounded part B of the plane such that the plane is covered by B's images under the various members of G;
(ii) there is a positive number δ such that $\|p - g(p)\| \geq \delta$ whenever $g \in G$ and p is a point of the plane for which $p \neq g(p)$.

The sets of plane isometries satisfying these conditions are called the *plane crystallographic groups* or the *wallpaper groups.* It is well known that there are (up to isomorphism) exactly 17 such groups (see [GS4] for many references related to this), and that each of them includes translations in two independent directions (whence isohedrality of a tiling implies periodicity).

In Exercise 7.4 we saw that a given prototile may be used in monohedral tilings having a variety of symmetry groups. Grünbaum and Shephard [GS3] called a polygon P an *all-purpose tile* if for each wallpaper group G there is a tiling $\mathcal{T}$ such that P is $\mathcal{T}$'s prototile and G is $\mathcal{T}$'s symmetry group. They proposed the following problem.

Problem 7.7

Does there exist an all-purpose tile? If not, which prototiles attain the maximum number of different symmetry groups?

They noted that a 30°-60°-90° triangle can be used to realize 14 of the 17 groups, and this triangle can be used together with a 15°-45°-120° triangle to realize the remaining 3 groups. They conjectured that no polygonal tile other than a 30°-60°-90° triangle can be used monohedrally to realize 14 or more of the wallpaper groups. It seems that the number of interesting unsolved problems about tilings of the plane is endless. As sources of additional such problems, we mention especially the book, *Tilings and Patterns* [GS4], and the papers by Grünbaum and Shephard [GS2, 3] and by Danzer, Grünbaum, and Shephard [DGS1, 2].

Before leaving the plane, we want to complete the discussion of equidissectability that was begun in Part One. The fact that a square fails to be k-equidissectable for odd k was first suspected by Richman and Thomas (see [Tho]) and established by Monsky [Mon1]. Later results of Kasimatis [Kas], S. Stein [Ste″2], and Monsky [Mon2] completed the proof of Theorem 7.5. An early form of Theorem 7.7 is due to Hales and Straus [HS], a sharper version to Kasimatis and Stein [KS]. Monsky's result was extended to higher dimensions by Mead [Mea], who showed that for the dissectability of a d-cube into k d-simplices of equal volume it is necessary and sufficient for k to be divisible by $d!$.

Perhaps the following should be regarded as the central problem in the theory of equidissectability of convex polygons.

Problem 7.8

If one is given a pair (n, k) and the coordinates of the successive vertices of a convex n-gon in the plane, how efficiently can P be be tested for k-equidissectability?

Because the property of k-equidissectability can be expressed in terms of polynomial equalities and inequalities involving the vertex-coordinates, it follows from Tarski's decision theory for real-closed fields [Tar2, CK] that algorithms for testing the property do exist. However, except for the special case treated in Exercises 8 and 12, it is not clear that the algorithms are really practical for this problem. (But see Renegar [Ren] for one of the best algorithmic implementations of ETR, the existential theory of the reals.) In particular, there are even trapezoids for which the question of equidissectability is unsettled.

The notion of tiling (and related definitions) can, of course, be formulated on the sphere and in higher-dimensional Euclidean spaces. Here is an attractive problem of L. Fejes Tóth [Fej]. (See Exercise 13 for the relationship between n and k.)

Problem 7.9

Let k great circles in general position (no three having a common point) divide a sphere into $n = k(k-1)+2$ spherical polygons of areas $\alpha_1, \dots, \alpha_n$. Must it be true that

$$\frac{\max\{\alpha_1\}}{\min\{\alpha_1\}} \to \infty \qquad \text{as } k \to \infty?$$

Fejes Tóth proves the existence of a constant k_0 such that

$$\frac{\max\{\alpha_1\}}{\min\{\alpha_1\}} > 7.43 \qquad \text{whenever } k > k_0.$$

Tilings of $\mathbb{R}^3$ have been much studied because of connections with crystallography. However, some of the simplest and most natural problems are still open. We shall merely state a few such problems, and refer to the survey article of Grünbaum and Shephard [GS1] for more details and additional references and problems.

Problem 7.10

Which tetrahedra tile 3-space monohedrally?

See Senechal [Sen] for a fascinating account of this problem's 2000-year history. The problem may eventually turn out to be barely tractable, but its difficulty is indicated by the fact that even the following is open.

Problem 7.11

Is there a tetrahedron that tiles $\mathbb{R}^3$ monohedrally but cannot do so periodically?

Danzer [Dan] has produced an aperiodic set that consists of four tetrahedra with associated matching conditions (analogous to the matching conditions for the Penrose rhombs). Problem 7.11 asks whether there is an aperiodic set that consists of a single tetrahedron without any matching conditions. It is unknown whether there is any convex 3-polytope that has this property. However, it is known for each dimension d that if $\mathbb{R}^d$ can be tiled by *translates* of a convex d-polytope P, then there is also a lattice tiling by translates of P (and hence P tiles periodically) (Venkov [Ven], McMullen [McM]). For examples of geometrically simple (though nonconvex) 3-dimensional bodies that tile by translation and yet do not admit lattice tilings (and for other interesting open problems), see Bezdek and Kuperberg [BK] and their references.

For $k \geq 5$, there is a great variety of space-fillers among the convex 3-polytopes with precisely k 2-faces. In fact, the variety is so great that there does not seem to be any hope of finding a usable intrinsic characterization of these polytopes. However, perhaps the following problem will prove to be barely tractable.

Problem 7.12

What is the largest k (if there is a largest) such that $\mathbb{R}^3$ can be tiled monohedrally by a convex polytope with k faces?

As we saw in Part One, 6 is the maximum number of edges for a polygon that tiles the plane monohedrally. The record for $\mathbb{R}^3$ is held at present by Engel [Eng′], who constructed a convex polytope with 38 2-faces that tiles 3-space, and even does so isohedrally.

As the dimension d increases, it becomes difficult to deal even with tilings by the most familiar of all d-dimensional geometric figures, the d-cube.

Problem 7.13

Is it true that in each tiling of $\mathbb{R}^d$ by congruent d-cubes, at least two of the tiles share a $(d-1)$-dimensional face?

For lattice tilings by congruent cubes, the answer is affirmative for all d (Hajos [Haj]), but without the lattice assumption an affirmative answer is known only for $d \leq 6$ (Perron [Per′]). S. Stein [Ste″1] surveys the interesting connections between cube-tiling and problems from number theory and algebra. Lawrence [Law] has an attractive purely graph-theoretic formulation that might, if accompanied by clever analysis and extensive computation, lead to further progress on Problem 7.13. Relative to a coloring of the edges of a graph, a *monochromatic odd cycle* represents a way of taking a walk along an odd number of edges, all of the same color, and returning to the starting vertex. Now with K_n denoting the complete graph on n vertices, consider the following assertion $L(n, c)$ (true for some pairs (n, c), false for others).

$L(n, c)$: For every way of coloring the edges of K_n with c colors, there is a monochromatic odd cycle or there is an edge e such that each way of changing e's color (to one of the remaining $c - 1$ colors) produces a monochromatic odd cycle.

Here is Lawrence's result.

Theorem 7.8. *For each $d \geq 2$, the answer to Problem 7.13 is affirmative if and only if the assertion $L(2^d, d)$ is true.*

See Corrádi and Szabó [CS1, 2] for alternative graph-theoretic formulations of Problem 7.13.

Exercises

10. (a) Assume that each nondegenerate p-hexagon (one with six sides) tiles the plane. Use this in conjunction with Theorem 7.7 and Exercise 7.7(a) to prove that every p-hexagon tiles the plane.

(b) Use Theorem 7.7 to prove that if $r = 1$ is indeed the only possibility in Problem 7.4, then the plane can be tiled by each convex polygon that can be surrounded by two successive zones formed from copies of itself.

11. Prove that each dissection of a convex n-gon P into $n-2$ or fewer triangles can be formed by adding $n-3$ segments so that each segment joins two of P's vertices and no two of the segments intersect except at a common endpoint.

12. Tell how to test a convex n-gon for k-equidissectability when $k \leq n-2$.

13. Suppose that k (≥ 1) great circles are used to dissect the sphere into n regions (the components of the complement of the union of the circles). Show that n must be even. Use Euler's theorem to show that $n = k(k-1)+2$ when the great circles are in general position.

14. (a) Investigate the truth of $L(n, c)$ for a few small values of n and c. In particular, show that $L(4, 2)$ is true but $L(4, 3)$ is false.

(b) Write a computer program that proves $L(8, 3)$.

8. PAINTING THE PLANE

Problem 8

What is the minimum number of colors for painting the plane so that no two points at unit distance receive the same color?

Problem 8 was posed in 1960–61 by M. Gardner [Gar1] and Hadwiger [Had2]. The four-color map problem was settled by Appel and Haken [AH] in 1976; see the books by Barnette [Bar] and Biggs, Lloyd, and Wilson [BLW] for a discusssion of its early history.

In Part One, a certain configuration of seven points was used to show very quickly that $t_2 \geq 4$. Now we want to outline the proofs of two less obvious results: $s_2 \geq 4$ and $u_2 \geq 6$. In the exercises, these inequalities are extended to $\mathbb{R}^d$ with the conclusions that $s_d \geq d+2$ and $u_d \geq 4d-2$.

The following proof is due to Raiskii [Rai]; a different proof of the result was given independently by Woodall [Woo].

Theorem 8.1. *If the plane is covered by three sets, at least one of them is a Δ-set; in other words, $s_2 \geq 4$.*

Outline of Proof. The proof uses the 7-point configuration of Part One in conjunction with the following strange lemma. The lemma's proof is not deep (it involves a use of "telescoping sums"), but it is omitted here.

Lemma. *If $f_1, \ldots, f_k$ are real-valued functions defined on $\mathbb{R}^2$ and $p_1, \ldots, p_k$ are points of $\mathbb{R}^d$ such that*

$$\sum_{i=1}^{k} [f_i(q) - f_i(q - p_i)] \geq 1$$

for each point q of $\mathbb{R}^2$, then at least one of the functions f_i is unbounded.

Now suppose that the plane is covered by three sets S_1, S_2, and S_3, none of which is a Δ-set, and assume without loss of generality that the sets are disjoint. For each i there is a positive number δ_i that is not realized as the distance between any two points of S_i. Let the set P_i be similar to the set $\{r, w, b, \overline{r}, w', b', \overline{r}'\}$ used in Part One, with the point r at the origin, but let P_i be formed from equilateral triangles of edge-length δ_i rather than 1. The following easily verified property of P_i is essential in the proof.

Among any three points of P_i there are two whose distance is δ_i.

For $1 \leq i \leq 3$, let $p_{i1}, \ldots, p_{i7}$ be the points of P_i, and define $f_{ij}(q)$ for each $q \in \mathbb{R}^2$ by setting $f_{ij}(q) = 0$ if $q + p_{ij} \in S_i$, and $f_{ij}(q) = 1$ otherwise. Note that for each fixed i and q, there are at most two zeros among the seven numbers $f_{ij}(q)$; for if there were three zeros then S_i would include three members of a translate of P_i and hence would include two points at distance δ_i. It follows that $\sum_{i=1}^{3}\sum_{j=1}^{7} f_{ij}(q) \geq 15$. Now fix the integer j, and note that by definition,

$$f_{ij}(q - p_{ij}) = 0 \quad \text{if and only if} \quad q \in S_i.$$

Hence (since the sets S_i are disjoint) the three numbers $f_{ij}(q - p_{ij})$ $(1 \leq i \leq 3)$ consist of a 0 and two 1s. Thus

$$\sum_{i=1}^{3}\sum_{j=1}^{7} f_{ij}(q - p_{ij}) = 14,$$

and for each point $q \in \mathbb{R}^2$, it is true that

$$\sum_{i=1}^{3}\sum_{j=1}^{7}[f_{ij}(q) - f_{ij}(q - p_{ij})] \geq 1.$$

Since each of the functions f_{ij} is bounded, this contradicts the lemma and completes the proof. □

Exercise 7 outlines a proof that for all d, $s_d \geq d + 2$. It is based on the immediate higher-dimensional analogue of the seven-point configuration used in $\mathbb{R}^2$ (the "spindle" of Moser and Moser [MM]). However, much stronger inequalities for large d have been obtained by using a variety of configurations in conjunction with the following powerful theorem of Larman and Rogers [LR].

If $\mathbb{R}^d$ contains a configuration of m points such that among each k of these points there are two at distance 1, then $s_d \geq m/k$.

This enabled Larman and Rogers [LR] and Larman [Lar] to show, respectively, that for all d,

$$s_d \geq d(d-1)/6 \quad \text{and} \quad s_d \geq (d-1)(d-2)(d-3)/178200,$$

and Frankl and Wilson [FW] to show that s_d grows exponentially with d. The results on asymptotic behavior may be summarized by saying that there are functions ξ_d and η_d such that $\xi_d \to 0$ and $\eta_d \to 0$ as $d \to \infty$, and such that for all d,

$$(1 + \xi_d)1.2^d \leq s_d \leq v_d \leq (3 + \eta_d)^d.$$

The proofs of these results are beyond the scope of our exposition.

For all $d \geq 2$, the best known lower bounds for t_d have in fact been established for s_d. Raiskii's bound of $d + 2$ is the best when $3 \leq d \leq 9$, but when $d \geq 10$ it is superseded by the quadratic bound of Larman and Rogers, which for larger values of d is overtaken first by the cubic bound of Larman and then by the exponential bound of Frankl and Wilson.

Recall that in the definition of t_d, the covering sets are unrestricted, while in the definition of u_d they are required to be closed. Of course, $t_d \leq m_d \leq u_d$, where in the definition of m_d the covering sets are required to be measurable. Falconer [Fal1] showed that $m_d \geq 5$, and Székely and Wormald [SW] made a comprehensive study of distance-realization in coverings of $\mathbb{R}^d$ by measurable sets.

The best lower bound on u_d that works for all d is $4d - 2$. It is due to Hadwiger [Had1], and comes from an extension of the proof of Theorem 8.2 to $\mathbb{R}^d$ (Exercise 11). This lower bound for u_d is not overtaken until $d = 25$ by the Larman–Rogers quadratic bound for s_d.

Theorem 8.2. *If the plane is covered by five closed sets, at least one of them includes two points at distance* 1; *that is,* $u_2 \geq 6$.

Outline of Proof. The argument uses some basic properties of compactness and connectedness, and also the following two lemmas.

Lemma. *If a circle is covered by two closed sets, and* $0 < \theta \leq \pi$, *then at least one of the sets includes two points at angular distance* θ.

Lemma. *If a square is covered by a finite number of closed sets, none of which intersects two opposite edges of the square, then there is a point that belongs to at least three of the sets.*

The first lemma is immediate from the fact that some point of the circle belongs to both sets. The second lemma, used here without proof, is a form of the Lebesgue tiling theorem that has played an important role in dimension theory (see [Tuc] for an elementary proof).

Using the two lemmas and some elementary properties of compactness and connectedness, we shall prove the following:

> *If a square* Q *of edge-length* 3 *is covered by five closed sets* $X_1, \ldots, X_5$ *then at least one of the sets includes two points at distance* 1.

Let us fix an arbitrary $\rho > 0$, and note that by the Heine-Borel theorem, each X_i is covered by the union Y_i of a finite number of disks of radius ρ centered in X_i. Since the X_is are compact and ρ is arbitrary, it will suffice to show that some Y_i

includes two points at unit distance. Note that Y_i has only a finite number of components, and each component is itself a closed set. Thus the collection of components of the various Y_is provides a covering of Q by finitely many closed sets.

Since the distance-function is continuous, each component realizes all distances between 0 and its own diameter. Thus if some component is of diameter at least 1, we have the desired conclusion. Suppose, on the other hand, that each component is of diameter less than 1, and let T be the square of edge-length 1 that is obtained by contracting Q toward its center. It follows from the second lemma that some point p of T belongs to at least three different components, and clearly these components are from three different sets Y_i—say, from Y_1, Y_2, and Y_3.

Now let C denote the circle of radius 1 centered at p. Then $C \subset Q$, and the proof is complete if X_1, X_2, or X_3 intersects C. In the remaining case, C is contained in $X_4 \cup X_5$, and, because $p \in T$, the desired conclusion follows from the first lemma. □

G. Wegner and T. Spieker have considered the analogue of Problem 8 in which *two* distances, δ and ϵ, are given, and we want to paint the plane so that two points receive different colors whenever the distance between them is either δ or ϵ. The minimum number of colors with which this can be accomplished depends on the ratio between δ and ϵ. They have examples of pairs of distances for which 6 colors are required, and they note that for any choice of two distances, no more than 49 colors are needed (see Exercise 13). Once again, a gap that begs to be closed!

We don't want to end the discussion of $\chi(\mathbb{R}^d)$ without mentioning the following theorem of Beckman and Quarles [BQ] and its curious relationship to Problem 8.

Theorem 8.3. *If $d \geq 2$ and $f : \mathbb{R}^d \to \mathbb{R}^d$ is a transformation that preserves unit distances, then f preserves all distances.*

There now exist short proofs of this theorem (e.g., [Elv]), and extensions to other spaces [Kuz, Len]. For references to several proofs, see Greenwell and Johnson [GJ], who raise the following problem.

Problem 8.4

For each dimension d, determine the set N_d of all dimensions n such that all distances are preserved by every unit–distance-preserving transformation of $\mathbb{R}^d$ into $\mathbb{R}^n$.

The restriction $d \geq 2$ is natural, because Theorem 8.3 fails when $d = 1$ (Exercise 8). The restriction $n \geq d$ is natural, because $\mathbb{R}^d$ contains a set of $d + 1$ points

$x_0, \ldots, x_d$ such that $\|x_i - x_j\|$ for all $i \neq j$ (Exercise 6), while $\mathbb{R}^n$ contains no such set when $n \leq d$. Nothing is known about Problem 8.4 beyond the following fact:

For each $d \geq 2$, N_d consists of all integers between d and some k_d with $d \leq k_d < \chi(\mathbb{R}^d) - 1$.

The relationship between k_d and $\chi(\mathbb{R}^d)$ was noticed by Dekster [Dek] and others. With $k = \chi(\mathbb{R}^d)$, let $X_1, \ldots, X_k$ be the disjoint sets in a 1-chromatic painting of $\mathbb{R}^d$, and let $p_1, \ldots, p_k$ be points of $\mathbb{R}^{k-1}$ such that the distance between any two of the p_is is equal to 1 (Exercise 6). Then define the transformation $f : \mathbb{R}^d \to \mathbb{R}^{k-1}$ by setting $f(x) = p_i$ whenever $x \in X_i$. It is clear that f preserves unit distances, and equally clear that it preserves *only* unit distances.

Since $d \leq k_d < \chi(\mathbb{R}^d)$, the number k_2 must be between 2 and 6. Not only is the precise value of k_2 unknown, but we do not even know whether $k_d = d$ for all $d \geq 2$. In other words, does there exist a pair (d, n) with $n > d$ such that all distances are preserved by every unit–distance-preserving transformation of $\mathbb{R}^d$ into $\mathbb{R}^n$?

Among other efforts to determine δ-chromatic numbers, we mention three directions of special interest. Chilakamarri [Chi3] extended the Euclidean result by showing that for an arbitrary 2-dimensional normed vector space, the 1-chromatic number is between 4 and 7. His paper contains other interesting results, and ends with a collection of unsolved problems.

Benda and Perles [BP] noted that for each d, $\chi(\mathbb{R}^d, 1) = \chi(\mathbb{A}^d, 1)$, where $\mathbb{A}^d$ is the subset of $\mathbb{R}^d$ consisting of the points whose coordinates are all algebraic numbers. (This follows from Tarski's theory of real-closed fields (see [CK]).) They also studied the 1-chromatic number of $\mathbb{Q}^d$, the subset of $\mathbb{R}^d$ consisting of the points whose coordinates are all rational (note that $\chi(\mathbb{Q}^d, 1) = \chi(\mathbb{Q}^d, \delta)$ for all rational $\delta > 0$). They showed by elementary number-theoretical arguments that $\chi(\mathbb{Q}^2, 1) = 2$ (discovered earlier by Woodall [Woo]—see Exercise 5), $\chi(\mathbb{Q}^3, 1) = 2$, and $\chi(\mathbb{Q}^4, 1) = 4$. (See [Chi1, 2, Fis, Joh, Zak 2, 3, 4] for other work in this direction.) The result for $\mathbb{Q}^4$ is especially interesting, because Zaks [Zak3] later showed that in each 1-chromatic painting of $\mathbb{Q}^4$ in four colors, each of the color classes must be dense in the space.

Although $\chi(\mathbb{Q}^d, 1)$ has been determined for $d \leq 4$, the following is open.

Problem 8.5

For $d \geq 5$, determine $\chi(\mathbb{Q}^d, 1)$.

A lower bound for $\chi(\mathbb{Q}^d, 1)$ is provided by $\omega(\mathbb{Q}^d, 1)$, defined as the maximum cardinality of a subset X of $\mathbb{Q}^d$ such that the distance between any two points of X is 1. It is well-known that $\omega(\mathbb{R}^d, 1) = d + 1$. Chilakamarri [Chi1] showed that for even d, $\omega(\mathbb{Q}^d, 1)$ is $d+1$ or d according as $d+1$ is or is not a perfect square; and for odd d, $\omega(\mathbb{Q}^d, 1)$ is $d-1$, d, or $d+1$ according to conditions that depend on whether $(d+1)/2$ is a perfect square and also on the solvability of a certain Diophantine equation. In particular, $\omega(\mathbb{Q}^5, 1) = 4$. However, $\chi(\mathbb{Q}^5, 1) \geq 6$ [Chi2].

Now let $\mathbb{S}^{d-1} = \{x \in \mathbb{R}^d : \|x\| = 1\}$, the unit sphere of $\mathbb{R}^d$. It is obvious that $\chi(\mathbb{S}^{d-1}, \delta) = 1$ for $\delta > 2$, and $\chi(\mathbb{S}^{d-1}, 2) = 2$.

Problem 8.6

Is it true that $4 \leq \chi(\mathbb{S}^{d-1}, \delta)$ for all $\delta < 2$?

Simmons [Sim] conjectured that the answer to Problem 8.6 is affirmative, and he proved this for $\delta \leq \sqrt{3}$. He conjectured that $\chi(\mathbb{S}^{d-1}, \sqrt{3}) = 4$ and proved that $\chi(\mathbb{S}^{d-1}, \sqrt{2}) = 4$. The number $\chi(\mathbb{S}^{d-1}, \sqrt{2})$ is of special interest, for it is the smallest number of colors that can be used to paint all the unit vectors in $\mathbb{R}^d$ in such a way that no two orthogonal vectors receive the same color.

Problem 8.7

Determine $\chi(\mathbb{S}^{d-1}, \sqrt{2})$ for $d \geq 4$.

It would not surprise us if the solutions of Problems 8.5–8.7 require a deep combination of methods from algebra, number theory, and combinatorics.

We end the discussion of δ-chromatic numbers by noting that for any set X on which a notion of distance is defined, the δ-chromatic number of X is equal to the supremum of the δ-chromatic numbers of the finite subsets of X (de Bruijn and Erdős [DE2]). In particular, to show that the entire plane can be painted in k colors so that no two friends have the same color, it would suffice to prove this for an arbitrary finite subset of the plane. Erdős, Harary, and Tutte [EHT] defined the *dimension* of a graph G to be the smallest d for which there exists an embedding of G's vertices in $\mathbb{R}^d$ such that two vertices are adjacent in G if and only if their images are at distance 1 in $\mathbb{R}^d$. In these terms, t_2 is the maximum of the chromatic numbers of the graphs of dimension 2. Purdy and Purdy [PP] described their use of a supercomputer in searching for minimal graphs of dimension 2.

There are many other covering problems that involve distance and are similar in spirit to the ones considered here. For example, a result of Erdős and Kakutani [EK] (based on the continuum hypothesis) implies that the real line $\mathbb{R}$ can be covered by countably many sets $S_1, S_2, \ldots$, none of which includes two different point-pairs with the same distance. (That is, if w, x, y, and z all lie in the same set

S_i, and if the distances wx and yz are equal, then $\{w, x\} = \{y, z\}$.) It is apparently unknown whether such a covering is admitted by $\mathbb{R}^d$ for any $d \geq 2$.

Exercises

5. (Woodall [Woo]) Prove that $\chi(\mathbb{Q}^2, 1) = 2$.

6. Prove that for each integer $d \geq 2$ and each positive real δ, the space $\mathbb{R}^d$ contains a set of $d+1$ points whose distance-set is $\{\delta\}$, and also contains a set P of $2d+3$ points such that among any three points of P there are two whose distance is δ.

7. (Raiskii [Rai]) Let s_d denote the minimum number of sets into which $\mathbb{R}^d$ can be decomposed so that in each set, at least one distance is omitted (i.e., none of the sets is a Δ-set). Prove that $s_d \geq d+2$.

8. Describe a one-to-one $f : \mathbb{R} \to \mathbb{R}$ such that f preserves all integral distances but does not preserve all distances.

9. (Hadwiger, Debrunner and Klee [HDK]) Prove that if a circle of radius 1 is covered by three closed sets, then at least one of the sets includes two points at distance $\delta = \sqrt{3}$. Show that this conclusion fails for every other positive value of δ.

10. Prove that if a square of edge-length 3 is covered by six closed sets, then at least one of the sets includes two points at distance 1 or two points at distance $\sqrt{3}$.

11. (Hadwiger [Had1]) Below are the d-dimensional analogues of the lemmas used in proving Theorem 8.2. The first lemma appears in [Had1], and the second can be found in many books on topology (e.g., [HW′]). Using these lemmas, prove that $u_d \geq 4d - 2$.

(1) If an m-dimensional Euclidean sphere is covered by $3m - 1$ closed sets, and $0 < \theta < \arccos(-1/m)$, then at least one of the sets includes two points at angular distance θ.

(2) Let $[0,1]^d$ denote the d-cube consisting of all points $x = (x_1, \ldots, x_d) \in \mathbb{R}^d$ such that $0 \leq x_i \leq 1$ for all i. For each i, let $F_i^- = \{x \in \mathbb{R}^d : x_i = 0\}$ and $F_i^+ = \{x \in \mathbb{R}_i : x_i = 1\}$. In any covering of $[0,1]^d$ by finitely many closed sets, none of which intersects both an F_i^- and the corresponding F_i^+, there are $d+1$ sets that have a point in common.

12. Use the method of Exercise 11 to prove that for an arbitrary d-dimensional normed vector space M with unit sphere $S = \{x \in M : \|x\| = 1\}$,

$$\overline{\chi}(M) \geq d + 1 + \chi(S, 1).$$

Using this inequality, show that

$$\overline{\chi}(M) \geq 6 \text{ when } d = 2, \quad \text{and} \quad \overline{\chi}(M) \geq d + 5 \text{ when } d \geq 3.$$

13. (G. Wegner) Show that for any choice of k distances $\delta_1, \ldots, \delta_k$, the plane can be painted in 7^k or fewer colors in such a way that no δ_i is realized as the distance between two points of the same color.

14. (Benda and Perles [BP], Chilakamarri [Chi1]) A famous theorem of Lagrange asserts that each positive integer is the sum of four squares (see [HW], or Part One of Section 20 in Chapter 2). Use this to show that when $d \geq 5$, any two points x and y of $\mathbb{Q}^d$ can be joined by a finite sequence $x = p_0, p_1, \ldots, p_k = y$ of points of $\mathbb{Q}^d$ such that $\text{dist}(p_i, p_{i-1}) = 1$ for $1 \leq i \leq k$. (It is known [BP, Chi1] that the conclusion fails for $d \leq 4$. Number-theoretic methods are used for $d = 1, 2$ by [BP] and for $d = 3, 4$ by Fischer [Fis] to provide a complete characterization of the possibilities for y when x is the origin.)

9. SQUARING THE CIRCLE

Problem 9

Can a circle be decomposed into finitely many sets that can be rearranged to form a square?

Tarski's circle-squaring problem is closely related to the famous Banach–Tarski paradox in three-dimensional space. This "paradox," discovered by Poland's Stefan Banach and Alfred Tarski in 1924 (and closely related to earlier work by Felix Hausdorff) sounds, on the face of it, absurd. Their result states (and proves!) that any two bodies in $\mathbb{R}^3$, say a ball the size of a pea and a ball the size of the sun, are equidecomposable. This assertion seems ridiculous; after all, the volumes of the pieces of the pea will add up to the volume of the pea, and volume is preserved under rigid motions, so the pieces can't be moved so that they combine to form the sun. But the notion of volume is a subtle one and, at the beginning of the twentieth century, mathematicians learned that not all sets of points can be measured (i.e., assigned a number in a way that conforms to our intuitive understanding of volume). The so-called nonmeasurable sets are sets far removed from physical reality and the reasoning underlying the preceding explanation of why the Banach–Tarski paradox cannot be true implicitly assumes that the pieces are measurable. Using nonmeasurable pieces, however, many weird and wonderful constructions are possible. (In fact, even without using nonmeasurable sets, it is possible to construct counterintuitive examples in a very explicit manner; see Exercise 5 for a famous example of a set that is congruent to half of itself.) See [Wag] for a complete discussion of the Banach–Tarski paradox and related mathematical topics. The BTP implies not only that a 3-dimensional ball is equidecomposable with a cube, but also that the ball is equidecomposable with any cube, no matter how large or small! A deep theorem of Banach, proven in 1923, asserts that no such paradoxical decompositions exist in the line or plane. He derived this from the following result (proof in [Wag, Chap. 10]) on the existence of an invariant measure in $\mathbb{R}^1$ and $\mathbb{R}^2$.

Theorem 9.2. *If $n = 1$ or 2, there is a function f (called a measure) assigning values in $[0, 1)$ to all bounded subsets of $\mathbb{R}^n$ and satisfying the following conditions:*

1. *f agrees with the notions of length (respectively, area) for those sets having area in the classical sense (i.e., intervals, polygons, circles);*
2. *f is finitely additive; that is, the measure of the union of a disjoint sequence of finitely many sets is the sum of the measures of the individual sets; and*
3. *f is invariant under rigid motions; that is, congruent sets receive the same measure.*

The existence of a measure as provided by the preceding theorem implies, by a formalization of the intuitive argument following the statement of the Banach–Tarski paradox, that no paradoxical decompositions exist for an interval or for a square or for a circle. The essential difference between the line and the plane, for which no paradoxes exist, and three- and higher-dimensional spaces, for which paradoxical decompositions do exist, is that the group of rigid motions of n-space is solvable if $n = 1$ or 2 and is highly nonsolvable (precisely, has free subgroups of rank 2) if $n \geq 3$.

Now, the motivation for Tarski's problem [Tar1] is clear. In 1925 he knew that a circle could not be equidecomposable with a square whose area was less than or greater than the circle's area. But could it be equidecomposable with a square of exactly the same area? This is the content of Problem 9, which, sixty years later, is still wrapped in mystery.

The situation for polygonal equidecomposability is well understood. By Theorem 9.2 equidecomposable polygons must have the same area and, by a variation of Theorem 9.1 due to Tarski (see Exercise 4), whenever two polygons have the same area the fact that they are congruent by dissection can be used to show that they are equidecomposable. In other words, the Wallace–Bolyai–Gerwien theorem is valid even if every last point is taken into account when constructing the decompositions.

Despite the Banach–Tarski paradox, a variation of Problem 9 can be formulated for 3-space. The third problem in the famous 1900 list of Hilbert was to show that the analogue of the computation of the area of a triangle by transforming it into a rectangle (see Exercise 1) is false for a regular tetrahedron. This was the first of the Hilbert Problems to be solved, as Max Dehn proved in 1900 that the regular tetrahedron is not congruent by dissection (i.e., transformable using subpolyhedra) with a cube. It had been known that some (nonregular) tetrahedra are congruent by dissection with a cube, and Dehn's result was extended by Sydler in 1965 to yield a necessary and sufficient condition for a polyhedron to be congruent by dissection with a cube. However, the Dehn–Sydler condition is abstract and not easy to apply in practice, and many unsolved problems remain; for example: Is there a purely geometric description of the tetrahedra that are equidecomposable with a cube? See [Bol′] for a lucid exposition of these topics.

Because of the Banach–Tarski paradox we know that if completely general sets are allowed as pieces, then any tetrahedron is equidecomposable with any cube, independent of their sizes. But the following question, which seems similar to Tarski's circle-squaring problem, is unsolved.

Problem 9.1

Is a regular tetrahedron equidecomposable to a cube using pieces that are measurable sets?

Only a few known results are relevant to Problem 9. One important theorem, discovered by Dubins, Hirsch, and Karush in 1963 [DHK], states that the circle is not equidecomposable with a square provided topologically simple pieces are used. More precisely, they proved that a circle is not "scissors-equidecomposable" with a square; that is, it is impossible to partition the circle into pieces each of which consists of the interior and boundary of a simple, closed curve. Recently Richard Gardner [Gar′1] has obtained additional negative results. He showed that the circle cannot be squared if the motions are restricted to a discrete group; a special case is that the circle cannot be squared if only rigid motions of the form $(x, y) \mapsto (x + p, y + q)$ are used where p and q are rational numbers. Perhaps it is possible to prove that the circle cannot be squared using only translations, but even this simple-sounding statement is unproven. Gardner has observed that his result about discrete groups implies that circle-squaring is impossible using translations and only three pieces. To see this, assume that one translation is the identity and the others translate by (a_1, a_2) and (b_1, b_2). These translations generate a discrete group unless $a_1 = b_1 = 0$ or $a_2 = b_2 = 0$. Assume the former. Then vertical lines are preserved by the translations. Banach's measure in $\mathbb{R}^1$ (Theorem 9.2) then implies that the vertical sections of the square have the same lengths as the vertical sections of a circle, which is absurd since these lengths for a square form a piecewise linear function, which $\sqrt{1 - x^2}$ is not.

In 1929 John von Neumann proved that paradoxes in the plane do exist (and hence the circle can be squared) provided one enlarges the set of transformations to include area-preserving shears, that is, linear transformations represented by 2×2 matrices of determinant ± 1. For a proof that the addition of the single shear given by $s(x, y) = (x + y, y)$ suffices, see [Wag, pp. 99–100].

Finally, we mention a one-dimensional version of Problem 9 (due to C. A. Rogers) that is unsolved[1] and that seems to be about as difficult as the circle-squaring problem.

Problem 9.2

Let A be the following union of infinitely many open intervals

$$\left(\frac{1}{3}, \frac{2}{3}\right) \cup \left(\frac{7}{9}, \frac{8}{9}\right) \cup \left(\frac{25}{27}, \frac{26}{27}\right) \cup \left(\frac{79}{81}, \frac{80}{81}\right) \cup \cdots.$$

A has total length equal to $\frac{1}{2}$. Is A equidecomposable to the interval $(0, \frac{1}{2})$?

---

[1] This problem has been solved by Laczkovich; see footnote on page 50. However, his methods do not solve all one-dimensional versions. A still unsolved case: Is A equidecomposable to an interval where A is the union of intervals of the form $\left(\frac{1}{\log n} - \frac{1}{n^2}, \frac{1}{\log n}\right)$, $n = 0, 1, 2, \ldots$?

Exercises

4 (a). Show that a disk D is equidecomposable with the disjoint union of a disk and a line segment whose length is less than the radius of the circle. (Hint: Use only two pieces to decompose D. Let one piece be any segment contained in a radius and congruent to the given segment together with the infinitely many segments that arise by repeatedly rotating the first segment about the center of D through an angle that is an irrational multiple of π; let the other piece be the remainder of the circle.)

(b). Show that a disk D is equidecomposable to the union of D and any finite number of line segments of any length. (The ideas used to solve this exercise form the heart of Tarski's proof that if two polygons are congruent by dissection then they are equidecomposable; it is simply a matter of "absorbing" the boundaries of the pieces of the dissection.)

5. (Sierpiński–Mazurkiewicz paradox, 1914) Consider the subset E of the plane consisting of all points corresponding to the complex numbers

$$a_0 + a_1 e^i + a_2 e^{2i} + \cdots + a_n e^{ni},$$

where n and the a_i are nonnegative integers and $a_n \neq 0$. Using de Moivre's formula ($e^{i\theta} = \cos\theta + i\sin\theta$) it is easy to see that this set consists of all points obtainable from $(1,0)$ by repeatedly translating one unit to the right and rotating one radian in the counterclockwise direction. Let A be the subset of E corresponding to points having a representation in which $a_0 = 0$, and let B correspond to points having a representation in which $a_0 > 0$.

(a) Show that $A \cup B = E$ and $A \cap B = \emptyset$. (Hint: For the latter, use the fact that e^i is a transcendental number, which is a consequence of Lindemann's theorem discussed in Section 22, Part Two.)

(b) Show that A is congruent to E (by rotation) and A is congruent to B by translation. It follows that A and B are congruent to each other, which justifies each being called a "half" of E. Hence the "paradox": E is a set that is congruent to half of itself.

10. APPROXIMATION BY RATIONAL SETS

Problem 10

Does the plane contain a dense rational set?

The history of this problem and its relatives can be traced up to about 1920 in Dickson's *History of the Theory of Numbers* [Dic]. The problem of finding all rational quadrilaterals was studied by the Hindu mathematicians Brahmegupta (born 598 A.D.) and Bhascara (born 1114 A.D.), and was in a sense solved by the famous German mathematician Kummer in 1848 [Kum], when he showed that the problem is equivalent to finding all rational solutions of a certain sixth degree polynomial equation in five variables. However, it is not clear from his result that every plane 4-set is rationally approximable.

In the 1940s, Stanislaw Ulam asked Paul Erdős whether the plane could contain a dense rational set. This was mentioned in a 1945 paper by Anning and Erdős [AE], which proved that each integral set in the plane is finite. (A set X is *integral* if all members of $\Delta(X)$ are integers)

Interest in rational approximability was revived in the late 1950s, when I.J. Schoenberg posed the problem of determining all rationally approximable finite subsets of the plane. The results in our Exercises 5 and (a special case of) 4 were obtained in 1959 by Besicovitch [Bes2], and in 1960 Mordell [Mor] proved the much deeper result that every plane 4-set is rationally approximable. In 1963, this result was sharpened in two ways. Almering [Alm] showed that for each rational 3-set $\{p, q, r\}$ in the plane, the set of all points s for which $\{p, q, r, s\}$ is rational is dense in the plane. And Daykin [Day1] showed that for the class of all parallelograms, the class of all quadrilaterals, and a certain class of hexagons, the members of the class that are rational and have rational area are dense in the class. Up to this point, all proofs of the rational approximability of arbitrary 4-sets had relied on a result of E. Nagell concerning the density of rational points on cubic curves. However, in 1966 Sheng [She] obtained sharper results by a more direct and elementary approach. His method was extended in 1969 by Ang, Daykin, and Sheng [ADS], but even their theorem does not settle the question of whether every 5-set in the plane is rationally approximable.

It was John Isbell who suggested that perhaps for m sufficiently large, each rational m-set (equivalently, each integral m-set) in the plane contains three collinear or four concyclic points. Examples of Harborth [Har1,2] showed that for each $m \geq 5$ there is an integral m-set in the plane such that no three of its points are collinear and no $m - 1$ of them are concyclic. An example of J. Leech (see Lagrange [Lag]) and Harborth and Kemnitz [HK] showed that $m = 6$ is also insufficient. The latter authors noted that whenever the side-lengths of a triangle are all rational, the triangle's area admits a unique representation in the form $q\sqrt{r}$ where q is rational and r is 1 or is a squarefree integer. Calling r the *characteristic* of the triangle, they showed that for each rational set in the plane, all triangles with vertices in the set have the same characteristic. This was used, in conjunction with a computer search, to show that 174 is the minimum diameter of an integral 6-set that has no three collinear and no four concyclic points. In the example $\{p_1, \ldots, p_6\}$ that realizes this minimum, 174 is the distance between p_1 and p_4, and the complete list of interpoint distances is as follows: [12]&[54] 85, [13]&[64] 158, [14] 174, [15]&[24] 131, [16]&[34] 68, [23]&[65] 87, [25] 136, [26]&[35] 127, [36] 170.

Guy [Guy] discusses Problem 10 in his chapter on Diophantine equations, and some other open problems presented there are also conveniently phrased in geometric language. For example: *Is there a point in the plane all of whose distances from the corners of a unit square are rational?*

Clearly a finite set X is rational if and only if there is a positive integer k such that the set $kX = \{kx : x \in X\}$ is integral (and see Exercise 6). However, it is clear from the following two results that the close connection between integral sets and rational sets breaks down in the case of infinite sets. Theorem 10.1 is Erdős's strengthening [Erd2] of the Anning–Erdős [AE] result mentioned above.

Theorem 10.1. *If an integral set X includes three noncollinear points p, q, and r, with distances $pq = j$ and $pr = k$, then the number of points in X is at most $4jk + 2j + 2k + 1$. Hence each infinite integral set in the plane is collinear.*

Proof. For each point $x \in X$, it is true that

$$|xp - xq| \leq pq = j.$$

Hence x lies on the line whose equation is $xp = xq$ or on one of the j hyperbolas whose equation is $|xp - xq| = i$, with $1 \leq i \leq j$. Similarly, x lies on the perpendicular bisector of the segment pr or on one of k hyperbolas with foci p and r. Each of the j hyperbolas in the first set has at most four points of intersection with each of the hyperbolas in the second set. The two lines have a single point of intersection, and the line in either set has at most two points of intersection with each hyperbola in the other set. □

The Anning–Erdős paper may have been the first to exhibit a rational set whose closure is a circle. Theorem 10.2 extends this by determining precisely which circles can be obtained. This appears to be new, but its proof is based on ideas of Sheng and Daykin [SD]. (See [HK″] for a different sort of result concerning rational points on circles.)

Theorem 10.2. *For each positive number ρ the following three conditions are equivalent:*

(i) *ρ^2 is rational;*
(ii) *on each circle of radius ρ there is a dense rational set;*
(iii) *on each circle of radius ρ there is a rational 3-set.*

Proof. It suffices, of course, to consider a circle of radius ρ centered at the origin, so that each point of C has polar coordinates (ρ, θ) for some angle θ with $-\pi < \theta \leq \pi$. We show first that (i) implies (ii). Let X denote the set of all points (ρ, θ) for which the number $(\tan \frac{1}{4}\theta)/\rho$ is rational. For the continuous mapping $f : (-\pi, \pi] \to (-1/\rho, 1/\rho]$ given by $f(\theta) = (tan\frac{1}{4}\theta)/\rho$, it is clear that f is one-to-one and hence has a continuous inverse. Since the rationals are dense in the interval $(-1/\rho, 1/\rho]$, it follows that the members of X are dense in the circle C. To show that the set X is rational, we use the trigonometric identities

$$\sin(\alpha \pm \beta) = \sin \alpha \cos \beta \pm \cos \alpha \sin \beta,$$

$$\sin \alpha \tan \frac{1}{2}\alpha = 1 - \cos \alpha,$$

and

$$\sin^2 \alpha + \cos^2 \alpha = 1.$$

When $\tan \frac{1}{4}\theta = a\rho$, it follows from the second identity with $\alpha = \frac{1}{4}\theta$ that

$$\cos^2 \frac{1}{2}\theta = (1 - a\rho \sin \frac{1}{2}\theta)^2,$$

whence by the third identity,

$$(1 + a^2\rho^2) \sin^2 \frac{1}{2}\theta = 2a\rho \sin \frac{1}{2}\theta.$$

Then $\sin \frac{1}{2}\theta = 0$ and $\cos \frac{1}{2}\theta = 1$, or

$$\sin \frac{1}{2}\theta = \frac{2a\rho}{1 + a^2\rho^2} \qquad \text{and} \qquad \cos \frac{1}{2}\theta = \frac{1 - a^2\rho^2}{1 + a^2\rho^2}.$$

We conclude that if ρ^2 is rational and $\tan \frac{1}{4}\theta$ is a rational multiple of ρ, then $\cos \frac{1}{2}\theta$ is rational and $\sin \frac{1}{2}\theta$ is a rational multiple of ρ. Now if (ρ, θ) and (ρ, ϕ) are two members of the set X and δ is the distance between them, then $\sin \frac{1}{2}(\theta - \phi) = \frac{\delta/2}{\rho}$ and hence, by the first identity, δ is the rational number

$$2\rho \sin \frac{1}{2}\theta \cos \frac{1}{2}\phi - 2\rho \cos \frac{1}{2}\theta \sin \frac{1}{2}\phi.$$

This shows that (i) implies (ii) in Theorem 10.2. Obviously (ii) implies (iii).

To complete the proof, we show that if C is a circle of radius ρ and there is a triangle with rational edge lengths whose vertices lie on C, then ρ^2 is rational. Let r, s, and t each denote half of one of the edge-lengths of the triangle, and let α (respectively, β, γ) denote half of the angle subtended at C's center by the edge of length r (respectively, s, t). Then $\gamma = \alpha + \beta$ or $\gamma = \pi - (\alpha + \beta)$, according as the triangle does not or does enclose the center of C. Also,

$$\sin \alpha = r/\rho, \qquad \sin \beta = s/\rho, \qquad \sin \gamma = t/\rho.$$

It then follows, with the aid of the first and third trigonometric identities listed earlier, that

$$\pm r\sqrt{\rho^2 - s^2} + \pm s\sqrt{\rho^2 - r^2} = t\rho,$$

where the two choices of $\pm$ do not depend on each other. Clearing of radicals in the usual way (twice transposing and squaring) yields an expression for ρ^2 as a rational linear combination of r, s, and t. □

Most of the natural higher-dimensional problems on rational approximability are untouched, though the reasoning of Anning and Erdős [AE] can be extended to show that in $\mathbb{R}^d$ for $d \geq 2$, each infinite integral set is collinear. Steiger [Ste] showed that for each $d \geq 2$, $\mathbb{R}^d$ contains arbitrarily large rational sets that do not lie in any hyperplane. In fact, "arbitrarily large" may be replaced by "infinite," as a consequence of Theorem 10.2 and Exercise 8. Aside from information obtained by repeated application of Exercise 8, the following problem seems to be open.

Problem 10.8

For which positive integers d and positive real numbers ρ is it true that spheres of radius ρ in $\mathbb{R}^d$

contain a dense rational set?

contain an infinite rational set that does not lie in any hyperplane?

In $\mathbb{R}^d$, the vertex-sets of regular simplices are precisely the sets X for which the distance-set $\Delta(X)$ has only one member. See Larman, Rogers, and Seidel [LRS] for some striking results on the structure of sets X in $\mathbb{R}^d$ for which $\Delta(X)$ has only two members. Equally striking is the following theorem of Graham, Rothschild, and Straus [GRS]: *For the existence of* $d+2$ *points in* $\mathbb{R}^d$ *such that the distance between any two of them is an odd integer, it is necessary and sufficient that* $d \equiv 14 \pmod{16}$.

Exercises

6. Show that a finite set $\{p_1, \ldots, p_m\}$ is rationally approximable if and only if for arbitrarily large positive integers n there is an integral set $\{q_1^n, \ldots, q_m^n\}$ such that $\|np_i - q_i^n\| < 1$ for all i.

7. Prove that if a set Y is rationally approximable then for each real number μ the set $\mu Y = \{\mu y : y \in Y\}$ is rationally approximable. (Thus it follows from Theorem 10.2 that each circle is rationally approximable.)

8. Suppose that x, y, and z are positive integers such that $x < y < z$ and $x^2 + y^2 = z^2$. Prove that if a sphere of radius ρ in $\mathbb{R}^d$ contains an infinite rational set that does not lie in any hyperplane, then in $\mathbb{R}^{d+1}$ the same is true of spheres of radius $\rho z^2/(2xy)$. Use this fact to show that for each rational $\rho > 0$ and each $d \geq 2$, each sphere of radius ρ in $\mathbb{R}^d$ contains an infinite rational set that does not lie in any hyperplane.

11. INSCRIBED SQUARES

Problem 11

Does every simple closed curve in the plane contain all four vertices of some square?

How should we assign a date to an unsolved problem? We could observe that if a problem is unsolved now then it has always been unsolved, but that's not very useful. Thus we choose to date each problem from its first formulation in print. Problem 11 was formulated by Toeplitz [Toe] in 1911, and thus appears to be the oldest of the geometric problems presented here. It was studied in 1913–1915 by Emch [Emc1, 2], who proved the existence of inscribed squares for convex Jordan curves that are polygonal or smooth. Later extensions of his or similar ideas [Zin, Chr] settled the general convex case, and more sophisticated methods [Sni, Jer, Gug, Str] later showed that each sufficiently smooth Jordan curve has an inscribed square (see also [GW′]). The sharpest of these results are the following two theorems of Stromquist [Str].

Theorem 11.7. *Each smooth simple closed curve in $\mathbb{R}^d$ admits an inscribed quadrilateral with equal sides and diagonals.*

Theorem 11.8. *Each locally monotone Jordan curve in R^2 admits an inscribed square.*

In Theorem 11.7, "smooth" means having a continuously turning tangent. When $d = 2$, the quadrilateral is a square. In Theorem 11.8, the requirement of local monotoneity amounts to saying that each point p of J admits a neighborhood in which J can be described can be described by a single-valued function $y = f(x)$ in some rectangular coordinate system whose orientation depends on p. This handles all convex Jordan curves and most Jordan curves that are piecewise smooth. Note, however, that as the theorems are stated here, none of 11.4, 11.5, and 11.7 subsumes either of the other two.

We don't know of any promising approach to finding inscribed squares in arbitrary Jordan curves. The problem's difficulty is illustrated by the fact that Ogilvy [Ogi1] published a proposed proof for the general case in 1950, but in his 1972 book [Ogi2] he conjectured that "there are some sufficiently nasty closed curves no four of whose points are the vertices of a square." There may even be a sense in which *most* Jordan curves fail to have inscribed squares, just as most continuous functions are nowhere differentiable and, as shown by Zamfirescu [Zam], most convex bodies have a variety of weird properties. At the other extreme, it is conceivable that the following has an affirmative answer. (A *continuum* is a set that is compact and connected.)

Problem 11.3

If K is a plane continuum whose complement is not connected, must K contain all four vertices of some square?

The proof of Theorem 11.3 (due to Meyerson [Mey1] and Kronheimer and Kronheimer [KK1]) can be extended to such continua. Meyerson [Mey1] showed also that for an arbitrary Jordan curve J, each point of J with at most two exceptions is a vertex of an inscribed equilateral triangle. He is responsible for Theorem 11.6 [Mey3] and for the following result [Mey1], which guarantees the existence of certain sorts of inscribed polygons even in simple closed curves situated in an arbitrary metric space. As an aid in understanding the theorem, focus first on its content when $n = 2$. In this case the theorem says that when a, b, and c are the sides of a triangle with $a \leq b \leq c$, then if $b = c$ every metric simple closed curve contains three points x, y, and z such that

$$xy : yz : zx = a : b : c;$$

and the existence of three such points is guaranteed only if $b = c$.

Theorem 11.9. *For each integer $n \geq 2$ and sequence $r_1, \ldots, r_n$ in $(0, 1]$, the following two conditions are equivalent:*

(i) *at least one r_i is equal to 1;*

(ii) *for each simple closed curve J in a metric space, there exist $n+1$ points $p_0, p_1, \ldots, p_n$ in cyclic order on J such that*

$$\text{dist}(p_{i-1}, p_i) = r_i \text{dist}(p_n, p_0) \qquad \textit{for } i = 1, 2, \ldots, n.$$

The existence part had been proved earlier [Mil] for the case in which all r_is are 1, and with $n = 3$ that case yields the existence of an inscribed rhombus for an arbitrary Jordan curve. Another paper of Meyerson [Mey4] contains an elegant proof by H. E. Vaughan that each Jordan curve admits an inscribed rectangle. However, these results do not prescribe the similarity class of the rhombus or rectangle.

The example in Exercise 5 appears in [Zak1], attributed to L. M. Kelly. It seems that even the following relative of Problems 11 and 11.1 is open.

Problem 11.4

Is either of the following statements true?

(a) For each quadrilateral Q there is a Jordan curve that does not admit any inscribed quadrilateral similar to Q.

(b) For each nonsquare quadrilateral Q there is a convex Jordan curve that does not admit any inscribed quadrilateral similar to Q.

Additional problems arise if one relaxes the similarity requirement, seeking inscribed figures that are only *affinely* equivalent to "nice" ones. See Grünbaum [Grü2] for the following problem, for references to papers that provide an affirmative answer when the curve is convex, and for other related problems.

Problem 11.5

Does each Jordan curve admit an inscribed affinely regular hexagon?

The proof of Theorem 11.1 is easily extended to establish an analogous result about circumscribed simplices in $\mathbb{R}^d$. However, most of the higher-dimensional results on inscribed or circumscribed figures require topological machinery that is too sophisticated to present here. A theorem of Kalisch and Straus [KS] sets forth conditions on a d-simplex S and a convex body C in $\mathbb{R}^d$ under which C's boundary admits an inscribed simplex homothetic to S. Specialized to $\mathbb{R}^2$, their result is as follows.

Theorem 11.10. *Suppose that T is a triangle in the plane, J is a convex Jordan curve, and at no point of J are there semitangents parallel to two edges of T. Then there is a homothet of T that is inscribed in J.*

Under the assumption that a smooth $(d-1)$-manifold M in $\mathbb{R}^d$ is the boundary of a convex body or of a region whose Euler characteristic differs from zero, Gromov [Gro1] proves that each d-simplex in $\mathbb{R}^d$ admits a homothet inscribed in M.

Both cubes and regular octahedra are natural 3-dimensional analogues of squares. It is known that each convex body in $\mathbb{R}^d$ admits a circumscribed cube [Kak, YY], while there is a 3-dimensional convex body whose boundary has no inscribed rectangular parallelotope and hence of course no inscribed cube [Bie]. The existence of inscribed regular octahedra has been discussed by several authors [Puc, Gug, HLM], but results are inconclusive. In particular, an affirmative solution of the following was claimed [Puc], but several steps in the proof are invalid [HLM].

Problem 11.6

If M is the boundary of a 3-dimensional convex body, must M admit an inscribed regular octahedron?

It is known [BR] that the boundary of each 3-dimensional convex body does admit an inscribed parallelotope. This implies that the boundary of each centrally symmetric 4-dimensional convex body admits an inscribed 4-parallelotope (see Exercise 7 for the essence of the simple argument). However, the following problem is open for all $d \geq 4$ and when central symmetry is assumed it is open for all $d \geq 5$.

Problem 11.7

For which d is it true that the boundary of each d-dimensional convex body admits an inscribed d-dimensional parallelotope?

It has been conjectured [BGKV] that for all sufficiently large d (probably $d \geq 4$ is sufficient), almost every d-dimensional convex body is such that its boundary *fails* to admit an inscribed d-dimensional parallelotope.

For additional references to related higher-dimensional results and problems, see [Grü2] and [HDK, pp. 49–50].

We would not want to leave the general subject of inscribed figures without stating a striking theorem of Lévy [Lév] and Hopf [Hop] on inscribed segments. It was motivated by Rolle's theorem, familiar from calculus, which asserts that if f is a differentiable real function on the interval $[0, 1]$, with $f(0) = f(1)$, then there exists $t_0 \in (0, 1)$ such that $f'(t_0) = 0$. Since $f'(t_0)$ is the limit of the slopes of segments that join points of f's graph near the point $(t_0, f(t_0))$, we might expect to find short horizontal segments inscribed in the graph. What can be said about the lengths of such segments? Here is the surprising answer, and part of the proof is outlined in Exercises 8 and 9.

Theorem 11.11. *For each positive number ϵ, the following three conditions are equivalent*:

(i) *for each differentiable real-valued function* f *on* $[0,1]$ *such that* $f(0) = f(1)$, *the graph of* f *has an inscribed horizontal segment of length* ϵ;
(ii) *for each plane continuum* K *and each segment* S *inscribed in* K, *there is a parallel inscribed segment whose length is* ϵ *times that of* K;
(iii) $1/\epsilon$ *is an integer.*

As was noted in [Mey4], Theorem 11.11 contributes to the 2-dimensional version of what we shall here call the *problem of the mountain picnic.* Suppose that we're planning a mountain picnic, taking tables and stools of various sizes. For each table or stool, the associated *endset* E is the set formed by the ends of its legs. We hope to be able to place all the endsets on the mountain so that each one is level. Will some reasonable assumption about the mountain ensure that this is possible?

In the 2-dimensional problem, each endset consists of just two points. A complete solution of the problem is outlined in Exercise 10.

In the 3-dimensional problem, the mountain's base is represented by a Jordan curve J in $\mathbb{R}^2$, the mountain itself by the graph (in $\mathbb{R}^3$) of a function $f : \mathbb{R}^2 \to \mathbb{R}$ such that $f \in F_J$, the set of all continuous functions on $\mathbb{R}^2$ that vanish outside J, or $f \in F_J^+$, the set of all nonnegative members of F_J. We assume for each table that the endset consists of the vertices of a cyclic quadrilateral and for each stool that the endset consists of the vertices of a triangle. In other words, we are dealing with 4-legged tables and 3-legged stools.

A *level set* of f is a set of the form $\{x : f(x) = \lambda\}$, where λ is a constant. We hope to be able to inscribe a congruent copy of each endset in some level set. Of course we could level all the endsets by going far from the mountain, out in the surrounding plain where the elevation (the value of f) is 0, but then it wouldn't be a mountain picnic. Thus we associate with each endset a *center*—the endset's circumcenter for the tables and an arbitrary interior point for the triangular stools, and we require that when an endset is leveled, the associated center should be over a point of the mountain.

Let us say that J is *good for square tables* if for each positive δ and each $f \in F_J^+$, some level set of f contains all four vertices of a square of edge-length δ whose center is inside J. Being *good for triangular stools* is similarly defined. The restriction to F_J^+ amounts to requiring that none of the mountain valleys dips below the level of the base. Note also that we are in effect identifying each table or stool with its endset, and ignoring the height of the table or stool. A consequence might be that leveling would require placing an endset on the mountain in such a way that part of the mountain sticks up through the middle of the table or stool—quite uncomfortable in practice, but viewed here with equanimity because it's only a mathematical mountain.

The following theorem is due to Fenn [Fen] in the case of tables, to Zaks [Zak1] and Kronheimer and Kronheimer [KK1] in the case of stools.

Theorem 11.12. *Each convex Jordan curve is good for all square tables and for all triangular stools.*

Meyerson [Mey2] shows that for square tables the convexity assumption cannot be abandoned, and remarks [Mey3] that the following problem is still open.

Problem 11.8

Is each convex Jordan curve good for all rectangular tables?

(He actually raised the problem for general cyclic quadrilateral tables, but there a negative answer follows from Exercise 5. See Exercise 11 for the relationship of Problem 11.8 to Problems 11.1 and 11.4.)

Kronheimer and Kronheimer [KK2] proved the following higher-dimensional relative of Theorem 11.12's result for stools.

Theorem 11.13. *For each d-simplex S in $\mathbb{R}$ and each compact d-dimensional submanifold M of $\mathbb{R}^d$ whose boundary is smooth at at least one point, there is a positive number σ such that for each $\tau \leq \sigma$ and each continuous nonnegative map $f : M \to \mathbb{R}$ that vanishes on the boundary, there is a simplex of diameter τ that is similar to S and is inscribed in a level set of f.*

It is unknown whether the smoothness condition can be dropped, and the extent to which nonnegativity of f is needed is also unsettled.

We end this section with a problem that is totally unrelated to Problem 11 except that it too concerns inscribed squares. See Graham [Gra2] for the background of this problem and its interesting relationship to Ramsey's theorem.

Problem 11.9

With L denoting the set of all points (x, y) in the plane such that x and y are both positive integers, suppose that W is a subset of L for which

$$\sum_{(x,y)\in W} \frac{1}{x^2 + y^2} = \infty.$$

Must W contain all four vertices of some square?

Exercises

7. Establish the following by modifying examples from Exercise 1.

For each $b > 0$ there is a Jordan curve J such that for each s with $0 < s \leq b$, there is a square of edge-length s that is inside J and inscribed in J.

8. Suppose that the convex body C in $\mathbb{R}^3$ is symmetric about the origin, and P is a plane intersecting C's interior but missing the origin. Show that $C \cup P$ contains one of the six 2-dimensional faces of a 3-dimensional parallelotope that is inscribed in C's boundary.

9. [Lév] Show that if the function $f : [0,1] \to \mathbb{R}$ is given by

$$f(x) = sin^2\left(\frac{\pi x}{\epsilon}\right) - xsin^2\left(\frac{\pi}{\epsilon}\right),$$

where ϵ is a positive number that is not the reciprocal of an integer, then there is no $x \in [0,1]$ such that $f(x) = f(x + \epsilon)$.

10. [Lév] Let f be a continuous real-valued function on $[0,1]$, with $f(0) = f(1)$, and let M denote the set of all $\epsilon > 0$ such that for all x with $0 \leq x < x + \epsilon \leq 1$, $f(x) \neq f(x + \epsilon)$. For $-1/2 \leq \delta \leq 1/2$, let f_δ denote the result of shifting f's graph by an amount δ to the right. Show that if $a \in M$ and $b \in M$, then (on the intervals common to the domains of the functions in question) either $f_{-a} < f$ and $f < f_b$, or $f_{-a} > f$ and $f > f_b$. Deduce from this that $a + b \in M$. Then, using the fact that $1 \notin M$, conclude that M omits each of the numbers $1/2, 1/3, 1/4, \ldots$.

11. [Mey4] Let F denote the set of all continuous functions $f : \mathbb{R} \to \mathbb{R}$ such that $f(t) = 0$ for all $t \leq 0$ and all $t \geq 1$. Let F^+ denote the set of all nonnegative members of F. Prove each of the following.

(a) For each $\epsilon > 0$, the following two conditions are equivalent:

(i) ϵ is the reciprocal of an integer;

(ii) for each $f \in F$ there exists t such that

$$0 \leq t \leq t + \epsilon \leq 1 \text{ and } f(t) = f(t + \epsilon).$$

(b) For each $\epsilon > 0$ and each $f \in F^+$ there exists t such that

$$0 \leq t \leq t + \epsilon \leq 1 \text{ and } f(t) = f(t + \epsilon).$$

(c) Whenever $0 \leq \delta \leq \epsilon$ and $f \in F$, there exists t such that

$$0 \leq t + \delta \leq 1 \text{ and } f(t) = f(t + \epsilon).$$

12. Suppose that C is a plane convex body with boundary J, and the origin 0 is interior to C. Prove the following:

(a) For each point $x \in C\backslash\{0\}$ there is a unique positive number $\mu(x)$ such that $\mu(x)x \in J$.

(b) A continuous function f on all of $\mathbb{R}^2$ is obtained by defining

$$f(x) = \begin{cases} 1 & \text{when } x \text{ is the origin } 0; \\ 1 & 1 - 1/\mu(x) \text{ when } x \in C\backslash\{0\}; \\ 0 & \text{when } x \notin C. \end{cases}$$

(c) For each positive level λ, the level sets of f are all homothets of J.

(d) If J is good (in the sense used earlier) for tables whose leg ends form a polygon P, then a polygon similar to P can be inscribed in J.

12. FIXED POINTS

Problem 12

Does each nonseparating plane continuum have the fixed-point property?

Our discussion of the fixed-point property is based in part on an article by Bing [Bin]. As he mentions, the fixed-point theorem that was proved by Brouwer [Bro1] in 1912 had been proved by Bohl [Boh] in 1904 for a more restricted class of mappings. However, Brouwer's methods have had more influence on subsequent developments in topology. It is interesting that Brouwer later lost faith in his own early work and, indeed, in many of the proofs that were then (and are still) accepted by most mathematicians. He developed new standards of mathematical definition and proof, so rigorous and "constructive" in nature that full acceptance of his ideas would mean abandoning large parts of mathematics as we know it. For details, see Brouwer [Bro2], van Stigt [vSt], and their references.

In proving Brouwer's theorem, we'll deal only with the case of topological disks. Two preliminary results are required, but they are of interest in their own right. The argument can be extended to show that topological d-balls (homeomorphs of the set $\{x \in \mathbb{R}^d : \|x\| \leq 1\}$) have the fixed-point property.

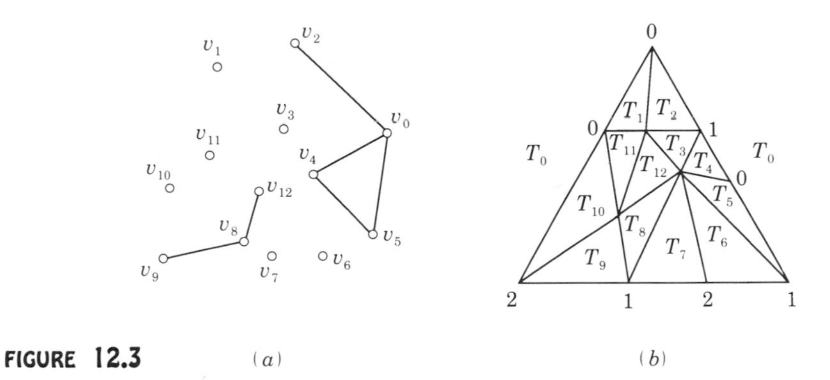

FIGURE 12.3

As the term is used here, a *graph* consists of a finite set of points (called *nodes*) together with a set of unordered pairs of distinct nodes (called *edges*). For each edge $\{x, y\}$, the edge and the node x (or y) are said to be *incident* to each other. The number of edges incident to a node is its *degree*. For example, the graph in Figure 12.3a has 11 vertices and 6 edges. The number of vertices of degree 0, 1,

2, and 3 is 6, 3, 3, and 1, respectively. (Figure 12.3 is from the book by Bondy and Murty [BM], which is a good general reference on graph theory.)

The first tool in proving Brouwer's theorem is the following lemma. Its easy proof, which is left to the reader (Exercise 5), uses one of the few general principles of combinatorial mathematics: If you can count something in two different ways, by all means do so!

Lemma 12.7. *In any graph, the number of nodes of odd degree is even.*

A *simplicial subdivision* of a triangle T is a finite collection of subtriangles with union T such that for any two intersecting members of the collection, the intersection is a vertex or an edge of both members. An associated *proper labeling* is an assignment of a label 0, 1, or 2 to each vertex in such a way that

(i) T's three vertices are labeled 0, 1, and 2, and
(ii) for $0 \le i < j \le 2$, each vertex on T's "ij" side is labeled i or j.

Relative to a proper labeling, a triangle in the subdivision is *distinguished* if its vertices have all three labels. In Figure 12.3b, the distinguished triangles are T_2, T_9, and T_{12}.

There are two results in the literature that are popularly called *Sperner's lemma,* even though they deal with quite different matters. Here is the one that we need [Spe].

Lemma 12.8 (Sperner's Lemma). *Every properly labeled simplicial subdivision of a triangle has an odd number of distinguished triangles.*

Proof. Construct a graph by associating a node with each of the subtriangles, and also a node p_0 with the large triangle T. Let two vertices be adjacent (i.e., joined by an edge) in the graph if the associated triangles intersect in a segment whose ends are labeled 0 and 1. Then each node adjacent to p_0 corresponds to a subtriangle having a 01 side in the 01 side of T, and the number of such subtriangles is odd (Exercise 6). A little reflection shows that, aside from p_0, each node of G is of degree 0, 1, or 2, and a node is distinguished (i.e., the vertex set of the associated subtriangle carries all three labels) if and only if it is of degree 1. Since the number of nodes of odd degree is even by Lemma 12.7, and since p_0 is of odd degree, there must be an odd number of nodes of degree 1 that are different from p_0. □

Theorem 12.9 (Brouwer Fixed-Point Theorem). *Topological disks have the fixed-point property.*

Proof. Because of Theorem 12.1, it suffices to consider a triangle T with vertices x_0, x_1, and x_2. Each point y of T has a unique expression in the form

$$y = \lambda_0(y)x_0 + \lambda_1(y)x_1 + \lambda_2(y)x_2$$

with nonnegative numbers $\lambda_i(y)$ whose sum is 1. (These numbers are the *barycentric coordinates* of y with respect to the triple (x_0, x_1, x_2). See [HY] for a good discussion of such coordinates.) Now consider an arbitrary continuous mapping f of T into T. Since each of the functions λ_i and each of the composite functions $\lambda_i f$ are continuous, the set

$$S_i = \{y \in T : \lambda_i(f(y)) \leq \lambda_i(y)\}$$

is closed for each i. Each point y of the intersection $S_0 \cap S_1 \cap S_2$ is fixed under f, because the barycentric coordinates of $f(y)$ are the same as those of y. (Use the fact that these coordinates are nonnegative and their sum is 1.)

To show that the set $S_0 \cap S_1 \cap S_2$ is not empty, it suffices to show that for each positive ϵ, T contains a triangle of diameter less than ϵ that intersects each of the sets S_i. To this end, consider a simplicial subdivision of T so fine that each of the subtriangles is of diameter less than ϵ (Exercise 8). Let a vertex p of the subdivision be labeled i if it belongs to S_i, being sure to use 0 or 1 for each vertex of the side x_0x_1 of T. Since this labeling is proper (Exercise 7), the desired conclusion follows from Lemma 12.8. □

Exercise 5 and Lemma 12.8 are the 1- and 2-dimensional cases of a d-dimensional lemma that leads to an easy proof of the d-dimensional version of Brouwer's fixed-point theorem. The d-dimensional lemma asserts that in each proper labeling of the vertices of a simplicial subdivision of a d-simplex (e.g., of a tetrahedron when $d = 3$), the number of distinguished d-simplices is odd. In the proof by mathematical induction, the argument for $d \geq 3$ is quite analogous to the one for $d = 2$, using the $(d-1)$-dimensional result to get started (see, e.g., [Tom]).

When $K \subset \mathbb{R}^m$ and $L \subset \mathbb{R}^n$, the *product* $K \times L$ is defined in an especially simple way as the set of all points $x = (x_1, \ldots, x_{m+n})$ of $\mathbb{R}^{m+n}$ for which the point $(x_1, \ldots, x_m)$ belongs to K and the point $(x_{m+1}, \ldots, x_{m+n})$ belongs to L. In particular, the product of K by $[0, 1]$ may be viewed as a sort of "prism" or "cylinder" constructed on the base K. Note that the d-cube $[0, 1]^d$ is homeomorphic with the product $[0, 1]^{d-1} \times [0, 1]$, and that an alternative proof of Brouwer's theorem in d dimensions would follow from this if the fixed-point property were preserved under product formation. However, not only is it not preserved in general, but perhaps even the following question of Bing [Bin] has a negative answer.

Problem 12.1

Suppose that the continuum K is contained in the plane or is 1-dimensional. If K has the fixed-point property, must the product $K \times [0, 1]$ have it?

Though the answer to Problem 12.1 is negative for high-dimensional K, several open problems (Bing [Bin], Brown [Bro]) ask what are the simplest K (in various senses) for which this happens.

The proof of Theorem 12.9 shows how coverings of a set by small subsets can play a role in the study of fixed points, and how associated graphs can be used to study such coverings. A different way of associating a graph with a covering is to take a node for each set and to join two nodes by an edge if and only if the corresponding sets intersect. The graph formed in this way is the *intersection graph* of the family of sets.

A continuum is said to be *snakelike* (respectively, *treelike*) if for each positive ϵ it admits a covering by open sets of diameter less than ϵ such that no three of the sets intersect and the intersection graph is a simple path (respectively, does not contain any circuit). Hamilton [Ham] proved that all snakelike continua have the fixed-point property. There were several attempts to do the same for treelike continua, but Bellamy [Bel] found a counterexample.

In connection with Problem 12, it should be added that each nonseparating plane continuum is the intersection of a decreasing sequence of topological disks. For $d \geq 3$, the intersection of a decreasing sequence of d-balls need not have the fixed-point property [Bin], but the case $d = 2$ is unsettled.

In several respects, the sharpest and most interesting results related to Problem 12 are Theorems 12.5 and 12.6 of Part One, both due to C. Hagopian. In order to explain his contributions more fully, let us say that a continuous mapping $f : K \to K$ *preserves arc-components* if for each $x \in K$ there is an arc in K that joins x to $f(x)$. A stronger condition is that f is a *deformation,* meaning that there is a continuous mapping h of the $K \times [0, 1]$ into K such that for each $x \in K$, $f(x, 1) = f(x)$ and $f(x, 0) = x$. (This is a precise formulation of the idea that K admits a continuous one-parameter family $f_1 : K \to K (0 \leq t \leq 1)$ of mappings such that $f_1 = f$ and f_0 is the identity.) Here are the main results of Hagopian [Hag1, 2, 3, 5].

Theorem 12.10. *If K is a nonseparating plane continuum then each mapping of K that preserves arc-components has a fixed point; thus K has the fixed-point property if K is arcwise connected.*

Theorem 12.11. *Suppose that K is a continuum that does not contain a simple closed curve. If K planar and arcwise connected then K has the fixed point property. If K is merely planar then K has the fixed-point property for arc-component-preserving maps; and if K is merely arcwise connected then K has the fixed-point property for deformations.*

An example of Young [You] shows that without the planarity requirement in Theorem 12.11, K may lack the fixed-point property. However, the following problem is open.

Problem 12.2

Must each tree-like continuum have the fixed-point property for arc-component–preserving maps?

Hagopian [Hag4] established an affirmative answer for a more restricted class of maps, and also [Hag3] for arc-component-preserving maps under the additional assumption that K does not contain uncountably many disjoint triods. (A *triod* is a homeomorph of the set formed by three closed segments issuing from a common vertex.)

We saw in Exercise 1 that if continua X and Y have the fixed-point property and the intersection $X \cap Y$ is just a point, then the union $X \cup Y$ also has the fixed-point property. What happens for more complicated intersections? It is known that if each of X, Y, and $X \cap Y$ is homeomorphic with a retract of a topological ball, then the same is true of $X \cup Y$ and hence $X \cup Y$ has the fixed-point property. However, the following problem is raised by Bing [Bin], who shows that thc answer is affirmative if K is not required to lie in the plane.

Problem 12.3

Does the plane contain a disk D and a continuum K such that K has the fixed-point property, $D \cap K$ is an arc, and $D \cup K$ lacks the fixed-point property?

Exercises

5. Prove that in any graph, the number of nodes of odd degree is even.

6. For points $p_0 < p_1 < \cdots < p_k$ in $\mathbb{R}$, suppose that p_0 is labeled 0, p_k is labeled 1, and each of the intermediate points p_i is labeled 0 or 1. Show that, among the

segments $[p_{i-1}, p_i]$ $(1 \leq i \leq k)$, there is an odd number for which one endpoint is labeled 0 and the other is labeled 1.

7. Show that in the proof of Theorem 12.9, the labeling is proper.

8. The proof of Theorem 12.9 uses the fact that for each positive ϵ the triangle T admits a simplicial subdivision in which each of the subtriangles is of diameter less than ϵ. Supply an essential ingredient in the proof of this by showing that in the barycentric subdivision of a triangle T the diameter of each of the six subtriangles is at most two-thirds the diameter of T. (Each triangle of T's barycentric subdivision has a vertex-set of the form

$$V(x, y, z) = \{x, \frac{1}{2}(x + y), \frac{1}{3}(x + y + z)\}$$

where (x, y, z) is a permutation of V's vertices.)

9. Show that the "$\sin \frac{1}{x}$ arc" A and the "$\sin \frac{1}{x}$ circle" C (in Figure 12.2) have the fixed-point property.

10. Let $\theta_1 > \theta_2 > \cdots$ be a sequence of numbers in the interval $(0, \pi/2)$, converging to 0. For each i, let p_i be the point of the plane whose polar coordinates are $(1, \theta_i)$, and let S_i denote the segment joining the origin O to p_i. Let X denote the union of these segments. Show that both X and its closure have the fixed-point property.

11. Among the specific continua mentioned here and in Part 1, which are snakelike and which are treelike?

REFERENCES

[AKW] J. Aldins, J. S. Kline, and S. Wagon, Problem 633, *Crux Mathematicorum,* 8 (1982) 120–121. [§9]

[Ale] R. Alexander, Lipschitzian mappings and total mean curvature of polyhedral surfaces, I, *Transactions of the American Mathematical Society,* 288 (1985) 661–678. [§3]

[Alm] J. H. Almering, Rational quadrilaterals, *Indagationes Mathematicae,* 25 (1963) 192–199. [§10]

[ADS] D. Ang, D. Daykin, and T. Sheng, On Schoenberg's rational polygon problem, *Bulletin of the Australian Mathematical Society,* 9 (1969) 337–344. [§10]

[AE] N. H. Anning and P. Erdős, Integral distances, *Bulletin of the American Mathematical Society,* 51 (1945) 598–600. [§10]

[AH] K. Appel and W. Haken, The solution of the four-color map problem, *Scientific American,* 237, No. 4 (Oct. 1977) 108–121. [§8]

[ABI] D. Avis, B. K. Bhattacharya, and H. Imai, Computing the volume of the union of spheres, *The Visual Computer,* 3 (1988) 323–328. [§3]

[AR] D. Avis and D. Rappaport, Computing the largest empty convex subset of a set of points, *Proceedings of the First Annual Symposium on Computational Geometry,* Association for Computing Machinery, New York, 1985, pp. 161–167. [§5]

[BBB] V. Bálint, A. Bálintová, M. Branická, P. Grešák, I Hrinko, P. Novotný, and M. Stacho, No midpoints in the subset, *Canadian Mathematics Bulletin,* to appear. [§5]

[BB] A. Bálintová and V. Bálint, On the number of circles determined by n points in the Euclidean plane, *Acta Mathematica Hungarica,* to appear. [§6]

[Bar] D. Barnette, *Map Coloring, Polyhedra, and the Four-Color Problem,* Mathematical Association of America, Washington, D.C., 1983 [§8]

[Bec] J. Beck, On the lattice property of the plane and some problems of Dirac, Motzkin, and Erdős in combinatorial geometry, *Combinatorica,* 3 (1983) 281–297. [§6]

[BQ] F. S. Beckman and D. A. Quarles, Jr., On isometries of Euclidean spaces, *Proceedings of the American Mathematical Society,* 4 (1953) 810–815. [§8]

[Bel] D. P. Bellamy, A tree-like continuum without the fixed-point property, *Houston Journal of Mathematics,* 6 (1979) 1–13. [§12]

[BP] M. Benda and M. Perles, Coloring of metric spaces, unpublished manuscript, Department of Mathematics, University of Washington, Seattle, 1976. [§8]

[Ber] M. Berger, Sur les cautistiques de surfaces en dimension 3, *Comptes Rendus de l'Académie Scientifique de Paris,* 311 (1990) 333–336. [§1]

[Bes1] A. S. Besicovitch, On Kakeya's problem and a similar one, *Mathematische Zeitschrift,* 27 (1928) 312–320. [§4]

[Bes2] ——, Rational polygons, *Mathematika,* 6 (1959) 98. [§10]

[BK] A. Bezdek and W. Kuperberg, Examples of space-tiling polyhedra related to Hilbert's problem 18, question 2, in *Topics in Combinatorics and Graph Theory, Essays in Honour of Gerhard Ringel,* R. Bodendiek and R. Henn, eds., Physica-Verlag, Heidelberg, 1990, pp. 87–92. [§7]

[Bez] K. Bezdek, Hadwiger–Levi's covering problem revisited, in *Progress in Discrete and Combinatorial Geometry,* J. Pach, ed., Springer, New York, to appear. [§1]

[BC] K. Bezdek and R. Connelly, Covering curves by translates of a convex set, *American Mathematical Monthly,* 96 (1989) 789–806. [§4]

[BDV] A. Bialostocki, P. Dierker, and B. Voxman, Some notes on the Erdős–Szekeres theorem, *Discrete Mathematics,* to appear. [§5]

[Bie] A. Bielecki, Quelques remarques sur la note précédente, *Annales Universitatis Mariae Curie-Skłodowska,* 8 (1954) 101–103. [§11]

[BR] A. Bielecki and K. Radziszewski, Sur les parallélépipèdes inscrits dans les corps convexes, *Annales Universitatis Mariae Curie-Skłodowska,* 8 (1954) 97–100. [§11]

[BLW] N. L. Biggs, E. K. Lloyd, and R. J. Wilson, *Graph Theory 1736–1936,* Clarendon Press, Oxford, 1976. [§8]

[Bin] R. H. Bing, The elusive fixed point property, *American Mathematical Monthly,* 76 (1969) 119–132. [§12]

[Bir] G. D. Birkhoff, *Dynamical Systems,* American Mathematical Society, Providence, 1927 (revised edition, 1966). [§1]

[BF1] K. Bisztriczky and G. Fejes Tóth, A generalization of the Erdős–Szekeres convex n-gon theorem, *Journal für die Reine und Angewandte Mathematik,* 395 (1989) 167–170. [§5]

[BF2] ——, Nine convex sets determine a pentagon with convex sets as vertices, *Geometriae Dedicata,* 31 (1989) 89–104. [§5]

[BF3] ——, Convexly independent sets, *Combinatorica,* 10 (1990) 195–202. [§5]

[BRW] W. Blaschke, H. Rothe, and W. Weitzenbock, Aufgabe 552, *Archiv der Mathematik und Physik,* 27 (1917) 82. [§2]

[BGKV] H. L. Bodlaender, P. Gritzmann, V. Klee, and J. van Leeuwen, Computational complexity of norm-maximization, *Combinatorica,* 10 (1990) 203–225. [§11]

[Boh] P. Bohl, Über die Bewegung eines mechanischen Systems in der Nähe einer Gleichgewichtslage, *Journal für die Reine und Angewandte Mathematik,* 127 (1904) 179–276. [§12]

[BKM] C. Boldrighini, M. Keane, and F. Marchetti, Billiards in polygons, *Annals of Probability,* 6 (1978) 532–540. [§1]

[Bol1] B. Bollobás, Filling the plane with congruent convex hexagons without overlapping, *Annales Universitatis Scientiarum Budapestinensis de Rolando Eötvös Nominatae, Sectio Mathematica,* 6 (1963) 117–123. [§9]

[Bol2] ——, Area of the union of disks, *Elemente der Mathematik,* 23 (1968) 60–61. [§3]

[Bol′] V. G. Boltianskii, *Hilbert's Third Problem,* trans. R. Silverman, Winston, Washington, D.C., 1978. [§7]

[BM] J. A. Bondy and U. S. R. Murty, *Graph Theory with Applications,* American Elsevier, New York, 1976. [§12]

[Bon] W. E. Bonnice, On convex polygons determined by a finite planar set, *American Mathematical Monthly,* 81 (1974) 749–752. [§5]

[Bor] K. Borsuk, Drei Sätze über die n-dimensionalen euklidische Sphäre, *Fundamenta Mathematicae,* 20 (1933) 177–190. [§4]

[BM′] P. Borwein and W. O. J. Moser, A survey of Sylvester's problem and its generalizations, *Aequationes Mathematicae,* 40 (1990) 111–135. [§6]

[Bou] G. Bouligand, Ensembles impropres et nombres dimensionnel, *Bulletin de la Société Mathématique de France,* 50 (1928) 320–344. [§3]

[Bro1] L. E. J. Brouwer, Über Abbildung von Mannigfaltigkeiten, *Mathematische Annalen,* 71 (1912) 97–115. [§12]

[Bro2] ——, *Brouwer's Cambridge Lectures on Intuitionism,* D. van Dalen, ed., Cambridge University Press, Cambridge, 1981. [§12]

[Bro′] R. F. Brown, The fixed-point property and Cartesian products, *American Mathematical Monthly,* 89 (1982) 654–678. [§12]

[But] G. J. Butler, *On the "equichordal curve" problem and a problem of packing and covering,* Ph.D. thesis, University College, London, 1968. [§2]

[CP] V. Capoyleas and J. Pach, On the perimeter of a point set in the plane, in *Discrete and Computational Geometry: Papers from the DIMACS Special year,* DIMACS Series in Discrete Mathematics and Theoretical Computer Science, American Mathematical Society, Providence, and Association for Computing Machinery, New York, 1991. [§3]

[CG] G. D. Chakerian and H. Groemer, Convex bodies of constant width, in *Convexity and its Applications,* P. M. Grüber and J.M. Wills, eds., Birkhäuser, Basel, 1983, pp. 49–96. [§4]

[CK] G. D. Chakerian and M. S. Klamkin, Minimal covers for closed curves, *Mathematics Magazine,* 46 (1973) 55–61. [§4]

[CK′] C. C. Chang and H. J. Keisler, *Model Theory,* North Holland, Amsterdam, 1973. [§8]

[Chi1] K. B. Chilakamarri, Unit-distance graphs in rational n-spaces, *Discrete Mathematics,* 69 (1988) 213–218. [§8]

[Chi2] ——, On the chromatic number of rational five-space, *Aequationes Mathematicae,* 39 (1990) 146–148. [§8]

[Chi3] ——, Unit-distance graphs in Minkowski metric spaces, *Geometriae Dedicata,* 37 (1991) 345–356. [§8]

[Chr] C. Christensen, Kvadrat indskrevet i konveks figur (Danish), *Matematisk Tidd-skrift B,* 1950 (1950) 22–26. [§11]

[Chv] V. Chvátal, A combinatorial theorem in plane geometry, *Journal of Combinatorial Theory B,* 18 (1975) 39–41. [§1]

[CEGSW] K. L. Clarkson, H. Edelsbrunner, L. J. Guibas, M. Sharir, and E. Welzl, Combinatorial complexity bounds for arrangements of curves and spheres, *Discrete and Computational Geometry,* 5 (1990), 99–160. [§6]

[Con] J. E. Connett, Trapped reflections?, *American Mathematical Monthly,* to appear. [§1]

[CS1] K. Corrádi and S. Szabó, A combinatorial approach for Keller's conjecture, *Periodica Mathematica Hungarica,* 21 (1990) 95–100.

[CS2] ——, Cube tiling and covering a complete graph, *Discrete Mathematics,* 85 (1990) 319–321.

[Cox1] H. S. M. Coxeter, A problem of collinear points, *American Mathematical Monthly,* 55 (1948) 26–28. [§6]

[Cox2] ——, Review of *Arrangements and Spreads* ([Grü5]), *Mathematical Reviews,* 46 (1973) 1059. [§6]

[CR] F. J. Craveiro de Carvalho and S. A. Robertson, Equichordal curves, *Proceedings of the Royal Society of Edinburgh,* 10A (1988) 85–91. [§2]

[CFG] H. T. Croft, K. J. Falconer, and R. K. Guy, *Unsolved Problems in Geometry,* Springer, New York, 1991. [Introduction]

[CS′] H. Croft and H. P. F. Swinnerton-Dyer, On the Steinhaus billiard table problem, *Proceedings of the Cambridge Philosophical Society,* 59 (1963) 37–41. [§1]

[CS″] J. Csima and E. T. Sawyer, A short proof that there exist $6n/13$ ordinary lines, manuscript, McMaster University, 1991. [§6]

[Cun] F. Cunningham, Jr., The Kakeya problem for simply connected and for star-shaped sets, *American Mathematical Monthly,* 78 (1971) 114–129. [§4]

[CS‴] F. Cunningham, Jr. and I. J. Schoenberg, On the Kakeya constant, *Canadian Journal of Mathematics,* 17 (1965) 946–956. [§4]

[DHH] G. B. Dantzig, A. J. Hoffman, and T. C. Hu, Triangulations (tilings) and certain block triangular matrices, *Mathematical Programming,* 31 (1985) 1–14. [§7]

[Dan] L. Danzer, Three-dimensional analogs of the planar Penrose tilings and quasicrystals, *Discrete Mathematics,* 76 (1989) 1–7. [§7]

[DGK] L. Danzer, B. Grünbaum, and V. Klee, Helly's theorem and its relatives, in *Convexity,* V. Klee, ed., Proceedings of Symposia in Pure Mathematics, vol. 7, American Mathematical Society, Providence, 1963, pp. 101–180. [§3]

[DGS1] L. Danzer, B. Grünbaum, and G. C. Shephard, Can all tiles of a tiling have five-fold symmetry?, *American Mathematical Monthly,* 89 (1982) 583–585. [§7]

[DGS2] ——, Equitransitive tilings, or how to discover new mathematics, *Mathematics Magazine,* 60 (1987) 67–89. [§7]

[Dav] R. O. Davies, Some remarks on the Kakeya problem, *Proceedings of the Cambridge Philosophical Society,* 69 (1971) 417–421. [§4]

[Day1] D. E. Daykin, Rational polygons, *Mathematika,* 10 (1963) 125–131. [§10]

[Day2] ——, Rational triangles and parallelograms, *Mathematics Magazine,* 38 (1965) 46–47. [§10]

[DE1] N. G. de Bruijn and P. Erdős, On a combinatorial problem, *Indagationes Mathematicae,* 10 (1948) 421–423. [§6]

[DE2] ——, A colour problem for infinite graphs and a problem in the theory of relations, *Indagationes Mathematicae,* 13 (1951) 369–373. [§§4 and 8]

[Deb] H. E. Debrunner, Tiling Euclidean d-space with congruent simplices, in *Discrete Geometry and Convexity,* J. E. Goodman, E. Lutwak, J. Malkevitch, and R. Pollack, eds., *Annals of the New York Academy of Sciences,* 440 (1985) 230–261. [§7]

[Dek] B. V. Dekster, Non-isometric distance 1 preserving mapping $E^2 \to E^6$, *Archiv der Mathematik,* 45 (1985) 282–283. [§8]

[DR1] D. W. DeTemple and J. M. Robertson, A billiard characterization of regular polygons, *Mathematics Magazine,* 54 (1981) 73–75. [§1]

[DR2] ——, Convex curves with periodic billiard polygons, *Mathematics Magazine,* 58 (1985) 40–42. [§1]

[Dic] L. E. Dickson, *History of the Theory of Numbers,* vol. II, Chelsea, New York, 1971 (reprint of 1923 edition). [§10]

[Dir1] G. A. Dirac, Collinearity properties of sets of points, *Quarterly Journal of Mathematics* (Oxford Series 2), 2 (1951) 221–227. [§6]

[Dir2] ——, Ovals with equichordal points, *Journal of the London Mathematical Society,* 27 (1952) 429–437. [§2]

[DEO] D. P. Dobkin, H. Edelsbrunner, and M. H. Overmars, Searching for empty polygons, Technical Report RUU-CS-88-11, Department of Computer Science, University of Utrecht, 1988. [§5]

[Dou] R. Douady, Applications du théorème des tores invariants, thèse 3ème cycle, Université de Paris VII, 1982. [§1]

[DHK] L. Dubins, M. Hirsch, and J. Karush, Scissor congruence, *Israel Journal of Mathematics,* 1 (1963) 239–247. [§9]

[Duf1] G. F. D. Duff, A smaller universal cover for sets of unit diameter, *Royal Society of Canada Mathematics Reports,* 2 (1980) 37–42. [§4]

[Duf2] ——, On universal covering sets and translation covers in the plane, *James Cook Mathematical News,* 2 (1980) 109–120. [§4]

[Ede] H. Edelsbrunner, *Algorithms in Combinatorial Geometry,* Springer, New York, 1987. [§6]

[Egg1] H. G. Eggleston, Figures inscribed in convex sets, *American Mathematical Monthly,* 65 (1958) 76–80. [§11]

[Egg2] ——, Minimal universal covers in E^n, *Israel Journal of Mathematics,* 1 (1963) 149–155. [§4]

[Ehr] E. Ehrhart, Un ovale a deux point isocordes?, *Enseignement Mathematiques,* 13 (1967) 119–124. [§2]

[Ell] P. D. T. A. Elliott, On the number of circles determined by n points, *Acta Academiae Scientiarum Hungaricae,* 18 (1967) 181–188. [§6]

[Elv] U. Elverling, The isometry problem, *The Mathematical Intelligencer,* 10, No. 4 (1988) 47. [§8]

[Emc1] A. Emch, Some properties of closed convex curves in a plane, *American Journal of Mathematics,* 35 (1913) 407–412. [§11]

[Emc2] ——, On the medians of a closed convex polygon, *American Journal of Mathematics,* 37 (1915) 19–28. [§11]

[Eng] A. Engel, Geometrical activities for the upper elementary school, *Educational Studies in Mathematics,* 3 (1971) 353–394. [§9]

[Eng′] P. Engel, Über Wirkungsbereichsteilungen von kubischer Symmetrie, *Zeitschrift für Krystallographie,* 154 (1981) 199–215. [§7]

[Erd1] P. Erdős, et al., Three point collinearity (Problem 4065), *American Mathematical Monthly,* 51 (1944) 169–171. [§6]

[Erd2] P. Erdős, Integral distances, *Bulletin of the American Mathematical Society,* 51 (1945) 996. [§10]

[Erd3] ——, Some unsolved problems, *Publications of the Mathematical Institute of the Hungarian Academy of Sciences,* 6 (1961) 221–254. [§6]

[Erd4] ——, Some combinatorial problems in geometry, in *Geometry and Differential Geometry,* (Proceedings of a Conference at the University of Haifa, 1979), Springer Lecture Notes in Mathematics, 792 (1979) 46–53. [§5]

[EGH] P. Erdős, P. M. Gruber, and J. Hammer, *Lattice Points,* Longman Scientific, Harlow, Essex, England (copublished with John Wiley, New York), 1989. [Introduction]

[EHT] P. Erdős, F. Harary, and W. T. Tutte, On the dimension of a graph, *Mathematika,* 12 (1965) 118–122. [§8]

[EK] P. Erdős and S. Kakutani, On non-denumerable graphs, *Bulletin of the American Mathematical Society,* 49 (1953) 457–461. [§8]

[EP1] P. Erdős and G. Purdy, Some extremal problems in geometry. III, *Congressus Numerantium,* 15 (1975) 291–308. [§6]

[EP2] ——, Some extremal problems in combinatorial geometry, in *Handbook of Combinatorics,* R. Graham, M. Grötschel and L. Lovász, eds., North Holland, Amsterdam, 1991. [§6]

[EP3] ——, *Extremal Problems in Discrete Geometry* (tentative title), Wiley, New York, to appear. [Introduction]

[ES1] P. Erdős and G. Szekeres, A combinatorial problem in geometry, *Compositio Mathematica,* 2 (1935) 463–470. [§5]

[ES2] ——, On some extremum problems in elementary geometry, *Annales Universitatis Scientiarum Budapestinensis de Rolando Eőtvős Nominatae, Sectio Mathematica,* 3 (1960/61) 53–62. [§5]

[Eve] H. Eves, *A Survey of Geometry,* vol. 1, Allyn and Bacon, Boston, 1963. [§9]

[ELR] G. Ewald, D. G. Larman, and C. A. Rogers, The directions of the line segments and of the r-dimensional balls on the boundary of a convex body in Euclidean space, *Mathematika,* 17 (1970), 1–20. [§1]

[Fal1] K. J. Falconer, The realization of distances in measurable subsets covering $\mathbb{R}^n$, *Journal of Combinatorial Theory A,* 31 (1981) 187–189. [§8]

[Fal2] ——, On the equireciprocal point problem, *Geometriae Dedicata,* 14 (1983) 113–126. [§2]

[FK] G. Fejes Tóth and W. Kuperberg, A survey of selected recent results in the theory of packing and covering, in *Progress in Discrete and Computational Geometry,* J. Pach, ed., Springer, New York, to appear. [§7]

[Fej] L. Fejes Tóth, On spherical tilings generated by great circles, *Geometriae Dedicata,* 23 (1987) 67–71. [§7]

[Fen] R. Fenn, The table theorem, *Bulletin of the London Mathematical Society,* 2 (1970) 73–76. [§11]

[Fis] K. G. Fischer, Additive k-colorable extensions of the rational plane, *Discrete Mathematics,* 82 (1990) 181–195. [§8]

[Fis′] S. Fisk, A short proof of Chvátal's watchman theorem, *Journal of Combinatorial Theory B,* 24 (1978) 374. [§1]

[FW] P. Frankl and R. M. Wilson, Intersection theorems with geometric consequences, *Combinatorica,* 1 (1981) 357–368. [§8]

[FW′] M. L. Fredman and B. Weide, On the complexity of computing the measure of $\bigcup[a_i, b_i]$, *Communications of the Association for Computing Machinery,* 21 (1978) 540–544. [§3]

[Fuj] M. Fujiwara, Über die Mittelkurve zweier geschlossenen konvexen Kurven in Bezug auf einen Punkt, *Tôhoku Mathematics Journal,* 10 (1916) 99–103. [§2]

[Gal] D. Gale, On Lipschitzian mappings of convex bodies, *Convexity,* V. Klee, ed., Proceedings of Symposia in Pure Mathematics, vol. 7, American Mathematical Society, Providence, 1963, pp. 221–223. [§3]

[Gal′] G. A. Galperin, On the existence of nonperiodic and nondense billiard trajectories in polygons and polyhedra, *Soviet Mathematics Doklady,* 29 (1984) 151–154. [§1]

[Gar1] M. Gardner, A new collection of brain teasers, *Scientific American,* 206 (Oct. 1960) 180. [§8]

[Gar2] ——, On tessellating the plane with convex polygon tiles, *Scientific American,* 221 (July 1975), 112–117. [§7]

[Gar3] ——, *Penrose Tiles to Trapdoor Ciphers,* W. H. Freeman, New York, 1989. [§7]

[Gar′1] R. J. Gardner, Convex bodies equidecomposable by locally discrete groups of isometries, *Mathematika,* 32 (1985) 1–9. [§9]

[Gar′2] ——, Chord functions of convex bodies, *Journal of the London Mathematical Society* (2), 36 (1987) 314–326. [§2]

[Gar′3] ——, Measure theory and some problems in geometry, *Atti Seminario Matematica e Fisica Universitá Modena,* 39 (1991) 39–60. [§2]

[GW] R. J. Gardner and S. Wagon, At long last, the circle has been squared, *Notices of the American Mathematical Society,* 36 (1989) 1338–1343. [§9]

[GN] D. Girault-Beauquier and M. Nivat, Tiling the plane with one tile, *Proceedings of the Sixth Annual Symposium on Computational Geometry,* Association for Computing Machinery, New York, 1990, pp. 128–138. [§7]

[GW′] C. H. Goldberg and D. B. West, Bisection of circle colorings, *SIAM Journal on Algebraic and Discrete Methods,* 6 (1985) 93–106. [§11]

[GP1] J. E. Goodman and R. Pollack, On the combinatorial classification of nondegenerate configurations in the plane, *Journal of Combinatorial Theory A,* 29 (1980) 220–235. [§5]

[GP2] ——, A combinatorial perspective on some problems in geometry, *Congressus Numerantium,* 32 (1981) 383–394. [§§5 and 6]

[GP3] ——, A theorem of ordered duality, *Geometriae Dedicata,* 12 (1982) 63–74. [§5]

[Gra1] R. L. Graham, The largest small hexagon, *Journal of Combinatorial Theory A,* 18 (1975) 165–170. [§4]

[Gra2] ——, Recent developments in Ramsey theory, *Proceedings of the International Congress of Mathematicians* (Warsaw, 1983), 1555–1567. [§11]

[GRS] R. L. Graham, B. L. Rothschild, and E. G. Straus, Are there $n+2$ points in E^n with odd integral distances?, *American Mathematical Monthly,* 81 (1974) 21–25. [§10]

[GJ] D. Greenwell and P. D. Johnson, Functions that preserve unit distance, *Mathematics Magazine,* 49 (1976) 74–79. [§8]

[Gro1] M. L. Gromov, Simplexes inscribed on a hypersurface (Russian), *Matematicheskii Zametki,* 5 (1969) 81–89. [§11]

[Gro2] ——, Monotonicity of the volume of the intersection of balls, in *Geometric Aspects of Functional Analysis,* J. Lindenstrauss and V. D. Milman, eds., Lecture Notes in Mathematics 1267, Springer, Berlin, 1987, pp. 1–4. [§3]

[Gru] P. Gruber, Convex billiards, *Geometriae Dedicata,* 33 (1990) 205–226. [§1]

[Grü1] B. Grünbaum, A generalization of theorems of Kirszbraun and Minty, *Proceedings of the American Mathematical Society,* 13 (1962) 812–814. [§3]

[Grü2] ——, Measures of symmetry for convex sets, in *Convexity,* V. Klee, ed., Proceedings of Symposia in Pure Mathematics, vol. 7, American Mathematical Society, Providence, 1963, pp. 233–270. [§11]

[Grü3] ——, Borsuk's problem and related questions, in *Convexity,* V. Klee, ed., Proceedings of Symposia in Pure Mathematics, vol. 7, American Mathematical Society, Providence, 1963, pp. 271–283. [§4]

[Grü4] ——, Arrangements of hyperplanes, in *Proceedings of the Second Louisiana Conference on Combinatorics, Graph Theory and Computing,* R. Mullin, K. B. Reid, D. P. Roselle, and R. S. D. Thomas, eds., Louisiana State University, Baton Rouge, 1971, pp. 41–106.

[Grü5] ——, *Arrangements and Spreads,* Regional Conference Series in Mathematics, vol. 10, American Mathematical Society, Providence, 1972. [§6]

[GS1] B. Grünbaum and G. C. Shephard, Tilings with congruent tiles, *Bulletin of the American Mathematical Society,* 3 (1980) 951–973. [§7]

[GS2] ——, Some problems on plane tilings, in *The Mathematical Gardner,* D. Klarner, ed., Prindle, Weber and Schmidt, Boston, 1981, pp. 167–196. [§7]

[GS3] ——, Is there an all-purpose tile?, *American Mathematical Monthly,* 93 (1986) 545–551. [§7]

[GS4] ——, *Tilings and Patterns,* W. H. Freeman, New York, 1987. [Introduction and §7]

[Gug] H. Guggenheimer, Finite sets on curves and surfaces, *Israel Journal of Mathematics,* 3 (1965) 104–112. [§11]

[GM] V. Guillemin and R. Melrose, An inverse spectral result for elliptical regions, *Advances in Mathematics,* 32 (1979) 128–148. [§1]

[Guy] R. K. Guy, *Unsolved Problems in Number Theory,* Springer, New York, 1981. [§10]

[GK] R. Guy and V. Klee, Monthly research problems, 1969–71, *American Mathematical Monthly,* 78 (1971) 1113–1122. [§1]

[Had1] H. Hadwiger, Überdeckung des euklidischen Raum durch kongruente Mengen, *Portugaliae Mathematicae,* 4 (1945) 238–242. [§8]

[Had2] ——, Ungelöste Probleme, Nr. 11, *Elemente der Mathematik,* 11 (1956) 60–61. [§3]

[Had3] ——, Ungelöste Probleme, Nr. 40, *Elemente der Mathematik,* 16 (1961) 103–104. [§8]

[HDK] H. Hadwiger, H. Debrunner, and V. Klee, *Combinatorial Geometry in the Plane,* Holt, Rinehart and Winston, New York, 1964. [§§8 and 11]

[HLM] H. Hadwiger, D. Larman, and P. Mani, Hyperrhombs inscribed to convex bodies, *Journal of Combinatorial Theory B,* 24 (1978) 290–293. [§11]

[Hag1] C. L. Hagopian, Uniquely arcwise connected plane continua have the fixed-point property, *Transactions of the American Mathematical Society,* 248 (1979) 85–104. [§12]

[Hag2] ——, The fixed-point property for deformations of uniquely arcwise connected continua, *Topology and its Applications,* 24 (1986) 207–212. [§12]

[Hag3] ——, Fixed points of arc-component-preserving maps, *Transactions of the American Mathematical Society,* 306 (1988) 411–420. [§12]

[Hag4] ——, Fixed points of tree-like continua, in *Fixed Point Theory and its Applications,* R. F. Brown, ed., Contemporary Mathematics, vol. 72, American Mathematical Society, Providence, 1988 pp. 131–137. [§12]

[Hag5] ——, Fixed points of plane continua, *Rocky Mountain Journal of Mathematics,* to appear. [§12]

[Haj] G. Hajos, Über einfach und mehrfache Bedeckung des n-dimensionalen Raumes mit einem Würfelgitter, *Mathematische Zeitschrift,* 47 (1942) 427–467. [§7]

[HS] A. W. Hales and E. G. Straus, Projective colorings, *Pacific Journal of Mathematics,* 99 (1982) 31–43. [§7]

[Hal] A. Hallstrom, Equichordal and equireciprocal points, *Bogazici University Journal of Science,* 2 (1974) 83–88. [§2]

[Ham] O. H. Hamilton, A fixed-point theorem for pseudo-arcs and certain other metric continua, *Proceedings of the American Mathematical Society,* 2 (1951) 173–174. [§12]

[Ham′] J. Hammer, *Unsolved Problems Concerning Lattice Points,* Pitman, London, 1977. [Introduction]

[Han] H. C. Hansen, A smaller universal cover for sets of unit diameter, *Geometriae Dedicata,* 4 (1975) 165–172. [§4]

[Han′1] S. Hansen, A generalization of a theorem of Sylvester on lines determined by a finite set, *Mathematica Scandinavica,* 16 (1965) 175–180. [§6]

[Han′2] ——, On configurations in 3-space without elementary planes and on the number of ordinary planes, *Mathematica Scandinavica,* 47 (1980) 181–194. [§6]

[Han′3] ——, *Contributions to the Sylvester–Gallai-Theory,* Doctoral dissertation, University of Copenhagen, 1981. [§6]

[Har1] H. Harborth, On the problem of P. Erdős concerning points with integral distances, *Annals of the New York Academy of Science,* 175 (1970) 206–207. [§10]

[Har2] ——, Antwort auf eine Frage von P. Erdős nach fünf Punkten mit ganzahligen Abständen, *Elemente der Mathematik,* 26 (1971) 112–113. [§10]

[Har3] ——, Konvexe Fünfecke in ebenen Punktmengen, *Elemente der Mathematik,* 34 (1978) 116–118. [§5]

[HK] H. Harborth and A. Kemnitz, Diameters of integral point sets, *Intuitive Geometry,* K. Böröczky and G. Fejes Tóth, eds., Colloquia Mathematica Societatis Janós Bolyai, vol. 48, 1987, pp. 255–266. [§10]

[HW] G. H. Hardy and E. M. Wright, *An Introduction to the Theory of Numbers,* 4th ed., Oxford, London, 1960. [§§1 and 8]

[Hay] T. Hayashi, On pseudo-central oval (Japanese), *Tôhoku Mathematics Journal,* 27 (1926) 197–202. [§2]

[Hee] H. Heesch, *Reguläres Parkettierungsproblem,* Westdeutscher Verlag, Cologne-Opladen, 1968. [§7]

[HK′] H. Heesch and O. Kienzle, *Flächenschluss. System der Formen lückenlos aneinanderschliessender Flachteile,* Springer, Berlin, 1963. [§7]

[Hel] E. Helly, Über Mengen konvexer Körper mit gemeinschaftlichen Punkten, *Jahresbericht der Deutschen Mathematischen Vereinigung,* 32 (1923) 175–176. [§3]

[HC] D. Hilbert and S. Cohn-Vossen, *Geometry and the Imagination,* trans. P. Nemenyi, Chelsea, New York, 1952. [§1]

[HH] M. D. Hirschhorn and D. C. Hunt, Equilateral convex polygons which tile the plane, *Journal of Combinatorial Theory A,* 39 (1985) 1–18. [§7]

[HY] J. G. Hocking and G. S. Young, *Topology,* Addison-Wesley, Reading, Mass., 1961. [§12]

[Hol] F. B. Holt, Reflecting paths of period $2 + 4k$ in right triangles, manuscript, Department of Mathematics, University of Washington, Seattle, 1990. [§1]

[Hop] H. Hopf, Über die Sehnen ebener Kontinuen und die Schleifen gescholossenen Wege, *Commentarii Mathematici Helvetici,* 9 (1936-37) 303–319. [§11]

[Hor] J. D. Horton, Sets with no empty 7-gon, *Canadian Mathematics Bulletin,* 26 (1983) 482–484. [§5]

[HK″] P. D. Humke and L. L. Krajewski, A characterization of circles which contain rational points, *American Mathematical Monthly,* 86 (1979) 287–290. [§10]

[HW′] W. Hurewicz and H. Wallman, *Dimension Theory,* Princeton University Press, Princeton, 1941. [§8]

[Jam] R. E. Jamison, Direction trees, *Discrete and Computational Geometry,* 2 (1987) 249–254. [§6]

[Jer] R. Jerrard, Inscribed squares in plane curves, *Transactions of the American Mathematical Society,* 98 (1961) 234–241. [§11]

[Joh] P. D. Johnson, Coloring abelian groups, *Discrete Mathematics,* 40 (1982) 219–223. [§8]

[Joh′] S. Johnson, A new proof of the Erdős-Szekeres convex n-gon result, *Journal of Combinatorial Theory A,* 42 (1986) 318–319. [§5]

[Kak] S. Kakutani, A proof that there exists a circumscribing cube around any closed bounded convex set in R^3, *Annals of Mathematics,* 43 (1942) 739–741. [§11]

[KKS] J. D. Kalbfleisch, J. G. Kalbfleisch, and R. G. Stanton, A combinatorial problem on convex regions, *Proceedings of the Louisiana Conference on Combinatorics, Graph Theory and Computing,* R. C. Mullin, K. B. Reid, and D. P. Roselle, eds., 1970, pp. 180–188. [§5]

[KS] G. K. Kalisch and E. G. Straus, On the determination of points in a Banach space by their distances from the points of a given set, *Anais Academii Brasililiensis Ciencias,* 29 (1958) 501–519. [§7]

[Kas] E. A. Kasimatis, Dissections of regular polygons into triangles of equal areas, *Discrete and Computational Geometry,* 4 (1989) 375–381. [§7]

[KS′] E. A. Kasimatis and S. K. Stein, Equidissections of polygons, *Discrete Mathematics,* 85 (1990) 281–294. [§7]

[Kel] J. B. Kelly, Power points, *American Mathematical Monthly,* 53 (1946) 395–396. [§2]

[KM] L. M. Kelly and W. Moser, On the number of ordinary lines determined by n points, *Canadian Journal of Mathematics* 10 (1958) 210–219. [§6]

[Kel′] P. J. Kelly, Curves with a kind of constant width, *American Mathematical Monthly,* 64 (1957) 333–336. [§2]

[KMS] S. Kerckhoff, H. Masur, and J. Smillie, Ergodicity of billiard flows and quadratic differentials, *Annals of Mathematics,* 124 (1986) 293–311. [§1]

[Ker1] R. B. Kershner, On paving the plane, *American Mathematical Monthly,* 75 (1968) 839–844. [§7]

[Ker2] ——, On paving the plane, *APL Technical Digest,* 8 (1969) 4–10. [§7]

[KK] B. Kind and P. Kleinschmidt, On the maximal volume of convex bodies with few vertices, *Journal of Combinatorial Theory A,* 21 (1976) 124–128. [§4]

[Kir] M. Kirszbraun, Über die zusammenziehenden und Lipschitzen Transformationen, *Fundamenta Mathematicae,* 22 (1934) 77–108. [§3]

[Kle1] V. Klee, An example related to the fixed-point property, *Nieuw Archief voor Wiskunde,* 8 (1960) 81–82. [§12]

[Kle2] ——, Is every polygonal region illuminable from some point?, *American Mathematical Monthly,* 76 (1969) 80. [§1]

[Kne1] M. Kneser, Einige Bemerkungen über das Minkowskische Flächenmaß, *Archiv der Mathematik,* 6 (1955) 382–390. [§3]

[Kne2] ——, Letter of May, 1968, to B. Bollobás. [§3]

[KS″] D. König and A. Szücs, Mouvement d'un point abandonné à l'intérieur d'un cube, *Rendiconti della Circolo Palermo,* 36 (1913) 79–90. [§1]

[KP] E. Kranakis and M. Pocchiola, Enumeration and visibility problems in integer lattices, *Proceedings of Sixth Annual Symposium on Computational Geometry,* Association for Computing Machinery, New York, 1990, pp. 261–270. [§6]

[KK′1] E. H. Kronheimer and P. B. Kronheimer, The tripos problem, *Journal of the London Mathematical Society,* (2), 24 (1981) 182–192. [§11]

[KK′2] ——, The n-dimensional simple tripos problem, *Journal of the London Mathematical Society,* (2), 31 (1985) 369–372. [§11]

[Kui] N. H. Kuiper, Double normals of convex bodies, *Israel Journal of Mathematics,* 2 (1964) 71–80. [§1]

[Kum] E. Kummer, *Journal für Mathematik,* 37 (1848) 1–20. [§10]

[Kup] W. Kuperberg, Packing convex bodies in the plane with density greater than 3/4, *Geometriae Dedicata,* 13 (1982) 149–155. [§7]

[Kuz] A. V. Kuzminyh, Mappings preserving unit distance (Russian), *Siberian Mathematics Journal,* 20 (1979) 417–421. [§8]

[Lac1] M. Laczkovich, Equidecomposability and discrepancy: a solution of Tarski's circle-squaring problem, *Journal für die Reine und Angewandte Mathematik,* 404 (1990) 77–117. [§9]

[Lac2] ——, Tilings of polygons with similar triangles, *Combinatorica,* 10 (1990) 281–306. [§7]

[Lag] J. Lagrange, Points du plan dont les distances mutuelles sont rationelles, Séminaire de Théorie des Nombres, Université de Bordeaux, Exposition 27, April 4, 1983. [§10]

[Lar] D. G. Larman, A note on the realization of distances within sets in euclidean space, *Commentarii Mathematici Helvetici,* 53 (1978) 529–535. [§8]

[LR] D. G. Larman and C.A. Rogers, The realization of distances within sets in euclidean space, *Mathematika,* 19 (1972) 1–24. [§8]

[LRS] D. G. Larman, C. A. Rogers, and J. J. Seidel, On the two distance problem, *Bulletin of the London Mathematical Society,* 9 (1977) 261–267. [§§8 and 10]

[LT] D. G. Larman and N. K. Tamvakis, A characterization of centrally symmetric convex bodies in E^n, *Geometriae Dedicata,* 10 (1981) 161–176. [§2]

[Las1] M. Lassak, Solution of Hadwiger's covering problem for centrally symmetric convex bodies in E^3, *Journal of the London Mathematical Society,* 30 (1984) 501–511. [§1]

[Las2] ——, Covering the boundary of a convex set by tiles, *Proceedings of the American Mathematical Society,* 104 (1988) 269–272. [§1]

[Law] J. Lawrence, Tiling $\mathbb{R}^d$ by translates of the orthants, in *Convexity and Related Combinatorial Geometry,* D. C. Kay and M. Breen, eds., Marcel Dekker, New York, 1982, pp. 203–207. [§7]

[Laz] V. F. Lazutkin, The existence of caustics for a billiard problem in a convex domain, *Mathematics of the USSR—Izvestija,* 7 (1973) 185–214. [§1]

[Len] H. Lenz, Der Satz von Beckmann–Quarles in rationaler Raum, *Archiv der Mathematik,* 49 (1987) 106–113. [§8]

[Lev] F. W. Levi, Überdeckung eines Eibereiches durch Parallelverschiebungen seines offenen Kerns, *Archiv der Mathematik,* 6 (1955) 369–370. [§1]

[Lév] P. Lévy, Sur une généralisation du théorème de Rolle, *Comptes Rendus de l'Académie Scientifique de Paris,* 198 (1934) 424–425. [§11]

[Lie] E. H. Lieb, Monotonicity of the molecular electronic energy in the nuclear coordinates, *Journal of Physics B: Atomic and Molecular Physics,* 15 (1982) L63–66. [§3]

[LS] E. H. Lieb and B. Simon, Monotonicity of the electronic contribution to the Born–Oppenheimer energy, *Journal of Physics B: Atomic and Molecular Physics,* 11 (1978) L537–542. [§3]

[Lov] L. Lovász, *Combinatorial Problems and Exercises,* North Holland, Amsterdam, 1979. [§5]

[Lub] A. Lubiw, Decomposing polygonal regions into convex quadrilaterals, *Proceedings of the First Annual Symposium on Computational Geometry,* Association for Computing Machinery, New York, 1985, pp. 97–106. [§7]

[Mas] H. Masur, Closed trajectories for quadratic differentials with application to billiards, *Duke Mathematical Journal,* 53 (1986) 307–314. [§1]

[Mat] J. N. Mather, Glancing billiards, *Ergodic Theory and Dynamical Systems,* 2 (1982) 397–403. [§1]

[McM] P. McMullen, Convex bodies which tile by translation, *Mathematika,* 27 (1980) 113–121.

[Mea] D. G. Mead, Dissection of the hypercube into simplices, *Proceedings of the American Mathematical Society,* 76 (1979) 302–304. [§7]

[Mel] E. Melchior, Über Vielseite der projektiven Ebene, *Deutsche Mathematik,* 5 (1940) 461–475.

[Mey1] M. D. Meyerson, Equilateral triangles and continuous curves, *Fundamenta Mathematicae,* 110 (1980) 1–9. [§11]

[Mey2] ——, Convexity and the table theorem, *Pacific Journal of Mathematics,* 97 (1981) 167–169. [§11]

[Mey3] ——, Balancing acts, *Topology Proceedings,* 6 (1981) 59–75. [§11]

[Mey4] ——, Remarks on Fenn's "The table theorem" and Zaks' "The chair theorem," *Pacific Journal of Mathematics,* 110 (1984) 167–169. [§11]

[MV] G. Michelacci and A. Volčič, A better bound for the eccentricities not admitting the equichordal body, *Archiv der Mathematik,* to appear. [§2]

[Mil] A. N. Milgram, Some metric topological invariants, *Reports of a Mathematical Colloquium* (2), 5–6 (1939–1948) 25–35, University of Notre Dame, South Bend, Indiana. [§11]

[Min] G. J. Minty, On the extension of Lipschitz, Lipschitz–Hölder continuous, and monotone functions, *Bulletin of the American Mathematical Society,* 76 (1970) 334–339. [§3]

[Mon1] P. Monsky, On dividing a square into triangles, *American Mathematical Monthly,* 77 (1970) 161–164. [§7]

[Mon2] ——, A conjecture of Stein on plane dissections, *Mathematische Zeitschrift,* 205 (1990) 583–592. [§7]

[Mor] L. J. Mordell, Rational quadrilaterals, *Journal of the London Mathematical Society,* 35 (1960) 277–282. [§10]

[MM] L. Moser and W. Moser, Solution to Problem 10, *Canadian Mathematics Bulletin,* 4 (1961) 187–189. [§8]

[MP] W. Moser and J. Pach, *Research Problems in Discrete Geometry,* McGill University, 1986 (new addition to be published by Academic Press). [Introduction and §5]

[Mot] T. S. Motzkin, The lines and planes connecting the points of a finite set, *Transactions of the American Mathematical Society,* 70 (1951) 451–464. [§6]

[MO] T. S. Motzkin and P. E. O'Neil, Bounds assuring sets in convex position, *Journal of Combinatorial Theory,* 3 (1967) 252–255. [§5]

[Niv] I. Niven, Convex polygons which cannot tile the plane, *American Mathematical Monthly,* 85 (1978) 785–792. [§7]

[Ogi1] C. S. Ogilvy, Square inscribed in an arbitrary simple closed curve, *American Mathematical Monthly,* 57 (1950) 423–424. [§11]

[Ogi2] ——, *Tomorrow's Math,* 2nd ed., Oxford, New York, 1972. [§11]

[ORo] J. O'Rourke, *Art Gallery Theorems and Algorithms,* Oxford, New York, 1987. [§1]

[OSV] M. Overmars, B. Scholten, and I. Vincent, Sets without empty convex 6-gons, Technical Report RUU-CS-88-12, Department of Computer Science, University of Utrecht, 1988. [§5]

[Pál1] J. Pál, Über ein elementäres Variationsproblem, *Danske Videnskabernes Selskabet Matematiske og Fysiske Meddelse,* 3 (1920), No.2. [§4]

[Pál2] ——, Ein Minimumproblem für Ovale, *Mathematische Annalen,* 83 (1921) 311–319. [§4]

[PP] L. Penrose and R. Penrose, Puzzles for Christmas, *New Scientist,* 25 December 1958, 1580–1581, 1597. [§1]

[Per] R. Perrin, Sur le problème des aspects, *Bulletin de la Société Mathématique de France,* 10 (1881–1882) 103–127. [§5]

[Per′] O. Perron, Über die lückenlose Erfüllung des Raumes mit Würfeln, *Mathematische Zeitschrift,* 46 (1940) 161–180. [§7]

[PC] C. M. Petty and J. M. Crotty, Characterization of spherical neighborhoods, *Canadian Journal of Mathematics,* 2 (1970) 431–435. [§2]

[Puc] C. Pucci, Sulla inscrivibilita di un ottaedro regolare in un insieme convesso limitato dell spazio ordinaria, *Atti della Academia Nazionale dei Lincei. Rendiconti Classe di Scienze Fisiche, Matematiche e Naturali,* 8 (1956) 61–65. [§11]

[Pur1] G. Purdy, A proof of a consequence of Dirac's conjecture, *Geometriae Dedicata,* 10 (1981) 317–321. [§6]

[Pur2] ——, Two results about points, lines and planes, *Discrete Mathematics,* 60 (1986) 215–218. [§6]

[Pur3] ——, Some combinatorial problems in the plane, in *Discrete Geometry and Convexity,* J. E. Goodman, E. Lutwak, J. Malkevitch, and R. Pollack, eds., Annals of the New York Academy of Sciences, 440 (1985) 65–68. [§6]

[PP] C. Purdy and G. Purdy, Minimal forbidden distance one graphs, *Congressus Numerantium,* 66 (1988) 165–171. [§8]

[Rad] J. Radon, Mengen konvexer Körper, die einen gemeinsamen Punkt enhalten, *Mathematischen Annalen,* 83 (1921) 113–115. [§3]

[Rai] D. E. Raiskii, Realization of all distances in a decomposition of the space R^n into $(n + 1)$ parts, *Mathematical Notes,* 7 (1970) 194–196; trans. from *Mathematicheskii Zametki,* 7 (1970) 319–323. [§8]

[Ram] F. P. Ramsey, On a problem of formal logic, *Proceedings of the London Mathematical Society,* (2), 30 (1929) 264–286. [§5]

[Rau] J. Rauch, Illumination of bounded domains, *American Mathematical Monthly,* 85 (1978) 359–361. [§1]

[RT] J. Rauch and M. Taylor, Penetration into shadow regions and unique continuation properties in hyperbolic mixed problems, *Indiana University Mathematics Journal,* 22 (1972) 277–285. [§1]

[Reh] W. Rehder, On the volume of unions of translates of a convex set, *American Mathematical Monthly,* 87 (1980) 382–384. [§3]

[Rei1] K. Reinhardt, *Uber die Zerlegung der Ebene in Polygone,* Inaugural-Dissertation, University of Frankfurt, R. Noske, Borna and Leipzig, 1918. [§7]

[Rei2] ——, Extreme Polygone mit gegebenen Durchmessers, *Jahresbericht der deutschen mathematischen Gesellschaft,* 31 (1922) 251–270. [§4]

[Ren] J. Renegar, The computational complexity and geometry of the first order theory of the reals, *Journal of Symbolic Computation,* to appear. [§7]

[Ren′] B. C. Rennie, The search for a universal cover, *Eureka,* 3 (1977) 61–63. Carleton-Ottawa Mathematics Association, Canada. [§4]

[Rog] C. A. Rogers, An equichordal problem, *Geometriae Dedicata,* 10 (1981) 73–78. [§2]

[Rog′1] J. C. W. Rogers, Remarks on the equichordal problem, manuscript, Polytechnic University of New York, 1989. [§2]

[Rog′2] ——, Asymptotic behavior of solutions of a nonlinear recursion relation arising in the study of the equichordal problem, research announcement, 1990. [§2]

[Ros] J. Rosenbaum, Power points, *American Mathematical Monthly,* 53 (1946) 395–396. [§2]

[RZ] M. Rosenfeld and J. Zaks, eds., *Convexity and Graph Theory,* Annals of Discrete Mathematics, vol. 20, North-Holland, Amsterdam, 1984. [§6]

[Rys] H. J. Ryser, *Combinatorial Mathematics,* Mathematical Association of America, Washington, D.C., 1963. [§6]

[SV] R. Schäfke and H. Volkmer, Asymptotic analysis of the equichordal problem, manuscript, University of Essen, 1990. [§2]

[Sch] D. Schattschneider, Tiling the plane with congruent pentagons, *Mathematics Magazine,* 51 (1978) 29–44. [§7]

[Sch′] I. J. Schoenberg, On a theorem of Kirszbraun and Valentine, *American Mathematical Monthly,* 60 (1953) 620–622. [§3]

[Sch″] O. Schramm, Illuminating sets of constant width, *Mathematika,* 35 (1988) 180–189. [§1]

[Sco] P. R. Scott, On the sets of directions determined by n points, *American Mathematical Monthly,* 70 (1970) 502–505. [§6]

[Sei] A. Seidenberg, A simple proof of the theorem of Erdős and Szekeres, *Journal of the London Mathematical Society,* 34 (1959) 352. [§5]

[Sen] M. Senechal, Which tetrahedra fill space?, *Mathematics Magazine,* 54 (1981) 227–244. [§7]

[She] T. K. Sheng, Rational polygons, *Journal of the Australian Mathematical Society,* 6 (1966) 452–459. [§10]

[SD] T. K. Sheng and D. E. Daykin, On approximating polygons by rational polygons, *Mathematics Magazine,* 38 (1965) 46–47. [§10]

[She′] T. Shermer, Recent results in art galleries, Technical Report TR 90-10, Department of Computer Science, Simon Fraser University, Burnaby, British Columbia, Canada, 1990. (To appear in *Proceedings of the IEEE* in 1992, in a special issue on computational geometry.) [§1]

[Sht] M. I. Shtogrin, Action centers of planigons, *Mathematical Notes of the Academy of Sciences USSR,* 44 (1988) 627–635; trans. from *Mathematicheskie Zametki,* 44(1988) 262–278. [§7]

[Sim] G. J. Simmons, The chromatic number of the sphere, *Journal of the Australian Mathematical Society,* Ser. A, 21 (1976) 473–480. [§8]

[SK] R. Sine and V. Kreĭnovič, Remarks on billiards, *American Mathematical Monthly,* 86 (1979) 204–206. [§1]

[Sni] L. S'nirelman, On certain geometrical properties of closed curves (Russian), *Uspehi Matematičeskih Nauk,* 10 (1944) 34–44. [§11]

[Spa] N. Spaltenstein, A family of curves with two equichordal points on a sphere, *American Mathematical Monthly,* 91 (1984) 423. [§2]

[Spe] E. Sperner, Neuer Beweis für die Invarianz der Dimensionzahl und des Gebietes, *Abhandlungen aus dem Mathematischen Seminar der Universität Hamburg,* 6 (1928) 265–272. [§12]

[Spi] P. G. Spirakis, The volume of the union of many spheres and point inclusion problems, *Proceedings of the Second Annual Symposium on Theoretical Aspects of Computer Science,* Lecture Notes in Computer Science 182, Springer, Berlin, 1985, pp. 328–338. [§3]

[Spr] R. Sprague, Über ein elementäres Variationsproblem, *Matematisk Tiddskrift,* 1936 (1936) 96–99.

[Ste] F. Steiger, Zu einer Frage über Mengen von Punkten mit ganzzahliger Entfernung, *Elemente der Mathematik,* 8 (1953) 66–67. [§10]

[Ste′] R. Stein, A new pentagon tiler, *Mathematics Magazine,* 58 (1985) 308. [§7]

[Ste″1] S. K. Stein, Algebraic tiling, *American Mathematical Monthly,* 81 (1974) 445–462. [§7]

[Ste″2] ——, Equidissections of centrally symmetric octagons, *Aequationes Mathematicae,* 37 (1989) 313–318. [§7]

[Sti] W. P. van Stigt, *Brouwer's Intuitionism,* North-Holland, Amsterdam, 1990. [§12]

[Str] W. Stromquist, Inscribed squares and square-like quadrilaterals in closed curves, *Mathematika,* 36 (1989) 187–197. [§11]

[Sud] V. N. Sudakov, Gaussian random processes and measures of solid angles in Hilbert space, *Soviet Mathematics Doklady,* 12 (1971) 412– 415. [§3]

[Sud′] G. Sudan, Sur le problème de rayon réfléchi, *Revue Roumaine de Mathématiques Pures et Appliquées,* 10 (1965) 723–733. [§1]

[Süs] W. Süss, Eibereiche mit ausgezeichneten Punkten, Sehnen, Inhalts und Umfangspunkt, *Tôhoku Mathematics Journal,* 25 (1925) 86–98. [§2]

[Syl] J. J. Sylvester, Mathematical Question 11851, *Educational Times,* 59 (1893) 98. [§6]

[SW] L. A. Székely and N. C. Wormald, Bounds on the measurable chromatic number of $\mathbb{R}^n$, *Discrete Mathematics,* 75 (1989) 343–372. [§8]

[ST] E. Szemerédi and W. T. Trotter, Extremal problems in discrete geometry, *Combinatorica,* 3 (1983) 381–392. [§6]

[Tar1] A. Tarski, Problème 38, *Fundamenta Mathematicae,* 7 (1925) 381. [§9]

[Tar2] ——, *A Decision Method for Elementary Algebra and Geometry,* 2nd ed., University of California Press, Berkeley, 1961. [§7]

[Tho] J. Thomas, A dissection problem, *Mathematics Magazine,* 41 (1968) 187–190. [§7]

[Thu] E. Thue Poulsen, Problem 10, *Mathematica Scandinavica,* 2 (1954) 346. [§3]

[Toe] O. Toeplitz, *Verhandlungen der Schweizerischen Naturforschendedn Gesellschaft in Solothurn* (August 1, 1911), 197. [§11]

[Tom] C. B. Tompkins, Sperner's lemma and some extensions, in *Applied Combinatorial Mathematics,* E.F. Beckenbach, ed., Wiley, New York, 1964, pp. 416–455. [§12]

[Tuc] A. W. Tucker, Some topological properties of disk and sphere, *Proceedings of the First Canadian Mathematical Congress,* Montreal, 1945, pp. 285–309. [§§8 and 12]

[Tur1] P. H. Turner, *Aspects of Convexity in Billiard Ball Dynamical Systems,* Ph.D. Dissertation, Department of Mathematics, Wesleyan University, Middletown, Connecticut, 1980. [§1]

[Tur2] ——, Convex caustics for billiards, in *Convexity and Related Combinatorial Geometry,* D. C. Kay and M. Breen, eds., Marcel Dekker, New York, 1982, pp. 85–106. [§1]

[Ung] P. Ungar, $2N$ noncollinear points determine at least $2N$ directions, *Journal of Combinatorial Theory A,* 33 (1982) 343–347. [§6]

[VW] J. van Leeuwen and D. Wood, The measure problem for rectangular ranges in d-space, *Journal of Algorithms,* 2 (1981) 282–300. [§3]

[Ven] B. A. Venkov, On a class of Euclidean polyhedra (Russian), *Vestnik Leningradskogo Universiteta. Seriya Matematiki, Fiziki, Himii* (Leningrad) 9 (1954) 11–31. [§7]

[Wag] S. Wagon, *The Banach–Tarski Paradox,* Cambridge University Press, New York, 1985. [§9]

[Wir] E. Wirsing, Zur Analytizität von Doppelspeichenkurven, *Archiv der Mathematik,* 9 (1958) 300–307. [§2]

[WW] J. A. Wiseman and P. R. Wilson, A Sylvester theorem for conic sections, *Discrete and Computational Geometry,* 3 (1988) 295–305. [§6]

[Woo] D. R. Woodall, Distances realized by sets covering the plane, *Journal of Combinatorial Theory,* 14 (1973) 187–200. [§8]

[YY] H. Yamabe and Z. Yujobô, On the continuous functions defined on a sphere, *Osaka Mathematics Journal,* 2 (1950) 112–116. [§11]

[Yan1] K. Yanagihara, On a characteristic property of the circle and the sphere, *Tôhoku Mathematics Journal,* 10 (1916) 142–143. [§2]

[Yan2] ——, Second note on a characteristic property of the circle and the sphere, *Tôhoku Mathematics Journal,* 11 (1917) 55–57. [§2]

[You] G. S. Young, Fixed-point theorems for arcwise connected continua, *Proceedings of the American Mathematical Society,* 11 (1960) 880–884. [§12]

[Zak1] J. Zaks, The chair theorem, in *Proceedings of the Second Lousiana Conference on Combinatorics, Graph Theory and Computing,* Louisiana State University, Baton Rouge, 1971, pp. 557–562. [§11]

[Zak2] ——, On four-colourings of the rational four-space, *Aequationes Mathematicae,* 37 (1989) 259–266. [§8]

[Zak3] ——, On the connectedness of some geometric graphs, *Journal of Combinatorial Theory B,* 49 (1990) 143–150. [§8]

[Zak4] ——, Uniform distances in rational unit-distance graphs, *Annals of Discrete Mathematics,* to appear. [§8]

[Zam] T. Zamfirescu, Baire categories in convexity, *Atti Seminario Matematica e Fisica Universitá Modena,* 39 (1991) 279–304. [§§1 and 11]

[ZK] A. Zemlyakov and A. Katok, Topological transitivity of billards in polygons, *Mathematical Notes of the Academy of Sciences USSR,* 18 (1975) 760–764; trans. from *Matematicheskii Zametki,* 18 (1975) 291–300. [§1]

[Zin] K. Zindler, Über konvexe Gebilde, *Monatshefte für Mathematik,* 31 (1921) 25–57. [§11]

[Zuc] L. Zuccheri, Characterization of the circle by equipower properties, preprint, University of Trieste, 1989. [§2]

CHAPTER 2
NUMBER THEORY

INTRODUCTION

Many unsolved problems in mathematics have famous (or infamous) reputations, but the adjective "notorious" is used more often to describe problems in number theory than those in other fields. This is because problems in number theory that sound so simple, involving only the most elementary notions of arithmetic, have proven to be exceptionally difficult. Because of the fame attached to certain unsolved problems (such as Fermat's last theorem or the existence of an odd perfect number), there regularly surface claims of a solution. But all such claims have turned out to be false, which adds to the notoriety of the problems. Also contributing to the mystique is the long history of some of these problems: Fermat's last theorem (it is a conjecture, not a theorem) goes back four centuries, while the search for perfect numbers began over two thousand years ago.

Several of the problems in this chapter can be understood by anyone knowing how to add and multiply. We begin with Fermat's problem, which asks whether two nth powers of positive integers can sum to another nth power (where $n > 2$). The next problem also deals with simple equations, a solution to which would yield a three-dimensional box that generalizes the famous 3-4-5 right triangle. And Problem 18, despite being posed in the twentieth century, involves the iteration of a function defined very simply in terms of multiplication by 3 and division by 2.

Prime numbers, those not divisible by any other number (except themselves and 1) are fundamental in number theory, and they are related to the issues raised in Problems 16–18. Problem 16, which has its origins in the search for perfect numbers, is tied to the question: Are there infinitely many primes of the form $2^n - 1$? Problem 17 is the famous Riemann hypothesis, which is equivalent to an assertion about the accuracy of the estimate $\pi(x) \approx x/\log x$, where $\pi(x)$ is the number of primes less than x. Problem 18, a more modern question related to the capabilities of high-speed computers, asks whether we can tell whether a number is prime in a reasonable amount of time. More precisely, the question is whether there is a polynomial time algorithm to determine whether a number is prime, by which is meant an algorithm for which there exists a polynomial function $p(n)$ such that the algorithm always halts in fewer than $p(n)$ steps for an input whose length (in this case, the number of digits) is n. The point is that the naive algorithms require an exponential amount of time, which, for large numbers, means a wait of months or years for an answer.

Algorithmic questions, a central theme of modern number theory, are also dealt with in Problems 13, 19, and 20. Problem 20 asks whether there is an algorithm that will decide, for certain equations, whether a solution in integers exists. In Problems 13 and 19, specific algorithms are defined, but it is not known whether they halt for every input.

The problem of listing interesting unsolved problems in number theory is one that definitely does not halt, and we hope the reader will gain a flavor of the field from the selection of old and new problems presented in this chapter. The reader interested in additional elementary unsolved problems in this area should consult the collections in the books by Guy [Guy] and Erdős and Graham [EG].

13. FERMAT'S LAST THEOREM

Problem 13

Do there exist positive integers x, y, and z and an integer $n \geq 3$ such that $x^n + y^n = z^n$?

It has been known since ancient times that two square numbers can add up to another square. The classical example is $3^2 + 4^2 = 5^2$, but there are infinitely many others, called Pythagorean triples. Pierre de Fermat (1608–1665) was the first to consider the generalization to arbitrary powers. In his copy of a Greek text by Diophantus he made his immortal comment that "It is impossible to separate a cube into two cubes, or a biquadrate [fourth power] into two biquadrates, or in general any power higher than the second into powers of like degree; I have discovered a truly remarkable proof, which this margin is too small to contain." To this day his claim has remained unproved, and Fermat's last "theorem" has become one of the most notorious unsolved problems about the integers. In number theory especially there are lots of easily-stated unsolved problems for which very little is known. But Fermat's last theorem is exceptional in that it has attracted the efforts of many mathematicians of the first rank, often with fruitful results. Although their work has not yet established Fermat's claim, it has given rise to a rich theory, in some cases opening up entirely new fields that have been central to the development of modern mathematics.

We list below some of the highlights in the history of Fermat's last theorem (FLT). First, some basic observations. If FLT is true for exponent n then it is true for all multiples of n, for if x, y, z satisfy Fermat's equation ($x^n + y^n = z^n$) with exponent kn, then x^k, y^k, z^k satisfy the equation with exponent n. It follows that to prove FLT it suffices to prove it for prime exponents and exponent 4, and this explains the interest in prime values of the exponent. Moreover, if x, y, z have a common divisor d and satisfy Fermat's equation with exponent n, then $\frac{x}{d}$, $\frac{y}{d}$, $\frac{z}{d}$ also satisfy the equation. Thus it suffices to consider *primitive* triples, by which we mean triples x, y, z having no common divisor.

1659	Fermat	FLT is true for $n = 4$.
1753	Euler	FLT is true for $n = 3$.
1825	Dirichlet, Legendre	FLT is true for $n = 5$.
1839	Lamé	FLT is true for $n = 7$.
1847	Kummer	FLT is true for $n < 37$ (using his powerful theorem that FLT is true whenever the exponent is a "regular prime").
1857	Kummer	FLT is true for $n \leq 100$ (using a criterion that works on some irregular primes).
1930–7	Vandiver	FLT is true for $n < 617$ (using a calculator and improvements to Kummer's criterion).
1953	Inkeri	If (x, y, z) is a primitive counterexample to FLT with exponent p (p prime), then $x > [(2p^3 + p)/\log(3p)]^p$. Hence by the 1991 result below, x has at least 17 million digits.
1954	D. H. Lehmer, E. Lehmer, and Vandiver	FLT is true for $n \leq 2500$ (using a high-speed computer).
1976	Wagstaff	FLT is true for $n \leq 125,000$.
1983	Faltings	For every $n \geq 3$, Fermat's equation has only finitely many primitive solutions.

1985	Granville, Heath-Brown	FLT is true for "almost all" exponents n.
1987	Tanner and Wagstaff	FLT is true for $n \leq 150,000$.
1991	Buhler, Crandall, and Sompolski	FLT is true for $n \leq 1{,}000{,}000$.

Many, many other mathematicians have had a hand in advancing the state of knowledge regarding Fermat's last theorem, and the literature on the subject is immense. Ribenboim's book [Rib1] contains over 500 references.

The definition of a regular prime is given in Part Two. It has not been proven that there are infinitely many regular primes (although there is good evidence to believe that there are), so the results above do not eliminate the possibility that FLT is false for all prime exponents past a certain point. Faltings's result shows that at least there are no other cases like the Pythagorean case, $n = 2$, since that is the only exponent greater than 1 for which Fermat's equation has infinitely many primitive solutions. To appreciate the striking 1985 result about almost all exponents, note that as soon as Fermat established the $n = 4$ case, it followed that FLT is true for all exponents that are multiples of 4, or 25% of the positive integers. To make this precise, let $N(x)$ denote the number of integers $n \in [2, x]$ such that FLT is true for exponent n. Then Fermat established that for any $\epsilon > 0$, $N(x)/x$ eventually stays greater than $\frac{1}{4} - \epsilon$.

Likewise, Euler's proof of the cubic case implied the truth of FLT for all exponents that are a multiple of 3 or a multiple of 4, and this yields that $N(x)/x$ eventually stays greater than

$$\frac{1}{3} + \frac{1}{4} - \frac{1}{12} - \epsilon = \frac{1}{2} - \epsilon$$

for any positive ϵ (the $\frac{1}{12}$ is subtracted to take into account numbers that are multiples of both 3 and 4). In short, we may say that FLT is true for, asymptotically, at least 50% of exponents. The phrase "almost all" means that the assertion is true for, asymptotically, 100% of the exponents, that is, $\lim_{x\to\infty} N(x)/x = 1$. Now, Wagstaff's 1976 result for primes up to 125,000 improved the limit's lower bound to 93% (essentially the same value follows from the 1987 improvement to 150,000). Further improvements along this line would raise the percentage, but still leave it short of 100%. The striking aspect of the 1985 result, whose proof, given in Part Two, is based on the deep work of Faltings, is that it shows that FLT is true for, asymptotically, 100% of the exponents. But keep in mind that the main interest is not in arbitrary exponents n, but in prime exponents p; because it has not yet been shown that FLT is valid for infinitely many primes, it is not known that the theorem is valid for even 1% of the possible prime exponents.

Combining Inkeri's estimate on the size of x in terms of n with the 1991 result about the necessary size of n shows that any counterexample must lead to immense numbers. Indeed, if x, y, z form a counterexample to FLT with exponent n, then x^n has at least 10^{13} digits!

Fermat was an exceptional mathematician, responsible for many important results in number theory. While the romantic notion that he possessed a remarkable proof cannot be disproved, it is, as noted by Ribenboim [Rib1, p. 2], "very difficult to understand today, how the most distinguished mathematicians could have failed to rediscover a proof, if one had existed." Historian André Weil, whose book [Wei] contains a detailed account of Fermat's achievements, summarizes the situation as follows: "For a brief moment perhaps, and perhaps in his younger days, he must have deluded himself into thinking that he had the principle of a general proof; what he had in mind on that day can never be known."

It should also be mentioned that even the best mathematicians can have incorrect intuition. For example, Euler thought that the exponent-three case of FLT, which he was the first to prove, could be generalized to: If $n \geq 3$, then fewer than n nth powers cannot sum to an nth power. Two hundred years later, in 1966, L. J. Lander and T. R. Parkin (see [Guy, pp. 79–80]) discovered by a computer search that $27^5 + 84^5 + 110^5 + 133^5 = 144^5$. And recently N. Elkies [Elk] has shown that there are infinitely many relatively prime triples of fourth powers that sum to a fourth power; the smallest counterexample, discovered by R. Frye who used various "Connection Machines" (see [Elk]), is

$$95800^4 + 217519^4 + 414560^4 = 422481^4.$$

For sixth or greater powers, Euler's conjecture is still unresolved.

Problem 13.1

Can five sixth powers sum to a sixth power?

Exercises

1. Suppose FLT is true. Show that it remains true even if negative integers are allowed for any of x, y, or z.

2. A *Pythagorean triple* is a set of 3 positive integers satisfying $x^2 + y^2 = z^2$.

(a) Prove that whenever r and s are positive, relatively prime integers having different parity and with $r > s$, then $(r^2 - s^2, 2rs, r^2 + s^2)$ is a primitive Pythagorean triple.

(b) Suppose (x, y, z) is a primitive Pythagorean triple. Show that exactly one of x, y is even.

(c) Suppose (x, y, z) is a primitive Pythagorean triple and y is even. Prove that there exist positive integers r and s as in (a) such that $x = r^2 - s^2$ and $y = 2rs$.

3. (Grünert, 1856) Prove that if x, y, and z are positive integers with $x < y$ and $x^n + y^n = z^n$, then $x > n$. (Hint: Use the factorization of $z^n - y^n$ into $(z - y)(z^{n-1} + z^{n-2}y + \cdots + y^{n-1})$.)

14. A PERFECT BOX

Problem 14

Does there exist a box with integer sides such that the three face diagonals and the main diagonal all have integer lengths?

Fermat's equation is not the only way to formulate interesting problems that stem from the classical Pythagorean relationships such as $3^2 + 4^2 = 5^2$. The 3-4-5 triangle can be viewed as half of a rectangle in which all the lengths obtained by joining vertices are integers. This leads to the question of whether the same type of object can exist in three dimensions, that is, whether there can be a rectangular box with integer sides such that the lengths obtained by joining all six vertices in pairs are also integers. There are four new lengths obtained in this way: the three diagonals of the faces, and the main diagonal that passes through the center of the box. More precisely, what is sought is a solution in positive integers (called a *perfect box*) to the following system of equations:

$$\begin{aligned} x^2 + y^2 &= a^2 \\ x^2 + z^2 &= b^2 \\ y^2 + z^2 &= c^2 \\ x^2 + y^2 + z^2 &= d^2. \end{aligned}$$

Here, x, y, z represent the lengths of the edges; a, b, c, the face diagonals; and d, the main diagonal. The problem can be viewed as trying to find seven related integer quantities: the lengths of the three edges, the three face diagonals, and the main diagonal. If we instead ask for only six of these to be integers, then three simpler problems arise, all of which have been solved.

Variation 1: The main diagonal is not required to have integer length.

Variation 2: One of the face-diagonals is not required to have integer length.

Variation 3: One of the edges is not required to have integer length.

In the first variation we seek three interconnected Pythagorean triangles, that is, three integers x, y, and z such that each pair forms the two smaller entries of a Pythagorean triple. Infinitely many solutions are known, the smallest of which is 44, 117, 240. As with all known solutions to any of the three variations, the seventh length, in this case the main diagonal, is not an integer: $\sqrt{73225} = 270.6\ldots$. Infinitely many solutions exist in the other variations as well. In variation 2, let the edge-lengths be 104, 153, 672. An example in variation 3 comes from the three edges-lengths 124, 957, $\sqrt{13{,}852{,}800}$; the four derived diagonals are all integers.

There are other unsolved Diophantine equations that arise from plane geometry. An integer-sided square cannot have an integer diagonal, since the diagonal-side ratio is the irrational number $\sqrt{2}$. But the following generalization is unsolved.

Problem 14.1

Can a square with integer sides contain a point whose distance from each corner is an integer?

A *Heron triangle* is one with rational sides and area. Finding these is eased by the use of Heron's formula for the area of a triangle with sides a, b, c: $A = \sqrt{s(s-a)(s-b)(s-c)}$, where s is one-half the perimeter; this formula was discovered in approximately the first century A.D. By this formula, the area of the triangle with sides 13, 14, 15 is 84, so this is an example of a Heron triangle. However, the following generalization with 7 unknowns is unsolved.

Problem 14.2

Is there a triangle with integer sides, medians, and area?

Turning from Diophantine problems to the theory of primes, we can get a different sort of unsolved problem related to the classical 3-4-5 triangle.

Problem 14.3

Are there infinitely many Pythagorean triples where the hypotenuse and one of the legs are prime numbers?

Exercises

1. (Brocard, 1895) Prove that if a perfect box exists, then two of its edges fail to be relatively prime.

15. EGYPTIAN FRACTIONS

Problem 15

Does the greedy algorithm always succeed in expressing a fraction with odd denominator as a sum of unit fractions with odd denominator?

Our current system of notation for numbers, using base ten, is by no means the only system possible. A variety of number representation schemes have been used by different civilizations, and the evolution continues even today as computers have brought base two (binary) and base sixteen (hexadecimal) notations into greater use. Hand-held calculators use still a different system, called binary-coded decimal, where a number is represented as a string of decimal digits, but each digit is represented internally in binary. The Babylonians used a base-sixty system, and the Romans used, naturally, Roman numerals. The even older system of the Egyptians is of interest because it differs so much from the base-n systems and raises interesting mathematical questions.

The Egyptians had special symbols for the unit fractions, which are the reciprocals of the integers: $\frac{1}{2}$, $\frac{1}{3}$, $\frac{1}{4}$, and so on. But they did not have symbols for fractions with numerators larger than 1 (except for $\frac{2}{3}$), and they represented such fractions as sums of unit fractions. Of course, this can be done in the obvious way by writing $\frac{4}{23}$, for example, as a sum of four summands, each equal to $\frac{1}{23}$, but there are generally shorter representations. One way of finding them is by the use of a so-called "greedy" algorithm, one that repeatedly chooses the largest unit fraction that will fit. More precisely, given a fraction $\frac{a}{b}$ where a and b are positive integers, choose a sequence of unit fractions $\frac{1}{x_1}$, $\frac{1}{x_2}$, and so on by letting x_1 be the least integer such that $\frac{1}{x_1} \leq \frac{a}{b}$, letting x_2 be the least integer different from x_1 such that $\frac{1}{x_1} \leq \frac{a}{b} - \frac{1}{x_1}$, and so on; for example, this technique applied to $\frac{4}{23}$ yields $\frac{1}{6} + \frac{1}{138}$; $\frac{41}{42}$ yields $\frac{1}{2} + \frac{1}{3} + \frac{1}{7}$. The terminology "greedy algorithm" is used because at each step we choose the desired object (in this case, a unit fraction differing from those already chosen) that eats up as much of the given fraction as possible. It is not obvious that this algorithm always terminates in a finite number of steps, but Fibonacci showed in 1202 A. D. that it does (details in Part Two). Thus any fraction can be represented as a sum of finitely many distinct unit fractions. Moreover, Fibonacci's proof shows that if the fraction is less than 1 then the number of unit fractions the algorithm produces is not greater than the numerator.

The Egyptians were aware of this technique, although the actual method they used was a combination of the greedy algorithm with other methods. For example, they preferred representing $\frac{2}{9}$ as $\frac{1}{6} + \frac{1}{18}$, while the greedy algorithm yields $\frac{1}{5} + \frac{1}{45}$. Much of our knowledge of Egyptian arithmetic comes from the Rhind papyrus which was discovered in 1858 and deciphered in 1877 with the aid of the famed Rosetta Stone. For more on these discoveries and the Egyptian system of arithmetic, in particular, division, where their fraction-representing scheme comes into play, see [RW]. See also [RS], who note: "Nowhere was greater skill and versatility shown than in the deployment of unit fractions. This was at once the glory and the straitjacket of Egyptian methodology."

Many interesting mathematical problems arise from the problem of representing a fraction as a sum of unit fractions. For instance, suppose we wish to use only even unit fractions (by which is meant the reciprocals of even integers), or only odd unit fractions. We will leave as an exercise (Exercise 5 in Part Two) the result that the greedy algorithm to represent any fraction as a sum of even unit fractions always terminates. The odd case, however, is quite a different matter. First observe that if a fraction has an even denominator when written in lowest terms then it cannot be represented as a finite sum of odd unit fractions (Exercise 1). But if the denominator is odd then it is not known whether the greedy algorithm always halts. All examples that have been tried do terminate; e.g., $\frac{4}{23} = \frac{1}{7} + \frac{1}{33} + \frac{1}{1329} + \frac{1}{2353659}$. As these numbers show, the greedy algorithm quickly leads to large integers, making it difficult to check the conjecture on a computer. Although the greedy algorithm using odd denominators is not well understood, it is at least known that every rational with an odd denominator can indeed be written as a sum of distinct odd unit fractions.

Another unsolved problem concerns the smallest number of summands necessary to represent a fraction between 0 and 1 as a sum of distinct unit fractions. Numbers of the form $\frac{2}{b}$ require two summands (provided b is odd), and some numbers of the form $\frac{3}{b}$ require three summands (e.g., $\frac{3}{7}$; see Exercise 2(d) in Part Two). But for fractions of the form $\frac{4}{b}$ it is not known that four summands are ever required. It has been conjectured by Erdős and Straus that three summands always suffice, and this has been verified for all denominators under 100,000,000.

Problem 15.1

Is it true that whenever $b > 1$ the fraction $\frac{4}{b}$ can be written as a sum of three or fewer distinct unit fractions?

Exercises

1. Show that if b is even and a is odd then $\frac{a}{b}$ is not a sum of finitely many odd unit fractions.

2. TRUE or FALSE: For any fraction with an odd denominator, the greedy algorithm that represents it as a sum of odd unit fractions yields at least as many terms as the ordinary greedy algorithm.

16. PERFECT NUMBERS

Problem 16

Is there an odd perfect number? Are there infinitely many even perfect numbers?

The search for perfect numbers is perhaps the oldest unfinished project of mathematics. A positive integer is perfect if it equals the sum of its divisors other than itself; for example, $6 = 1+2+3$ and $28 = 1+2+4+7+14$. The ancient Greeks knew the first four perfect numbers and by 1914 eight more had been discovered. The advent of modern computers brought new life into the search: the thirteenth perfect number was discovered in 1952, and now 31 such numbers are known. Here are the first nine:

$$6$$
$$28$$
$$496$$
$$8{,}128$$
$$33{,}550{,}336$$
$$8{,}589{,}869{,}056$$
$$137{,}438{,}691{,}328$$
$$2{,}305{,}843{,}008{,}139{,}952{,}128$$
$$2{,}658{,}455{,}991{,}569{,}831{,}744{,}654{,}692{,}615{,}953{,}842{,}176.$$

These numbers grow quickly—the largest known perfect number has over 130,000 digits—and this list immediately leads to the two unsolved problems of this section:

(1) Does the list have infinitely many entries?

(2) Does an odd number ever occur?

It might seem that the two questions are similar but there are fundamental differences between even and odd perfect numbers that make the questions quite different.

The even perfect numbers are closely connected to primes of the form $2^n - 1$, which are called Mersenne primes; they can arise only when the exponent is prime

(Exercise 1). The first few examples are 3, 7, 31, 127, 8191. Note that these primes, when multiplied by the next smaller power of 2, yield perfect numbers: $3 \cdot 2 = 6$, $7 \cdot 4 = 28$, $31 \cdot 16 = 496$, and so on. This is true for all Mersenne primes, a fact known to the ancient Greeks (Proposition 36 in Book IX of Euclid's *Elements*; see [Hea]). Moreover, *all* even perfect numbers arise in this way, that is, are the product of a Mersenne prime with the next smaller power of 2; this was stated by many prior to Euler, but the earliest known proof is Euler's. The discussion is eased by the introduction of the function $\sigma(n)$, defined to be the sum of the divisors of n (including both 1 and n). In terms of σ, a number n is perfect if and only if $\sigma(n) = 2n$. A fundamental fact about this function (Exercise 2) is that whenever a and b are relatively prime integers, $\sigma(ab) = \sigma(a)\sigma(b)$.

Theorem 16.1 (Euclid–Euler Formula). *For every Mersenne prime $2^n - 1$, the even number $2^{n-1}(2^n - 1)$ is perfect. Moreover, all even perfect numbers have this form.*

Proof. If $2^n - 1$ is prime then it has only two divisors, itself and 1, whence $\sigma(2^n - 1) = 2^n$. And the only divisors of a power of 2 are itself and the smaller powers of 2, whence $\sigma(2^{n-1}) = 1 + 2 + 4 + \cdots + 2^{n-1} = 2^n - 1$. Now $\sigma(2^{n-1}(2^n - 1)) = \sigma(2^{n-1})\sigma(2^n - 1) = 2^n(2^n - 1)$, as required. Conversely, suppose r is an even perfect number; write r as $2^{n-1}m$ where $n > 1$ and m is odd. Since $2^n m = \sigma(2^{n-1}m) = (2^n - 1)\sigma(m)$, we have

$$\sigma(m) = m + \frac{m}{2^n - 1}.$$

This implies that $m/(2^n - 1)$ is an integer, whence m and $m/(2^n - 1)$ are both divisors of m. The expression for $\sigma(m)$ then implies that $m/(2^n - 1)$ is equal to 1 and m has no other divisors. In other words, $m = 2^n - 1$ and m is prime. □

The preceding theorem shows that infinitely many even perfect numbers exist if and only if infinitely many Mersenne primes exist. This question is just one of dozens of unsolved problems regarding the existence of infinitely many primes of a certain form. In the case of Mersenne primes there are some reasons for believing that infinitely many do exist. They turn up at fairly regular intervals and one can provide heuristic arguments that this regularity should continue forever (see Part Two). Here are some other famous unsolved problems about prime numbers.

Problem 16.1 (Twin Prime Conjecture)

Are there infinitely many twin primes, that is, primes, such as 17 and 19, or 227 and 229, that differ by 2?

Problem 16.2

Are there infinitely many primes of the form $n^2 + 1$?

Problem 16.3 (P. Erdős)

Are there arbitrarily long arithmetic progressions of primes? In other words, is it true that for every n there exist n primes that form an arithmetic progression?

The question about odd perfect numbers is different, since the theory of these numbers is not closely connected with the theory of primes or other parts of number theory. Several results are known; for example, an odd perfect number must have at least 50 digits and at least eight distinct prime factors. But the situation is delicate as is shown by Descartes's discovery (1638) that $N = 22021 \cdot 3^2 \cdot 7^2 \cdot 11^2 \cdot 13^2$ is tantalizingly close to being perfect. Since, for p prime, $\sigma(p) = p + 1$ and $\sigma(p^2) = 1 + p + p^2$, and since the following relation holds (Exercise 4(a)):

$$(22021 + 1)(1 + 3 + 3^2)(1 + 7 + 7^2)(1 + 11 + 11^2)(1 + 13 + 13^2) = 2N,$$

it seems that N is an example of an odd perfect number. But there is a flaw in the reasoning, and N is not perfect (Exercise 4(b)). Nevertheless, this near-miss shows how fragile the problem is; perhaps, as Descartes believed, some relatively simple combination of a dozen or so primes does yield an odd perfect number.

Closely related to perfect numbers are the amicable pairs, numbers m, n such that the sum of the divisors of each (excluding the number itself) equals the other. In terms of σ the definition is $\sigma(m) = m + n = \sigma(n)$. The classical example, discovered by the Pythagoreans, is the pair 220, 284 (Exercise 3). The next discovered pair is 17296 and 18416 (14th century, Ibn al-Banna, Morocco; see [BH]). Both perfect and amicable numbers have been surrounded by much numerological speculation. For example, Saint Augustine suggested that God took six days to create the heavens and earth because 6 is a perfect number and hence this would signify the perfection of the work. Amicable numbers have often been associated with the perfect friendship of two persons. Indeed, in 1007 in Madrid it was suggested that these numbers could be used as a love-potion: serve the smaller number to the object of your affections while eating the larger number yourself (for example, 220 and 284 grains of rice) [Dic1, p. 39]. See [BBC, Chap. 2] for more lore on perfect numbers as well as proofs of several elementary facts about them.

Amicable pairs are similar to perfect numbers, in that the following is the central unsolved problem.

Problem 16.4

Are there infinitely many amicable pairs?

By 1866, 65 pairs of amicable numbers were known, three discovered before Euler, 59 discovered by Euler, and three discovered after Euler. Most of these are quite large, so it was a surprise when a sixteen-year-old Italian, N. Paganini, announced in 1866 his discovery of the amicable pair 1184 and 1210. Dickson, in 1913, proved that the five smallest amicable pairs are $(220, 284)$, $(1184, 1210)$, $(2620, 2924)$, $(5020, 5564)$, and $(6232, 6368)$. Several thousand amicable pairs are now known [BH, LM, tR1, 2].

Exercises

1. Show that if n is composite then so is $2^n - 1$.

2. Show that whenever a and b are relatively prime integers, $\sigma(ab) = \sigma(a)\sigma(b)$. (Hint: First show $\sigma(a)\sigma(b) \leq \sigma(ab)$ and then show $\sigma(ab) \leq \sigma(a)\sigma(b)$.)

3. Use the result of the preceding exercise to compute $\sigma(220)$ and $\sigma(284)$ and verify that these numbers form an amicable pair; do the same for 17296 and 18416 (whose prime factorizations are $2^4 \cdot 23 \cdot 47$ and $2^4 \cdot 1151$, respectively).

4 (a). Verify equation (1), using only a hand calculator.

(b). Find the flaw in the argument purporting to show that Descartes's number N is an odd perfect number.

17. THE RIEMANN HYPOTHESIS

Problem 17

Do the nontrivial zeros of the Riemann zeta function all have real part $\frac{1}{2}$?

Deciding the status of the Riemann hypothesis[1] is considered by many to be the single most important unsolved problem in mathematics. To understand it fully requires some knowledge of complex function theory, for it is an assertion about the imaginary numbers for which the Riemann zeta function equals zero. However, there are equivalent versions that make no mention of complex numbers, and we shall discuss some of these. More complete treatments, but still at a relatively elementary level, can be found in [Rib3, Rie, Wag5].

The main importance of Riemann's zeta function (which will be defined in Part Two) is its connection with the function $\pi(x)$, the number of primes less than or equal to x (it is convenient to allow noninteger values of x in the definition of $\pi(x)$). A study of tables of primes led Gauss to observe that, near an integer x, the "probability" that a number is prime is about $1/\log x$. Of course, primeness is not a random phenomenon (see Section 16, Part Two); Gauss's observation was that the number of primes in an interval of size D near x, where D is neither too large nor too small, is about $D/\log x$. For example, there are 75 primes between 1,000,000 and 1,001,000, and

$$\frac{1{,}000}{\log 1{,}000{,}000} = 72.38\ldots.$$

This observation led Gauss (at the age of 15) to conjecture that $\pi(x)$, for an integer x, is well approximated by a definite integral that counts integers by assigning to each number x a probability of $1/\log x$ that x will be counted. This can easily be made precise by the following definition.

Definition 17.1. The logarithmic integral[2] of x, denoted $\text{li}(x)$, is

---

[1] Named after its formulator, Bernhard Riemann (1826–1866), a German mathematician known to generations of calculus students because of his definition of the definite integral using approximations by rectangles.

[2] There is a singularity at $t = 1$. Thus, to be precise, $\text{li}(x)$ is defined to be the limit as $\epsilon \to 0$ of $\int_0^{1-\epsilon} \frac{1}{\log t}\,dt + \int_{2+\epsilon}^{x} \frac{1}{\log t}\,dt$. One can think of $\text{li}(x)$ as being just the integral from 2 to x, since the omitted part, that is, $\text{li}(2)$, is less than 1.05.

$$\int_2^x \frac{1}{\log t} dt.$$

One way to say that one function is well approximated by another is to say that their quotient approaches 1. Precisely, $f(x)$ and $g(x)$ are called *asymptotic* if

$$\lim_{x \to \infty} \frac{f(x)}{g(x)} = 1.$$

We may now state the celebrated prime number theorem, proved in 1896 almost 100 years after Gauss conjectured it.

Theorem 17.2 (Prime Number Theorem). *The function* $\pi(x)$ *is asymptotic to* $\text{li}(x)$.

Because $\text{li}(x)$ is not too hard to compute for large x (details in Part Two; a rough approximation is given by $\frac{x}{\log x}$ (Exercise 1)), this theorem allows one to very quickly get an approximation to, say, the number of primes under a billion (10^9). The theorem predicts that it will be near $\text{li}(10^9)$, which equals 50,849,234. In fact, there are 50,847,534 primes under a billion, so the relative error in this approximation is only 0.003%. The prime number theorem can be interpreted as saying that the relative error in the approximation tends to zero; in other words, the relative error is, asymptotically, 0%. However, the prime number theorem does not say anything about the speed with which the relative error gets small. One can compute $\text{li}(10^{100})$, for example, and get $4.36197\ldots \times 10^{97}$, but Theorem 17.2 does not say explicitly how close this is to the true value of $\pi(10^{100})$. The function $\pi(x)$ has been computed for various values up to 10^{16}; these values lead (see Table 17.1) to the guess that the error in the approximation is roughly on the order of $\sqrt{x}$. Subtler considerations lead to a slightly more complicated conjecture, which is equivalent to the Riemann hypothesis. More precisely, RH is equivalent to an affirmative answer to the following question.

Problem 17.1

Is there a constant c such that, for any x, $\pi(x)$ is within $c\sqrt{x}\log x$ of $\text{li}(x)$?

Some bounds on the error in the use of $\text{li}(x)$ are known, but they do not come close to solving Problem 17.1. In particular, it has not been proved that there are constants c and d, with $d < 1$, such that $|\pi(x) - \text{li}(x)| < cx^d$. The Riemann hypothesis would imply that the error is bounded by cx^d for any $d > \frac{1}{2}$. Returning to the specific example of 10^{100}, all that is currently known is that the error in using the

x	$\pi(x)$	$\text{li}(x) - \pi(x)$	Relative error
100	25	5	0.2
10,000	1,229	17	0.014
1,000,000	78,498	130	0.0017
10^8	5,761,455	754	0.00013
10^{10}	455,052,511	3,104	6.82×10^{-6}
10^{12}	37,607,912,018	38,263	1.02×10^{-6}
10^{14}	3,204,941,750,802	314,890	9.83×10^{-8}
10^{16}	279,238,341,033,925	3,214,632	1.15×10^{-8}

TABLE 17.1
The relative error in the use of the logarithmic integral to approximate the number of primes less than x approaches zero as x gets large.

logarithmic integral to approximate $\pi(10^{100})$ yields an error less than 3×10^{95}; in other words, $\pi(10^{100}) = (4.36197 \pm .03) \times 10^{97}$. The Riemann hypothesis, however, implies that the error is at most 10^{51} (this is not simply a consequence of the bound in Problem 17.1, but follows from additional work). Therefore, under RH, if $\text{li}(10^{100})$ is computed to 50 decimal places—which is not terribly difficult—the result will agree with $\pi(10^{100})$ to about 47 decimal places.

As far as lower bounds on the error are concerned, more is known. In 1914 Littlewood showed that there is a positive constant c such that, for infinitely many values of x, $|\pi(x) - \text{li}(x)| > c\sqrt{x} \log\log\log x / \log x$. Thus the order of the approximation is never, in a consistent fashion, much better than $\sqrt{x}$. However, just as a stopped clock is right two times each day, $\text{li}(x)$ occasionally does yield $\pi(x)$ exactly (see remarks concerning Littlewood's result in Part Two).

To give another equivalent version of RH we need to introduce the Möbius function, $\mu(n)$, which is defined on positive integers as follows. If n has the form $p_1 p_2 \cdots p_r$ where the p_i are distinct primes, then $\mu(n)$ is defined to be $+1$ or -1 according as r is even or odd ($\mu(1)$ is taken to be $+1$); and $\mu(n) = 0$ if n is not of this form and hence is divisible by the square of some prime. This function has many applications in combinatorics and number theory. The connection with RH arises by considering by how much the $+1$'s can dominate the -1's (and vice versa) as values of $\mu(n)$. To be precise, let $M(x)$ be the sum of $\mu(n)$ for all $n \leq x$. Most values of μ will be zero, and there will be a lot of cancellation between the $+1$'s and -1's. What we are interested in is the rate of growth of $|M(x)|$. It is known that $|M(x)|$ can be greater than $\sqrt{x}$, and it is suspected that for any $c > 0$ there are values of x for which $|M(x)| > c\sqrt{x}$. The Riemann hypothesis is equivalent to an affirmative answer to the following question.

Problem 17.2

Is it true that for every $\epsilon > 0$ there is a constant c such that $|M(x)| < cx^{\frac{1}{2}+\epsilon}$?

Exercise

1. Show that $\mathrm{li}(x)$ is asymptotic to $x/\log x$. (Hint: Use l'Hôpital's rule.)

18. PRIME FACTORIZATION

Problem 18

Is there a polynomial-time algorithm for obtaining a number's prime factorization?

Every positive integer N can be written uniquely in the form $p_1^{e_1} p_2^{e_2} \cdots p_r^{e_r}$, where the p_i are distinct prime numbers. But in practice it is very difficult to determine this prime factorization. Since the length of N, when written in base 2, is $\lfloor \log_2 N \rfloor + 1$, what we are seeking is an algorithm that determines the prime factorization of N in time bounded by a power of $\log_2 N$ (or, equivalently, by a power of $\log N$). This problem has become one of great importance because of cryptographic applications that assume there is no polynomial-time algorithm for factoring.

If N is prime then N is its own prime factorization. Thus it is natural to break Problem 18 into two components.

Problem 18.1 (Primality Testing)

Is there a polynomial-time algorithm that determines whether a number is prime?

Problem 18.2 (Factoring)

Is there a polynomial-time algorithm that, given a composite number N, computes a nontrivial (i.e., other than 1 or N) divisor of N?

If there are polynomial-time algorithms for primality testing and factoring, then they can be combined to obtain the prime factorization of N. First test N for primality. If N is prime we're done; if not, use the other algorithm to obtain a nontrivial divisor d of N. Then repeat, replacing N by each of d and N/d. This procedure will eventually terminate having found all prime divisors of N, and the number of times each divides N. Each of the two algorithms will have been used $O(\log N)$ times, so the running time is polynomially bounded.

If running time is not a consideration, there is a straightforward algorithm to obtain a number's factorization. Roughly speaking, one simply tries all possible divisors. This can be done systematically by first reducing N by all the 2s that divide it, then by all the 3s, then by all the 5s, then by all the 7s, and so on. Once all the

2s are divided out of N no other even number will divide N, which is why subsequent trial divisors are odd. Similarly, 9, 15, and other multiples of 3 can be skipped as trial divisors once N is reduced by 3s. In fact, we need only consider primes as trial divisors, though in practice one would not want to take the time to test trial divisors for primality. Since a composite number must have at least one nontrivial divisor less than its square root, the algorithm can terminate when the trial divisor is greater than the square root of the current (possibly reduced) value of N, declaring this value prime. This algorithm does not run in polynomial-time, however, because its main loop is executed roughly $\sqrt{N}$ times, and $\sqrt{N}$ is not bounded by any polynomial in $\log N$. It would take years to prove a 36-digit number prime by this technique, even on today's fastest computers, which can do a billion operations in a second. The trial divisor method is over 2,000 years old; modern primality-testing algorithms are much, much faster.

Despite the sophisticated algorithms that have been developed to attack these problems, factoring large numbers is still very time-consuming. For example, the fastest known factoring algorithm, programmed on today's fastest computers, might require centuries to factor an arbitrary 200-digit number. Numbers with fewer than 80 digits can be handled, and the 100-digit barrier has just been broken, but if the problem is as inherently difficult as researchers believe, further progress will be very slow. On the other hand, the best primality-testing algorithms, while not quite polynomially bounded, work very well for numbers of moderate size: primality of a 200-digit number can be determined in less than ten minutes. Thus there are numbers proved to be composite, but for which no nontrivial divisors are known. Examples: $2^{523} - 1$ is not prime, but no divisors of this 158-digit number are known; $2^{445} - 1$ is divisible by the primes 31, $2^{89} - 1$, and 2671, but, at press time, no prime factors are yet known for the number that remains after these primes are divided out, a number having 103 digits and known to be composite. See [BLSTW] for the limits of what is known along these lines.

The reader might well wonder how it can be known that a number is composite without knowing any of its divisors. This turns out to be not too difficult because of Fermat's little theorem, which states that if N is prime, then for any number b with $1 < b < N$, $b^{N-1} \equiv 1 \pmod{N}$. So if, say, $N > 2$ and 2^{N-1} fails to be congruent to $1 \pmod{N}$ (where $N > 2$), N cannot be prime. Unfortunately the converse to Fermat's little theorem is false (Exercise 1), so this cannot be used to prove that a number is prime. Nevertheless, the great majority of composite numbers fail this test (i.e., for most composite N, $2^{N-1} \not\equiv 1 \pmod{N}$). Thus if one is willing to live with a small chance of error, the Fermat little theorem test with $b = 2$ is a fast way to determine primality. It might seem that the computation of large powers of $2 \pmod{N}$ would be slow, but there is a very efficient way of doing this (Exercise 2).

One of the main ideas behind fast primality-testing algorithms is a strengthening of the Fermat little theorem approach that is quite simple and has many ramifications. As just pointed out, if N is prime and $1 < b < N$ then $b^{N-1} \equiv 1 \pmod{N}$. And Exercise 2 shows how to efficiently compute this power by first computing b^m where m is the odd integer so that $N-1 = 2^s m$, and then squaring s times. Let's call this sequence of $s+1$ numbers—the sequence: $b^m, b^{2m}, \ldots, b^{N-1}$ (all mod N)—N's *b-sequence.* Now, if $*$ denotes a number that is not congruent to $\pm 1 \pmod{N}$, there are five[1] forms the b-sequence of N can take. The first two will be called "type 1" and the other three, "type 2" (note that the type of a b-sequence can be determined without computing the last term):

b^m	b^{2m}	b^{4m}			$\cdots$			b^{N-1}	
$+1$	$+1$	$+1$	$+1$	$+1$	$+1$	$+1$	$\cdots$	$+1$	type 1
$*$	$*$	$*$	$\cdots$	$*$	-1	$+1$	$\cdots$	$+1$	type 1
$*$	$*$	$*$	$\cdots$	$*$	$+1$	$+1$	$\cdots$	$+1$	type 2
$*$	$*$	$*$	$*$	$*$	$*$	$*$	$\cdots$	$*$	type 2
$*$	$*$	$*$	$*$	$*$	$*$	$*$	$\cdots$	$*$	type 2

The point of this classification is that if N is prime then not only must b^{N-1} be $+1$ (Fermat's little theorem) but, because ± 1 are the only numbers (mod N) that square to 1 (Exercise 3), the path to $+1$ must be of type 1. Consider 561, which masqueraded as a prime (see Exercise 1) under the Fermat little theorem test. Since $560 = 2^4 \cdot 35$, the 2-sequence has 5 terms, which are: 263, 166, 67, 1, 1. If 561 was prime, it would not be possible for 67^2 to be congruent to $1 \pmod{561}$. In other words, the 2-sequence is of type 2 so the given number must be composite.

Even with this more complicated test, a composite can masquerade as a prime as far as a few choices of b are concerned (examples: 2047, with 2-sequence 1, 1; 1373653, with 2-sequence 890592, -1, 1 and 3-sequence 1, 1, 1), but no composite has a type 1 b-sequence for too many bs. The exact meaning of "too many" is the key point in formulating a precise test. It seems that all odd composite numbers are shown to be composite by some b less than $2(\log N)^2$; numerical checking has verified this for numbers up to two trillion [Wag4]. Moreover, this assertion is a consequence of the extended Riemann hypothesis, a notorious unproved hypothesis involving number theory and complex analysis. Under the assumption that this upper bound on witnesses for compositeness is universally valid, the following algo-

[1] In fact, the final row can never arise for odd N, prime or composite. A proof using elementary facts about the order of an integer mod N was found by John Ewing.

rithm (due to Gary Miller) determines primality of an odd integer N in polynomial time.

For each b less than $2(\log N)^2$:
If N's b-sequence is of type 2, then output COMPOSITE, else
Output PRIME.

In short, if the b-sequence test for all b less than $2(\log N)^2$ fails to determine that N is composite, then N is declared prime.

Miller's algorithm is simple and elegant, but it has the unfortunate drawback that it is not known if it always tells the truth. There is as yet no guarantee that the assumption underlying the algorithm is valid, so if it outputs "PRIME" one cannot be certain that the input is, in fact, prime. But a variation worked out by Michael Rabin adds a lot of power to Miller's approach. Rabin observed that, without any unproved assumptions, if N is composite (and odd) then N's b-sequence is of type 2 for at least one-half of the choices of b in $[2, N-1]$. This was later improved by Rabin and, independently, L. Monier, to three-quarters of the choices. These results can be turned into a probabilistic algorithm as follows. Given an odd integer N choose, say, 100 random bs in $[2, N-1]$ and compute the 100 b-sequences for N. If any one of these sequences is of type 2 then N is definitely composite. If they are all of type 1, then N is almost certainly prime. For if N is composite, the probability that this will not be discovered by at least one of the 100 b-sequences is less than 4^{-100}, a number so small that it is unimaginable (except to mathematicians) that the input fails to be prime. Indeed, one can argue that some accepted mathematical theorems have such complicated proofs that the probability of an error is significantly greater than the probability that a "probable prime" in the above sense is not really prime (see [Pom2, 3] for a discussion, and criticism, of this view).

There may simply be no polynomial-time algorithm (probabilistic or deterministic) for factoring; any result along these lines would be a major achievement. Primality-testing seems to be quite different, however, and the recent breakthroughs in both theory and practice are as much an improvement over the classical trial divisor method as todays computers are faster than paper-and-pencil arithmetic.

Exercises

1. The number $561 = 3{\cdot}11{\cdot}17$, and so is not prime. Show that the number satisfies Fermat's little theorem with $b = 2$ (i.e., $2^{560} \equiv 1 \pmod{561}$). (Hint: Show that 2^{560} is congruent to 1 mod 3, mod 11, and mod 17). The number 561 is particularly bad, since $b^{560} \equiv 1 \pmod{561}$ for every number b that is relatively prime to 561.

2. Explain why the following algorithm for computing $X = b^m$ works.

   1. Take the base 2 representation of m, replace each 0 by an "S", and replace each 1, except the leftmost 1, by "SM".

2. Set $X = b$.
3. Work from left to right through the string of Ss and Ms. At each S replace X by X^2; at each M replace X by bX.

Note that this technique requires fewer than $2\lceil \log_2 m \rceil$ multiplications (some additional work is needed to get the base 2 representation of m), many fewer than the less sophisticated method of repeatedly multiplying by b. If this technique is used to form $b^m \pmod{N}$ where b^m is greater than the overflow limit of the computer, one would follow each S step and each M step by a reduction of $X \pmod{N}$.

3. Prove that if p is prime and $a^2 \equiv 1 \pmod{p}$ then a is congruent to either $+1$ or $-1 \pmod{p}$.

19. THE $3n + 1$ PROBLEM

Problem 19

Is every positive integer eventually taken to the value 1 by the $3n + 1$ function?

One of the fascinating aspects of mathematics is the continual discovery of interesting and difficult problems involving the most elementary concepts. The $3n+1$ problem, first considered in the 1950s, is a notorious recent example. Consider the following function (called the "$3n + 1$ function"), which takes positive integers to positive integers:

$$f(n) = \begin{cases} n/2, & \text{if } n \text{ is even} \\ 3n + 1, & \text{if } n \text{ is odd.} \end{cases}$$

The $3n + 1$ problem asks what happens when f is applied repeatedly, starting with an arbitrary positive integer.

The best way to get a feeling for this question is to go immediately to your computer and write a program to examine the results of iterating f. A program to do this in, say, BASIC, would be only two or three lines long. Or get a calculator (and lots of paper) and keep track of the results for various starting values. As an example, here is the result of 30 iterations with starting value 1,001:

$$1001,\ 3004,\ 1502,\ 751,\ 2254,\ 1127,\ 3382,\ 1691,$$

$$5074,\ 2537,\ 7612,\ 3806,\ 1903,\ 5710,\ 2855,\ 8566,$$

$$4283,\ 12850,\ 6425,\ 19276,\ 9638,\ 4819,\ 14458,\ 7229,$$

$$21688,\ 10844,\ 5422,\ 2711,\ 8134,\ 4067,\ 12202.$$

At first the results seem haphazard: sometimes the output of f is larger than the input, sometimes smaller. But there are definitely some patterns. For example, an odd step (i.e., an application of $f(n) = 3n + 1$) is always followed by an even step; this is simply because $3n + 1$ is even whenever n is odd. More mysterious is the pattern in the long-term behavior of the iterations (the iteration sequence starting from n is called the *trajectory* of n). Though not evident in the case of the 30 sample iterations just given, all trajectories eventually reach 1, and thereafter produce the loop $1, 4, 2, 1, 4, 2, \ldots$. Well, at least all trajectories ever examined—and three trillion have been checked so far—reach this loop. The $3n+1$ conjecture is the assertion that this is *always* the case. This question, which sounds so simple, is apparently very difficult and has frustrated the efforts of many mathematicians. Paul Erdős has commented, "Mathematics is not yet ready for such problems."

In addition to the verification of the conjecture for all starting values up to 3×10^{12}, there are some heuristic arguments in favor of the conjecture, discussed below and in Part Two.

A fundamental result of computer science is that there does not exist an algorithm that, given an arbitrary program, decides whether the program halts for all of its inputs. Anyone who has written a complex computer program is aware of how difficult it can be to tell whether the program has an infinite loop. The $3n+1$ problem shows that even programs whose logic is transparent can lead to difficulties. A program that computes trajectories and halts upon reaching 1 is only a few lines long. But the assertion that such a program halts for all inputs is equivalent to the $3n+1$ conjecture.

Note that there are two ways in which the program just mentioned might fail to halt. There might be a starting value whose trajectory grows without bound (a *divergent trajectory*); or there might be a *nontrivial loop,* that is, a value m other than 1, 2, or 4 such that the trajectory of m eventually returns to m, after which the loop is repeated indefinitely.

Another way of stating the conjecture is to start from 1 and work backwards, using the inverse of the $3n+1$ function, as illustrated in Figure 19.1. The $3n+1$ conjecture is equivalent to the assertion that every positive integer appears somewhere in this diagram.

Although there is nothing truly random about the trajectories—they are completely determined once the starting value is specified—the apparent randomness can be used to formulate a heuristic argument that sheds a little light on why the sequences are expected to return to 1. The idea is to consider a random walk designed to approximate the behavior of the trajectories.

Since a random integer is equally likely to be even or odd, consider the random process that takes a starting value x and repeatedly multiplies by $\frac{3}{2}$ or $\frac{1}{2}$, with equal probability. We use $\frac{3}{2}$ because an odd step in the $3n+1$ trajectories is always followed by an even step; the replacement of $3n+1$ by $3n$ simplifies the analysis and does not have a significant effect (for one could replace $3n$ in the argument that follows by $3.5n$, which is greater than $3n+1$ if $n > 2$). This random process produces sequences of the form:

$$x \to d_1 x \to d_2 d_1 x \to d_3 d_2 d_1 x \to \cdots,$$

where each d_i is either $\frac{3}{2}$ or $\frac{1}{2}$.

In order to apply the theory of random walks, we need an additive process rather than a multiplicative one. This is easily arranged by taking logarithms; it is convenient to use logarithms to the base 2, which we denote simply by log. So consider the random walk that starts at $\log x$ and at each step adds $0.584\ldots$ or -1 with equal probability ($0.584\ldots = \log \frac{3}{2}$; $-1 = \log \frac{1}{2}$). Let $S(r)$ denote the total change after r steps. Because the steps are assumed to be independent, $\lim_{r\to 1} S(r)/r$ is, with probability 1, equal to the expected step-size.[1] But this expected value is just the weighted average of the steps: $\frac{1}{2}(0.584\ldots) + \frac{1}{2}(-1) = -0.207\ldots$. This means that, with probability 1, $S(r) \to -\infty$. Hence, for the random walk with starting value $\log x$, it is almost certainly true that there will be leftward motion totalling at least $-\log x$, which corresponds to the original multiplicative random process getting from x to below 1.

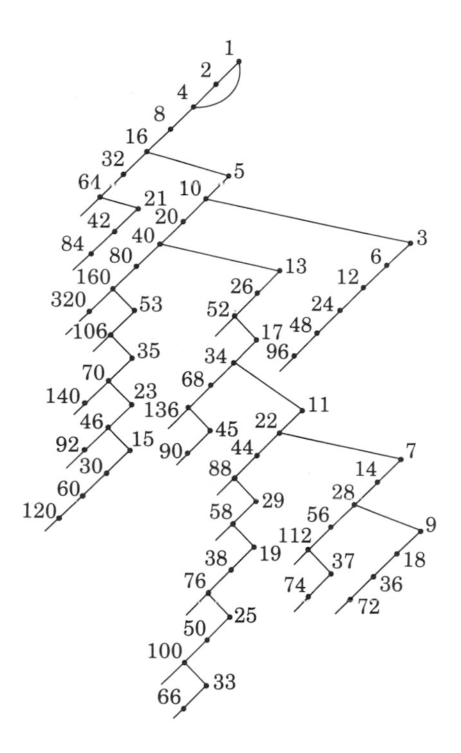

FIGURE 19.1
The $3n+1$ conjecture states that all positive integers appear in the continuation of this diagram. An integer m branches to $2m$ and, if $(m-1)/3$ is odd, to $(m-1)/3$. Where is 27?

If the "3" in the $3n+1$ function is replaced by a "5" or larger odd number, the corresponding random walk diverges to infinity. For in these cases the average step

---

[1] This is a consequence of the strong law of large numbers; see any text on probability theory.

is at least $\frac{1}{2}(\log(\frac{5}{2}) - 1)$, which is positive. Thus one expects that, in general (see Exercise 3), the trajectories for the "$\alpha n + 1$ Problem" will diverge to infinity when α is an odd integer greater than 3. But this has not been proved in even one case. The following problem gives a specific example, for which some computation has been done. The trajectory of 3, using the $7n + 1$ function, grows as expected and a computation of 36,500 odd steps yields values beyond 10^{9000} that, quoting Crandall [Cra], show "no apparent tendency to return."

Problem 19.1

Does the starting value 3 in the $7n + 1$ problem lead to a divergent trajectory?

Of course, heuristic arguments involving random walks do not prove anything about the deterministic $3n + 1$ trajectories. But there is evidence (discussed at the end of Part Two) that the trajectories do behave, to some degree, like random walks, so this sort of analysis at least offers a glimmer of insight into why the $3n + 1$ conjecture is believed to be true.

Exercises

1. Define the *stopping time* of a positive integer n to be the least positive integer r such that $f^r(n) < n$. Show that the $3n+1$ conjecture is equivalent to the assertion that all integers greater than 1 have finite stopping time.

2. Write a computer program that accepts a positive integer n and computes its stopping time and trajectory. Verify that the trajectory of 1,001 hits 1 in 142 iterations of f. Use this program to locate the position of 27 in the extension of Figure 19.1. (Some shortcuts for computing stopping times will be discussed in Part Two.)

3. Consider the analogue of the $3n + 1$ problem with the "3" replaced by a "5." One might expect that, except for the loop starting at 1—1, 6, 3, 16, 8, 4, 2, 1—all trajectories are divergent. But this conjecture is false. Why?

4. One can just as well consider negative starting values in the $3n + 1$ function. What is a reasonable conjecture about the behavior of trajectories with negative starting values?

20. DIOPHANTINE EQUATIONS AND COMPUTERS

Problem 20

> Is there an algorithm that decides, given a polynomial P with integer coefficients, whether there are rational numbers q_i such that $P(q_1, q_2, q_3, \ldots) = 0$?

In the present section *polynomial* will mean a polynomial, in one or more variables, having integer coefficients. A Diophantine equation is one of the form $P(x_1, \ldots, x_n) = 0$ where P is a polynomial and solutions are sought in integers (or, perhaps, in $\mathbb{N}$ or $\mathbb{N}^+$, the set of positive integers). Some notorious unsolved Diophantine equations are discussed in Sections 13 and 14. When Hilbert posed his famous problem list in 1900 he included a very general question about such equations, asking for a method powerful enough to decide the solvability of all Diophantine equations.

Hilbert's Tenth Problem

> To devise a process according to which, given a polynomial P, it can be determined by a finite number of operations whether the equation $P = 0$ is solvable in integers.

In modern terms, this problem asks for a computer program that accepts arbitrary polynomials as input, does a finite amount of computation, and outputs either a "Yes" or a "No" according as a solution in integers does or does not exist. If there were such a program, then when the output is "Yes" the computer could, by a systematic search that is guaranteed to halt, find positive integers for which the polynomial is zero (see Exercise 1). Unfortunately, as was proved by Yuri Matijesivič in 1970, no such program exists!

The solution to Hilbert's tenth problem was one of the most dramatic results of twentieth-century mathematics. Gödel's famous theorem of 1931 showed that for any reasonable axiom system (i.e., one that is consistent (there is no proof of a contradiction from the axioms) and recursive (the axioms are recognizable by a computer program)) that is strong enough to do arithmetic, there are statements that can be neither proved nor disproved from the axioms. Gödel's sentences involve a clever use of self-reference and are somewhat unnatural, more of interest to logicians than to mathematicians (see [NN]). But the proof that no Diophantine-equation–solving computer program exists is related to an important strengthening of Gödel's theorem: *Given a consistent and recursive system of axioms for arithmetic, there is a polynomial P in 9 variables such that $P = 0$ has no solution in positive integers,* but yet this fact cannot be proved[1] from the axioms! In other words, there are undecidable Diophantine equations: they have no solutions, but their unsolvability is unprovable from the usual axioms of arithmetic.

There are many other remarkable consequences of the solution to Hilbert's tenth problem. For example, there is a polynomial P such that the nonnegative values of this polynomial (when all possible natural numbers are substituted for the variables) is exactly the set of primes. Indeed, the polynomial can be written down in a few lines [JSSW; Rie, p. 43]. Another surprising consequence is that there is a single polynomial $P(n, x, y, z, x_1, \ldots, x_9)$ such that a solution in positive integers to $P = 0$ exists if and only if there are positive integers x, y, z, and n (with $n > 2$) satisfying $x^n + y^n = z^n$. In other words, Fermat's last theorem, which is generally thought of as a family of infinitely many Diophantine equations, one for each exponent, is equivalent to a single Diophantine equation (albeit with more unknowns). This result is based on a definition of exponentiation (x^n) in terms of $+$ and $\times$, a crucial step in the solution of Hilbert's tenth problem. Yet another surprising consequence is that there is a "universal" polynomial, that is, a polynomial $Q(a, x_1, \ldots, x_9)$ such that for any polynomial P, there is an integer n, which depends on P, such that $P = 0$ is solvable in integers if and only if $Q(n, x_1, \ldots, x_9) = 0$ is solvable in integers. Prior to Matijasevič's work it was believed by many that such a universal polynomial ought not to exist. For example, G. Kreisel observed [Dav, p. 266] that "it is not altogether plausible that all (ordinary) Diophantine equations are uniformly reducible to those in a fixed number of variables of fixed degree." However, Q, the universal polynomial, does exist and can even be written down [Jon2]. See [Dav, DH, DMR] for more on Hilbert's tenth problem, a fundamental result that combines elements of number theory, logic, and computer science.

The extension of Hilbert's tenth problem to the domain of rational numbers has not been solved, however. In other words, it is not known whether there is an algorithm that can decide the solvability in rational numbers of all polynomial equations, for example, the system of equations corresponding to the existence of a rational box (Section 14) or the system of two equations $x^m + y^m = 1$ and $xyw = 1$, where m is fixed. A solution in rationals to this latter system corresponds to a nontrivial solution in integers to Fermat's last theorem with exponent m ($xyw = 1$ guarantees that neither x nor y equals 0). Note that a finite system of equations is

---

[1] See Part Two for a discussion of the natural question: If a sentence cannot be proved, how do we know it's true?

equivalent to a single equation since the polynomials P, Q, $R, \ldots$ are simultaneously equal to zero for certain rational values of the unknowns if and only if the single equation $P^2 + Q^2 + R^2 + \cdots = 0$ has a rational solution. Note also that an equation of the form $R = Q$ is equivalent to one of the form $P = 0$ (let $P = R - Q$).

In order to compare various versions of the problem let's use the term X-*algorithm* to refer to an algorithm that decides the solvability of polynomial equations with integer coefficients where the variables are allowed to take on values in X. The 1970 breakthrough was in the context of $\mathbb{N}$-algorithms, and stated that no $\mathbb{N}$-algorithm exists. It follows that no $\mathbb{N}^+$-algorithm exists either (Exercise 2(a)). This implies that no $\mathbb{Z}$-algorithm exists, thus solving Hilbert's problem in the form he stated it. This implication is proved by observing that a $\mathbb{Z}$-algorithm, if it existed, could be transformed into an $\mathbb{N}$-algorithm as follows. Suppose a polynomial equation $P(x, y, \ldots) = 0$ with n unknowns is given. Let Q, a polynomial in $4n$ variables be defined to be $P(a^2 + b^2 + c^2 + d^2, e^2 + f^2 + g^2 + h^2, \ldots)$. If $P = 0$ has a solution in $\mathbb{N}$, then, because of Lagrange's theorem that every nonnegative integer is a sum of four squares (see [HW, Chap. 20]), there is a solution to $Q = 0$ in $\mathbb{Z}$; and a solution to $Q = 0$ in $\mathbb{Z}$ clearly implies that $P = 0$ has a solution in $\mathbb{N}$. Thus a $\mathbb{Z}$-algorithm could be transformed into an $\mathbb{N}$-algorithm. It is also possible to transform an $\mathbb{N}$-algorithm into a $\mathbb{Z}$-algorithm (Exercise 3; see also Exercise 2). In any event, because of the negative solution to Hilbert's tenth problem for $\mathbb{N}$, neither type of algorithm exists.

The preceding discussion shows that because there is no $\mathbb{N}$-algorithm and because $\mathbb{N}$ can be defined in $\mathbb{Z}$ using only existential quantifiers ($\exists$, "there exists") ranging over $\mathbb{Z}$ and symbols for $+$, $\times$, $=$, 0, 1, there cannot be a $\mathbb{Z}$-algorithm. More precisely, the definition is:

$$x \in \mathbb{N} \text{ if and only if } \exists a, b, c, d\ (x = a \times a + b \times b + c \times c + d \times d).$$

This same argument could be used to prove that no $\mathbb{Q}$-algorithm exists provided one could find an existential definition of $\mathbb{Z}$ (or $\mathbb{N}$) in $\mathbb{Q}$. But it is not known if such a definition exists.

Problem 20.1

Is there a sentence S with a parameter x that uses only existential quantifiers (ranging over $\mathbb{Q}$), the operations $+$ and $\times$, the equality relation, the constants 0 and 1, and variables and parentheses, such that $S(x)$ is true when a rational number q is substituted for x if and only if q is an integer? In short, is $\mathbb{Z}$ existentially definable in $\mathbb{Q}$?

Exercises

1. Suppose it is known that the polynomial equation $P(x, y) = 0$ has a solution in integers. Construct an algorithm that finds a solution.

2 (a). Given a polynomial P, construct another polynomial Q such that $P = 0$ has a solution in $\mathbb{N}$ if and only if $Q = 0$ has a solution in $\mathbb{N}^+$.

(b). Given a polynomial P, construct another polynomial Q such that $P = 0$ has a solution in $\mathbb{N}^+$ if and only if $Q = 0$ has a solution in $\mathbb{N}$.

3. Given a polynomial P, construct another polynomial Q such that $P = 0$ has a solution in $\mathbb{Z}$ if and only if $Q = 0$ has a solution in $\mathbb{N}$.

NUMBER THEORY: Part 2

13. FERMAT'S LAST THEOREM

Problem 13

Do there exist positive integers x, y, and z and an integer $n \geq 3$ such that $x^n + y^n = z^n$?

There is a lot of colorful history surrounding Fermat's last theorem. Monetary prizes have been offered, the most famous of which is the Wolfskehl Prize offered in the will of P. Wolfskehl of Darmstadt, Germany, in 1908. The prize was a substantial one, 100,000 Marks, but inflation has reduced its current value to about 10,000 Deutschmarks. See [Rib1, §1.7] for an account of such prizes, which have the unfortunate effect of encouraging hundreds of amateurs to submit "solutions." For more on the history and mathematics of FLT, see [Edw2, 3, Rib1, 2, Wag3, Wei].

Kummer's work on the connection between regular primes and FLT yielded the first comprehensive result on the problem. Regularity can be defined in several ways, the simplest of which is in terms of the Bernoulli numbers (defined in Section 24, Part Two). An odd prime p is regular if it does not divide the numerator of any of the Bernoulli numbers $B_2, B_4, \ldots, B_{p-3}$. The first irregular prime is 37, which divides the numerator of

$$B_{32} = -\frac{37 \cdot 683 \cdot 305065927}{510}.$$

Roughly speaking, the importance of the regular primes is that in certain generalizations of the integers that are relevant to FLT, unique factorization into (generalized) primes does not hold. But if p, which is used to form the generalized integers, is a regular prime, then a version of unique factorization does hold, and this was used by Kummer to prove FLT for such exponents. For more on Bernoulli numbers and regular primes, a good place to start is [IR, Chap. 15].

It is not known whether there are infinitely many regular primes, but a heuristic argument (see [Wag′]) suggests that roughly 60.653% $(= 1/\sqrt{e})$ of the primes ought to be regular; Wagstaff's computations, combined with more recent computations by Tanner and Wagstaff [TW1] yielded that 8,398 of the 13,847 odd primes under 150,000 are regular, or 60.649%. Further statistical tests confirm the hypothesis that the regular primes occur quite regularly, so there is good reason to believe that FLT is true for infinitely many prime exponents. This has not yet been proved, although it is known that FLT is true for infinitely many pairwise relatively prime exponents (Exercise 4(b)).

Kummer also began the investigation into the case of an exponent that is an irregular prime. He developed criteria for FLT that worked in some irregular cases, including all the irregular primes less than 100 $(37, 59, 67)$. Kummer's proofs here were not entirely complete, but Vandiver proved that all the essential ideas were correct. Vandiver, along with D. H. and Emma Lehmer, continued the development of criteria for the irregular case, and in 1954 they found a condition especially well suited to machine computation. Their condition worked on all the irregular primes less than 2,500, and when Wagstaff checked the primes to 125,000 in 1976 he found that their condition worked on all the irregular primes. Moreover, a recent computation by Buhler, Crandall, and Sompolski has shown that the condition is satisfied by all irregular primes less than one million. Thus the 1954 condition (see [Rib1, p. 202]) has never failed to work on an irregular prime; if one could prove that it held for every irregular prime, one would have proved FLT.

Wagstaff's computations took over a year of CPU time on a large computer. The criterion used to verify FLT in the irregular case is rather complicated, but only 10% of the computing time was spent on this aspect of the problem. The more time-consuming part of the calculation (90%) was determining which primes are irregular. Rather than using the definition directly, this is done by checking a complicated set of equations. Jonathan Tanner, in a 1985 undergraduate honors thesis at Harvard, showed how these equations could be improved slightly, and his technique has been used on the primes between 125,000 and 150,000 [TW1]. Tanner's improvement works best on primes congruent to 1 modulo 4; for such a prime near 125,000, Tanner's method uses one-half the time of previous techniques.

Researchers have found it fruitful to divide FLT into two cases. Here we consider only prime exponents and primitive solutions. The *first case* asserts that Fermat's equation has no solution in which each of x, y, z is relatively prime to p; the *second case* asserts that the equation has no solution in which p divides at least one member of the relatively prime triple x, y, z. The first case has been of particular interest. For example, in 1823, well before Kummer's work and using more elementary techniques, Sophie Germain was able to show that the first case holds for all exponents under 100. She proved this by showing that for any odd prime p, the first case holds for p if any of $2p + 1$, $4p + 1$, $8p + 1$, $10p + 1$, $14p + 1$, $16p + 1$ is also prime. This result and its generalizations underlie a recent breakthrough by Adleman, Fouvry, and Heath-Brown [HB2], who used deep theorems on the distribution of primes to prove that the first case of FLT is true for infinitely many primes. It is still not known, however, whether there are infinitely many primes p such that $2p+1$ is also a prime. Thanks to recent computations of A. Granville and M. B. Monagan (extended by Wagstaff, Tanner, and Coppersmith [Cop]), it is now known that the first case holds for all prime exponents less than 10^{17} [Rib2, TW2].

It is unlikely that elementary techniques will be enough to solve FLT (assuming that it is true!). As observed by Edwards [Edw2, p. ix], no elementary proof has ever been found that works for all exponents less than 37. (For a short, elementary proof of the cubic case, see [MW]; the case of fourth powers is even simpler, and can be proved using only the characterization of Pythagorean triples in Exercise 1; see [Rib1, §3.2].) However, elementary techniques continue to contribute small pieces to the FLT puzzle. This is well illustrated by Filaseta's result in Exercise 4, proved when he was a graduate student, and the extension (Theorem 13.2) by Heath-Brown [HB1] and Granville [Gra′], independently (the latter also a graduate student). The reader familiar with the Euler ϕ-function will be able to follow the complete proof, given below. We emphasize, however, that the main tool underlying these results is the deep theorem of Gerd Faltings, that for any $n \geq 3$, Fermat's equation does not have infinitely many primitive solutions.

Theorem 13.2. *Fermat's last theorem is true for almost all exponents.*

Proof. We must show that

$$\frac{F(x)}{x} \to 0 \qquad \text{as } x \to 0,$$

where $F(x)$ is the number of positive integers $n \leq x$ such that FLT is false for exponent n. Let $p_1, p_2, \ldots$ enumerate the odd primes.

Since

$$\prod_{i=1}^{\infty}\left(1-\frac{1}{p_i}\right)=0$$

(Exercise 5) it is possible, given $\epsilon > 0$, to choose r such that

$$\prod_{i=1}^{r}\left(1-\frac{1}{p_i}\right)<\frac{\epsilon}{4}.$$

Let $P = p_1 p_2 \cdots p_r$, and use Filaseta's corollary to Faltings's theorem (Exercise 4(a)) to choose B so that FLT is true for all multiples of p_1 or p_2 or ... or p_r that are greater than B. Then let $x > \max\{P, \frac{2B}{\epsilon}\}$ be arbitrary, and let m be the unique integer such that $mP \leq x < (m+1)P$.

We then have:

$$\begin{aligned} F(x) &< B + |\{n < x : \gcd(n,P) = a\}| \\ &= B + (m+1)\,|\{n \in \{1,\dots,P\} : \gcd(n,P) = 1\}| \\ &= B + (m+1)\,\phi(P) \\ &= B + (m+1)\,P\prod_{i=1}^{r}\left(1 - \frac{1}{p_i}\right) \end{aligned}$$

Therefore $\frac{F(x)}{x} < \frac{B}{x} + \frac{\epsilon(m+1)P}{4mP} < \frac{\epsilon}{2} + \frac{2\epsilon}{4} = \epsilon$ □

For an interesting connection between Fermat's last theorem, Newton's method for solving equations, and continued fractions, see [McC].

Exercises

4 (a). (M. Filaseta, 1984 [Fil]) Deduce the following corollary to Faltings's theorem that for every $n \geq 3$ Fermat's equation has only finitely many primitive solutions: For every odd prime p there is an integer B such that FLT is true whenever the exponent is a multiple of p that is greater than B. (Hint: If not, then Fermat's equation has a solution for infinitely many exponents that are multiples of p. Show how these lead to infinitely many primitive solutions in the case of exponent p.)

(b). Use Exercise 4(a) to construct infinitely many integers, no two of which have a common divisor, such that FLT is true whenever one of these integers is the exponent.

5. Prove that

$$\prod_{i=1}^{\infty}\left(1 - \frac{1}{p_i}\right) = 0.$$

(Hint: Use the fact that the harmonic series diverges.)

14. A PERFECT BOX

Problem 14

Does there exist a box with integer sides such that the three face diagonals and the main diagonal all have integer lengths?

Variations on this problem have been considered at least since 1719, when Halcke discovered the triple $(44, 117, 240)$, which has all face-diagonals integers (see [Dic2, p. 497]). Brocard, in 1895, thought he had proved that no perfect box exists, but his proof was erroneous since he had assumed that the side-lengths are pairwise relatively prime (see Exercise 1). If a perfect box does exist, then the four numbers $(xd)^2, (xc)^2, (cd)^2, (bc)^2$ are such that their differences are all perfect squares. No such sequence of four squares is known to exist, so anyone who is searching for a perfect box might first search for such a sequence.

Problem 14.4

Do there exist four distinct nonzero squares such that each of the six differences they yield is also square?

If only three squares are wanted with all differences being square, then an example arises from any solution to Variation 2, where a face-diagonal, say $y^2 + z^2$, is irrational. The squares z^2, $x^2 + z^2$, d^2 have all differences square.

If instead we consider squares all of whose sums in pairs are square, then asking for three such squares is just a restatement of Variation 1, and so again $(44, 117, 240)$ is an example. A natural generalization is whether there exist longer sequences of numbers with all sums of pairs square. No such sequence of length 4 is known.

Negative results in this area are rare. One approach is to examine the classes of known solutions to Variations 1–3 and prove that the seventh quantity is definitely not an integer. See [Lee] for some examples of this.

The three 6-parameter variations to the problem all have relatively small solutions, but it is known that this is not the case for a perfect box. If a perfect box exists, then all of its sides must be longer than one million, as verified by a computer search carried out by I. Korec [Kor1, 2].

There are also lots of variations to Problems 14.1–3. For example, it is known (J. A. H. Hunter, 1967) that a unit square contains a point having rational distances from three of the corners (see [Guy, Problem D19]). The smallest example is given in Exercise 2. The restriction to integer-sided squares in Problem 14.1 is essential, since there are triangles such that no point is a rational distance from each of its vertices (Exercise 3). However, given a rational-sided triangle, there are lots of points having rational distance from each of the triangle's vertices; indeed, as proved by Almering [Alm], the set of such points is dense in the plane. This result implies an older result of Mordell on *rational quadrilaterals* (quadrilaterals with rational sides and rational diagonals): Given any four points in the plane, there is a rational quadrilateral with vertices arbitrarily close to the given points (Exercise 5). (These results are also discussed in Part Two of Section 10.) Regarding Problem 14.2, H. G. Eggleston [Egg] has given an elementary proof that no isosceles triangle can have integer sides and integer medians. H. Schubert conjectured in 1905 that no triangle could have integer sides, integer area, and two integer medians, but this was disproved by R. H. Buchholz and R. L. Rathbun (independently) in 1987 (see Exercise 6). For diverse results concerning Heron triangles whose area is a perfect square (e.g., sides 9, 10, 17, area 36), see [Mel].

There are analogous problems in higher dimensions. It is known that *rational tetrahedra,* that is, tetrahedra having rational sides, face areas, and volume, exist (thus generalizing Heron triangles); the smallest known example is given in Exercise 7. Moreover, any triangle in the plane can be approximated arbitrarily closely by a Heron triangle (Exercise 8). This leads to the following question.

Problem 14.5

Can every tetrahedron be approximated arbitrarily closely by a rational tetrahedron?

By arguments similar to those in Exercise 8, it can be shown that every tetrahedron can be approximated by one with rational sides. See [Dic2, Chaps. 4 and 5], [Guy, Chap. D], and [BG] for many more problems and results regarding Diophantine equations that arise from geometrical considerations.

Exercises

2. Show that there is a point inside a square of side-length 53 having distances 25, 51, and 52 from three of the corners.

3. Show that there exists a triangle such that no point has a rational distance from all three vertices. In fact, such a triangle can be found such that two of its sides are one unit long. (Hint: Use the fact that the rationals are countable while

the reals are not. A variation of this problem appeared as problem A-4 in the 1990 Putnam exam.)

4. Prove that the area of a Heron triangle must be divisible by 6. (Hint: Decompose the triangle into two right triangles and use Exercise 13.2.)

5. Use Almering's result about triangles to deduce that given any four points in the plane there is a rational quadrilateral with vertices arbitrarily close to the given points.

6. (R. H. Buchholz, R. L. Rathbun, 1987) Show that the triangle with sides 52, 102, and 146 is a Heron triangle such that two of its medians have integer length.

7. (a) Show that if $\triangle ABC$ is acute, then a tetrahedron exists with each of its faces congruent to $\triangle ABC$.

(b) (J. Leech) Let $\triangle ABC$ have sides $a = 148$, $b = 195$, and $c = 203$ and let T be the tetrahedron obtained from this triangle as in (a). Show that each face of T has integer area and that T has integer volume. (Hint: Use the following formula for the volume of a tetrahedron in which all faces are congruent:

$$V = \sqrt{\frac{(a^2+b^2-c^2)(a^2+c^2-b^2)(b^2+c^2-a^2)}{72}}$$

This formula can be derived using calculus (the volume of a solid can be obtained by integrating its cross-sectional areas) and elementary geometry of triangles.)

8. Show that any triangle in the plane can be approximated arbitrarily closely by triangles with rational sides and rational area. (Hint: First show that any right-angled triangle can be approximated by a right-angled triangle with rational sides. For an arbitrary triangle, break it into two right triangles.)

15. EGYPTIAN FRACTIONS

Problem 15

Does the greedy algorithm always succeed in expressing a fraction with odd denominator as a sum of unit fractions with odd denominator?

The fact that each fraction can be represented "greedily" as a sum of distinct unit fractions was proved by Fibonacci. We shall indicate the proof in the case that the given fraction is less than 1; for the rest see Exercise 4. The greedy algorithm proceeds by choosing the least positive integer m such that $\frac{1}{m} \leq \frac{a}{b} < \frac{1}{m-1}$. Note that as the algorithm proceeds, it will never be the case that the least m for which this happens has already been used. For if this ever happened, it would mean that at some stage $\frac{1}{m} + \frac{1}{m} < \frac{1}{m-1}$, which implies $m < 2$ contradicting the assumption that the given fraction is less than 1. Now, the inequality defining m yields that the remainder, $\frac{a}{b} - \frac{1}{m}$, has a numerator $(ma - b)$ that is smaller than a. Therefore, as the algorithm continues, the numerators will form a decreasing sequence of positive integers, whence the remainder will eventually be zero, as desired.

Although the greedy algorithm is easy to handle theoretically (at least in the case where there is no restriction on the unit fractions), its behavior in practice may be quite poor. For example, when applied to the modest fraction $\frac{5}{121}$ it yields the following representation:

$$\frac{1}{25} + \frac{1}{757} + \frac{1}{763309} + \frac{1}{873960180913} + \frac{1}{1{,}527{,}612{,}795{,}642{,}093{,}418{,}846{,}225}.$$

This dramatically illustrates one of the drawbacks of the algorithm: the denominators in the representation grow very rapidly (Exercise 6). However, there are better algorithms; an algorithm due to Bleicher keeps all denominators less than b^2. Applying his algorithm to the extreme example above yields the representation

$$\frac{5}{121} = \frac{1}{25} + \frac{1}{1225} + \frac{1}{3477} + \frac{1}{7081} + \frac{1}{11737}.$$

And ad hoc techniques yield even better representations (Exercise 7). See [Ble] for a comprehensive survey of Egyptian fraction algorithms; a good elementary introduction to Egyptian fractions can be found in [BBC]. In 1950 P. Erdős [Erd] proved that $\frac{b-1}{b}$ requires at least $\lfloor \log\log b \rfloor$ terms in its representation as a sum of distinct unit fractions. Erdős conjectured that, for some constant c, any fraction $\frac{a}{b}$ between 0 and 1 can be written as a sum of $c \log\log b$ distinct unit fractions. The best result known is due to M. Vose [Vos] who proved that a constant c exists so that any

$\frac{a}{b}$ between 0 and 1 can be written as a sum of at most $c\sqrt{\log b}$ distinct unit fractions. Note that Erdős's result implies that some fraction with a 175-digit denominator requires at least 5 terms. In fact, as shown by Bleicher [BBC], $\frac{21}{23}$ requires five unit fractions (it equals $\frac{1}{2} + \frac{1}{3} + \frac{1}{15} + \frac{1}{115} + \frac{1}{230}$).

The question of whether all rationals with odd denominator can be expressed as a sum of distinct odd unit fractions was first raised as a problem in the *American Mathematical Monthly,* and solved affirmatively by Breusch and Stewart [BS].

The following problem has led to some interesting work.

Problem 15.2

For which sets S of unit fractions is it true that each fraction can be written as a sum of distinct members of S?

R. Graham [Gra] proved several results including:

1. S has this property if for some constant c, S contains all primes and all squares beyond c; and
2. if S is the set of reciprocals of squares, then any positive rational except those between $0.6449\ldots$ $(= \pi^2/6 - 1)$ and 1 can be written as a sum of distinct fractions in S (see Exercise 7).

Graham also obtained a generalization of the result of Breusch and Stewart concerning odd unit fractions.

There are several unsolved problems concerning the representation of the number 1 as a sum of distinct unit fractions.

Problem 15.3

Is it true that whenever the integers greater than 1 are partitioned into two sets, one of them has the property that 1 can be represented as a sum of finitely many distinct reciprocals from the set?

Perhaps much more is true; it may be that, for any partition, one of the sets is rich enough to represent all rationals (using finitely many distinct reciprocals from the set). At least we know this is true for the partition into evens and odds (Exercise 5). See [EG, Guy] for more unsolved problems regarding Egyptian fractions. An extensive bibliography appears in [Guy], and an even longer list (almost 300 items) is obtainable from Paul Campbell of Beloit College.

Exercises

3 (a) Show that if b is odd then $\frac{2}{b}$ can be written as a sum of two distinct unit fractions.

(b) Show that if b is even then $\frac{3}{b}$ can be written as a sum of two distinct unit fractions.

(c) Show that if b is not divisible by 3 but is divisible by some prime that is congruent to $-1 \pmod 6$ then $\frac{3}{b}$ is a sum of two distinct unit fractions.

(d) (N. Nakayama, 1940) Show that if all primes dividing b are congruent to $+1 \pmod 6$ then $\frac{3}{b}$ is not a sum of two unit fractions.

4. Show that the greedy algorithm for representing a positive rational greater than 1 as a sum of distinct unit fractions always terminates.

5. Show that the greedy algorithm for representing any fraction as a sum of even unit fractions terminates.

6. Suppose that for a positive rational $\frac{a}{b} < 1$ the greedy algorithm outputs, in order, the reciprocals of the integers $m_1, m_2, \ldots, m_k$. Let $d = m_1^2 - m_1$. Prove that $m_2 > d, m_3 > d^2, m_4 > d^4, \ldots, m_k > d^{2^{k-2}}$.

7. Obtain shorter Egyptian fraction representations of $\frac{5}{121}$ by (a) writing it as $\frac{1}{121} + \left(\frac{2}{11}\right)^2$ and multiplying together two different representations of $\frac{2}{11}$; (b) finding a representation of $\frac{5}{11}$ and multiplying each denominator by 11; (c) by finding an integer n such that $121n$ has divisors $\{d_i\}$ adding up to $5n$; it follows that

$$\frac{5n}{121n} = \sum \frac{1}{121n/d_i}.$$

8. Show that if $\frac{a}{b}$ is between $\frac{\pi^2}{6} - 1$ and 1 then $\frac{a}{b}$ is not a finite sum of reciprocals of distinct squares. (Hint: See Section 24.)

16 PERFECT NUMBERS

Problem 16

Is there an odd perfect number? Are there infinitely many even perfect numbers?

As discussed in Section 18, the best known techniques for determining the primality of an arbitrary number are quite time-consuming, and are feasible only for numbers of fewer than 1,000 digits. But the largest Mersenne prime known, discovered by D. Slowinski in 1985 using three hours on a CRAY computer, is $2^{216,091} - 1$, a number of over 65,000 digits. The reason that such large primes can be found is that there is a special test for primality of numbers of the form $2n - 1$ due to E. Lucas who, in 1876, verified by hand that $2^{127} - 1$ is prime. Let $M(p)$ denote $2^p - 1$. The test goes as follows. First define the sequence $\{L_k\}$ by setting $L_0 = 4$ and letting $L_{k+1} = L_k^2 - 2$; this sequence begins: 4, 14, 194, 37634, 1,416,317,954. Then $M(p)$ is prime if and only if $M(p)$ divides L_{p-2} (see [Rie, Chap. 4] or [Ros′] for a proof). Examples: $M(5) = 31$ which divides 37634; $M(6) = 63$ which does not divide L_4. This test is very efficient since only about p multiplications are needed to decide whether $2^p - 1$ is prime (see [Ros, p. 183] for a more detailed analysis of the efficiency of the Lucas test).

The 31 known Mersenne primes are $M(p)$ for the following values of p: 2, 3, 5, 7, 13, 17, 19, 31, 61, 89, 107, 127, 521, 607, 1279, 2203, 2281, 3217, 4253, 4423, 9689, 9941, 11213, 19937, 21701, 23209, 44497, 86243, 110503, 132049, 216091. The first 30 of these numbers are known to be the first 30 Mersenne primes, but not all values of p under 216091 have been checked so it is possible (but unlikely, as explained below) that another Mersenne prime is hiding below $M(216091)$. When $M(11213)$ was proved prime by D. B. Gillies of the University of Illinois in 1963, the university celebrated by asserting "$2^{11213} - 1$ is prime" in the space beside the postmark on its metered mail. The Mersenne prime $M(21701)$ is noteworthy in that it was discovered by two California high school students, Curt Noll and Laura Nickel, in 1978. Although it is not evident from the first few Mersenne primes, now that 31 are known their regularity is striking. For example, if $M(p)$ is prime then there is a prime q less than $2p$ such that $M(q)$ is prime (with two exceptions, $p = 127$ and $p = 4423$). This regularity is made more apparent by graphing the base-two logarithms of the values of p leading to Mersenne primes $M(p)$.

The near-linearity of the data in Figure 16.1 shows that $\log_2 p$ is increasing very regularly for those p yielding Mersenne primes. The average change in $\log_2 p$ is about 0.5766; another interpretation is that if $M(p)$ and $M(q)$ are successive

Mersenne primes, then the ratio $\frac{q}{p}$ is, on average, approximately equal to $2^{0.5766} = 1.4716\ldots$. The surprising thing is that using facts about the form of potential factors of $M(p)$ one can actually formulate a heuristic argument that implies that the expected ratio is $2^{1/e^{\gamma}} = 2^{1/e^{0.5772\ldots}} = 1.4757\ldots$ (see Section 24 for the definition of γ). The idea of performing a statistical analysis on the Mersenne primes is due to Gillies, the discoverer of the 21st, 22nd, and 23rd Mersenne primes, and has been refined by S. Wagstaff (see [Sch, §3.5]). These ideas can be used to predict roughly where the next Mersenne prime should be. Using the theory's ratio 1.4757 leads to the value $1.4757 \cdot 216091 = 318885$ as the vicinity where the next p yielding a Mersenne prime will be found.

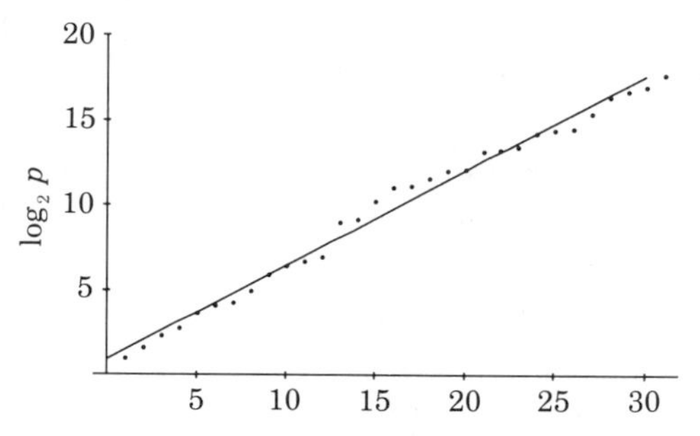

FIGURE 16.1
A comparison of the actual values of $\log_2 p$ for which $M(p)$ is prime (dots), and the values predicted by a heuristic statistical argument (line).

It must be emphasized that the reasoning leading to a justification of the regularity in the Mersenne primes is completely unproven. Such heuristic arguments abound in number theory, but one must always be a little wary. For an extreme example, it is known that the function $\frac{1}{\log x}$ is a good approximation to the density of the primes near x (see Section 17, Part One); indeed it is a proved theorem that the ratio of $\pi(x)$, the number of primes less than x, to $\int_0^x \frac{1}{\log t}\,dt$ converges to 1 as x approaches infinity (Theorem 17.2). This can be loosely interpreted as saying that the "probability" that an integer near x is prime is approximately $\frac{1}{\log x}$. This leads to accurate results when used conservatively, but primeness is a deterministic, not random, phenomenon and a too-literal use of such heuristics lead to absurdities. For example, one could argue that the probability that both x and $x+1$ are prime is $\frac{1}{\log x \log(x+1)}$; integrating this quantity from 2 to ∞ leads to the conclusion that there are infinitely many integers x such that x and $x+1$ are both prime, when in reality there is only one such integer.

Another famous sequence of numbers is the sequence, F_n, of Fermat numbers: 3, 17, 257, 65537, 4294967297, 18446744073709551617, ... These are the values of the expression $F_n = 2^{2^n} + 1$. Fermat conjectured that they are all prime, but each Fermat number whose prime/composite nature is known is composite, except for the first five: it is now known that F_5–F_{21} are composite. The most recent computation [YB] used a special technique for Fermat numbers to establish the compositeness of F_{20}, a 315,653-digit number. This computation took 10 days on a CRAY 2, all for an ultimate output of a single bit, the answer "composite"; this prompted the programmers to quip: "Never have so many circuits labored for so many cycles to produce so few output bits."

Problems 16.1–16.3 are special cases of more general conjectures, some of which are very far-reaching. A famous theorem on primes is Dirichlet's theorem (1837) that any arithmetic progression for which the first term and common difference are relatively prime contains infinitely many primes. This is equivalent to the $r = 1$ case of the following problem, which L. E. Dickson conjectured to have an affirmative answer.

Problem 16.5 (L. E. Dickson, 1904)

Suppose $a_1 + b_1x, a_2 + b_2x, \ldots, a_r + b_rx$, where $a_i, b_i \in \mathbb{Z}$ and $b_i \geq 1$, satisfy the following condition:

There is no integer $n > 1$ such that for each integer k, n divides the product $(a_1 + b_1k)(a_2 + b_2k) \cdots (a_r + b_rk)$.

Then there exist infinitely many natural numbers m such that each $a_i + b_im$ is prime.

The condition on the functions in Problem 16.4 is necessary, for otherwise the desired numbers m do not exist (Exercise 5). One can weaken the conclusion by requiring only one value of m yielding prime values of the r functions, but the resulting conjecture is equivalent to the ostensibly stronger one (see [Rib3]). Dickson's conjecture implies both the twin prime conjecture and a positive solution to Problem 16.3. For the first, just consider the two linear functions x and $x+2$, which are easily seen to satisfy the condition in Dickson's conjecture. For the second, see [Rib3]. A different strengthening of the twin prime conjecture arises from more quantitative aspects of prime distributions. Heuristic arguments can be used to formulate a conjecture about the number of twin primes in the interval $[1, x]$ (see [Rib3, Zag]). The resulting estimate is surprisingly accurate, thus adding plausibility to the twin prime conjecture. For example, the conjecture predicts that there will be 166 twin pairs between 10^{15} and $10^{15} + 150,000$; in fact, there are 161 such pairs.

Dickson's conjecture can be generalized to higher-degree polynomials in a way that would yield an affirmative answer to Problem 16.2. The strongest conjecture along these lines is Schinzel's *hypothesis H,* a proof of which would solve Problem 14.3 as well (see [Rib3]).

Problem 16.3 is a special case of a notorious problem of P. Erdős and P. Turán. Erdős, a Hungarian mathematician who spends most of his time travelling to mathematics departments throughout the world, is known for offering financial rewards for solutions to many of his problems. Solving even a $10 Erdős problem is quite prestigious, and can be very difficult. Erdős has offered $3,000 for the solution to the following.

Problem 16.6

Is it true that whenever $\{a_n\}$ is a sequence of positive integers such that $\sum_{n=1}^{\infty} \frac{1}{a_n}$ diverges, then the sequence $\{a_n\}$ contains arbitrarily long arithmetic progressions?

Because the sum of the reciprocals of the primes diverges (see Theorem 17.3), an affirmative answer to Problem 16.6 would settle Problem 16.3. An important intermediate result has been obtained by G. Szemerédi who proved that if a set of integers A has positive upper density (meaning: there is some $c > 0$ such that for n sufficiently large, the fraction of integers in $[1, n]$ that are in A is greater than c), then A contains arbitrarily large arithmetic progressions (see [GRS, §2.5]). Unfortunately, the primes do not have positive upper density (Exercise 17.4), so Szemerédi's theorem does not settle Problem 16.3. The largest known arithmetic progression of primes has 19 terms [Pri]. See [ME] for a computational approach to Problem 16.6.

The theory of odd perfect numbers is less comprehensive than for even ones, but there are several results about the form an odd perfect number must have, if one exists. If N is an odd perfect number, then the following results are known.

L. Euler	N has the form $p^a M^2$ where p is prime and $p \equiv a \equiv 1 \pmod 4$.
R. Steuerwald, 1937	In Euler's result M cannot be squarefree.
D. Suryanarayana and P. Hagis, 1970	The sum of the reciprocals of the prime divisors of N lies between 0.596 and 0.694.

P. Hagis, 1975, 1983	N has at least 8 distinct prime factors; if N is not divisible by 3 then N has at least 11 prime factors.
J. Condict, 1978, undergraduate thesis at Middlebury College	N has a prime factor larger than 300,000.
P. Hagis, 1981	The second-largest prime factor of N is at least 1,000.
G. L. Cohen, 1987	The largest prime power dividing N is greater than 1020.
R. P. Brent and G. L. Cohen, 1988	$N > 10^{160}$.
R. P. Brent, G. L. Cohen, H. J. J. te Riele, 1989	$N > 10^{300}$.

In 1959 Wirsing (see [Wag2]) obtained an upper bound on the function $V(x)$, the number of perfect numbers (odd or even) less than x, a corollary to which is that the sum of the reciprocals of the odd perfect numbers is finite. Of course, by the Euclid–Euler formula this is clear for even perfect numbers. Some interesting results relating odd perfect numbers to computability have been obtained by C. Pomerance. Consider the set of k for which there is an odd perfect number having exactly k distinct prime factors. The only specific information known about this set is Hagis's result that the set contains none of 1, 2, 3, 4, 5, 6, 7. Because numbers with k distinct prime factors can be arbitrarily large, it is not clear that this set is computable. However, Pomerance [Pom1] proved that it is, by finding an easily-computed bound on odd perfect numbers with k distinct prime factors. His work was motivated by a conjecture of C. W. Anderson, who observed that an odd perfect number exists if and only if 2 is in the set R, defined to be the set of values (all rational) of $\frac{\sigma(n)}{n}$, for $n = 1, 2, 3, \ldots$. Given a rational number q it is not at all clear how to determine in finite time whether q lies in R; this leads to the following question, which Anderson conjectured had an affirmative answer.

Problem 16.7 (C. W. Anderson)

Is R a computable set?

For more on the preceding results concerning odd perfect numbers see [BC1, BCT, Hag1, 2, Pom1, Wag2] and the references cited in those papers.

Writing in 1877, Sylvester [Syl1] felt that the set of conditions that an odd perfect number has to satisfy make it unlikely that such exists: "the existence of [an odd perfect number]—its escape, so to say, from the complex web of conditions which hem it in on all sides—would be little short of a miracle." Descartes's belief that replacing the primes in his near-miss $22021 \cdot 3^2 \cdot 7^2 \cdot 11^2 \cdot 13^2$ by other primes will lead to an odd perfect number is false, both because of Steuerwald's result and the necessity of having at least 8 prime factors. Nevertheless, no heuristic argument is known that predicts the nonexistence of an odd perfect number, and there is a very real possibility that some product of dozens, or perhaps hundreds, of primes leads to one. On the other hand, perhaps there is a proof that none exists; according to Sylvester [Syl2], "Whoever shall succeed in demonstrating their absolute nonexistence will have solved a problem of the ages comparable in difficulty to that which previously to the labours of Hermite and Lindemann environed the subject of the quadrature of the circle."

Exercises

5. Show that if the condition on the functions $a_i + b_i x$ in Problem 16.5 is not satisfied, then there cannot be infinitely many values of m yielding prime values of the s functions.

6. (S. C. Root, 1969) Use the primality testing algorithm described in Exercise 18.6 to find out how many terms of the following arithmetic progression are prime: first term $=$ 53297929, common difference $=$ 9699690. Do you notice anything special about the form of the difference, and can you explain it?

7. (G. Nocco, 1863) Show that a perfect number cannot have only two prime divisors. (Hint: Use the formula for the sum of a geometric series to obtain a contradiction from the assumption that $2 = \frac{\sigma(p^a q^b)}{p^a q^b}$.)

17. THE RIEMANN HYPOTHESIS

Problem 17

Do the nontrivial zeros of the Riemann zeta function all have real part $\frac{1}{2}$?

For integers $s > 1$ we may define $\zeta(s)$ to be

$$\sum_{n=1}^{\infty} \frac{1}{n^s}.$$

This series converges for any real value $s > 1$ (see Section 24), so we may view ζ as a function from $(1, \infty)$ to $\mathbb{R}$. However, the real power of the zeta function arises when it is considered as a function defined on all complex numbers (except $s = 1$) and taking values in $\mathbb{C}$. The construction of this extension is complicated, but it can be done (see [Edw1]) and the result, which, for real $x > 1$, agrees with the definition using series, is called the Riemann zeta function. Actually, it is not difficult to see how to define $\zeta(s)$ when the real part of s is positive. The following identity is easy to verify for real $s > 1$ (Exercise 2).

$$\zeta(s) = \frac{1}{(1 - 2 \cdot 2^{-s})} \sum_{n=1}^{\infty} \frac{(-1)^{n+1}}{n^s}. \tag{1}$$

The nice thing about (1) is that even though the left-hand side does not make sense if $s \leq 1$—the series defining $\zeta(s)$ diverges for such s—the series on the right is convergent for any positive real s. This allows us to use the right side of (1) to define $\zeta(s)$ for positive real s (except $s = 1$, for then $1 - 2^{1-s} = 0$). So, for example, one can check that $\zeta(\frac{1}{2}) = 1.4\ldots$ (Exercise 3).

The preceding technique also works for complex values of s, provided the real part of s is positive and $s \neq 1$. We omit the details, which can be easily worked out by writing s as $a + bi$ and n^s as $e^{s \log n}$, and using $e^{it} = \cos t + i \sin t$. This leads to another formulation of RH that does not mention complex numbers.

Problem 17.3

Is it true that $a = \frac{1}{2}$ is the only positive real number for which there exist values of b making the following two series vanish:

$$\sum_{n=1}^{\infty} \frac{(-1)^{n+1}}{n^a} \cos(b \log n),$$
$$\sum_{n=1}^{\infty} \frac{(-1)^{n+1}}{n^a} \sin(b \log n)? \tag{2}$$

Problem 17.3 is equivalent to RH because a pair of reals (a, b) (where $a > 0$) for which both series above equal 0 corresponds to a complex number $a + bi$ for which $\zeta(a + bi) = 0$.

The construction of an extension of ζ to all of $\mathbb{C}\backslash\{1\}$ is a refinement of the preceding ideas, in that a formula is found that agrees with ζ for s having positive real part, but which is defined in all of $\mathbb{C}\backslash\{1\}$. This was done in 1859 by Riemann, who realized the subtle connections between the zeros of the zeta function and the distribution of primes. The theory of analytic continuation (see any text on complex analysis, e.g., [Ahl]) guarantees that any two analytic extensions of the original function on $(1, \infty)$ are equal.

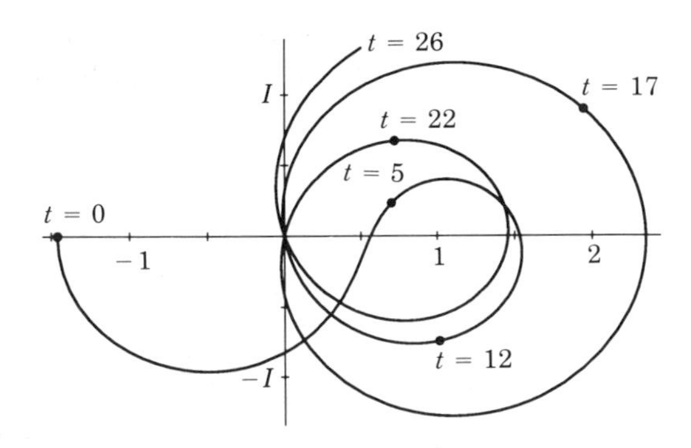

FIGURE 17.2
The values of the zeta function on the critical line $\frac{1}{2} + it$, for $0 \le t \le 26$, showing the first three zeros.

The Riemann zeta function turns out to be zero whenever s is a negative even integer, but it has other zeros as well. The precise location of these additional zeros, called the *nontrivial zeros,* is what the Riemann hypothesis is concerned with. Riemann conjectured that all nontrivial zeros have real part $\frac{1}{2}$, and it is this assertion that is the original formulation of the celebrated Riemann hypothesis. Millions of zeros have been computed, and they all have real part $\frac{1}{2}$. Figure 17.2 shows $\zeta(\frac{1}{2} + it)$ for $0 \le t \le 26$, with the first three zeros visible. Indeed, van de Lune, te Riele, and Winter have computed the first 1,500,000,000 zeros, and they all lie on the so-called critical line, the line in $\mathbb{C}$ consisting of complex numbers with real part $\frac{1}{2}$. See [Wag5] for a sketch of some of the ideas behind such computations, and a discussion of whether this numerical work yields good evidence for RH. A more detailed discussion of zeta-function computations can be found in [Edw1], while a deeper discussion of the question of numerical evidence in favor of the Riemann hypothesis may be found in [Odl]. A good introduction to the theory of the zeta function may be found in [Pat]; for more advanced topics, see [Ivi].

It is not immediately clear what the zeta function would have to do with the distribution of the primes. To point out one connection, recall the proof (Exercise 13.5) that

$$\prod_{i=1}^{\infty}\left(1-\frac{1}{p_i}\right)=0,$$

where $\{p_i\}$ is the set of primes. The proof can be neatly summed up by the formula

$$\frac{1}{\sum_{n=1}^{\infty}\frac{1}{n}}=\prod_{i=1}^{\infty}\left(1-\frac{1}{p_i}\right).$$

However, neither side of this formula is meaningful as it stands. One way around the difficulty is to consider partial sums and products, as was done in the solution of Exercise 13.5. Another way is to consider the sth powers of n and p_i, where $s>1$. This yields

$$\frac{1}{\sum_{n=1}^{\infty}\frac{1}{n^s}}=\prod_{i=1}^{\infty}\left(1-\frac{1}{p_i^s}\right), \tag{3}$$

which can be proved using the ideas of the solution to Exercise 13.5 (and a bit more work to justify some manipulations with series), and which has the advantage that both sides are meaningful, that is, are definite real numbers. The importance of formula (3), which was known to Euler, is that it relates the primes, a central topic in number theory, to the function $\zeta(s)$, which does not explicitly mention primes and can be studied by the methods of analysis.

The vanishing infinite product of Exercise 13.5 can be used to prove that the series of reciprocals of primes diverges.

Theorem 17.3. *The series $\sum_{i=1}^{\infty}\frac{1}{p_i}$ is divergent.*

Proof. In this proof all sums and products range over all prime values of p. Because $\prod(1-\frac{1}{p})=0$, the series $\sum\log(1-\frac{1}{p})$ diverges to $-\infty$. Assume that the series of prime reciprocals converges. Then the series $\sum -\log(1-\frac{1}{p})-\frac{1}{p}$ diverges to ∞.

But, using the Maclaurin series for $\log(1-x)$, the general term of this series equals

$$\frac{1}{2p^2}+\frac{1}{3p^3}+\frac{1}{4p^4}+\cdots,$$

which is less than

$$\frac{1}{2p^2}+\frac{1}{3p^3}+\frac{1}{4p^4}+\cdots=\frac{1}{2p(p-1)}.$$

This last is the general term of a convergent series (it's less than $1/(2(p-1)^2)$), which is the desired contradiction. □

Theorem 17.3 gives some information on the magnitude of the set of primes. For example, since the series of square reciprocals converges we may conclude that the primes are, in a sense, more numerous than the squares even though there are infinitely many of each. For a discussion of the effect of the zeros of ζ on the distribution of primes, see [Rie, Chap. 2] (see also [Wag6]). For example, the prime number theorem (Theorem 17.2) is equivalent to the assertion that $\zeta(s)$ has no zeros of the form $1+it$. Note that one consequence of the prime number theorem is that the primes do not have positive upper density (Exercise 4).

One aspect of the data in Table 17.1 stands out, namely that $\text{li}(x)$ always overestimates $\pi(x)$. For many years it was believed that this was always the case, but in 1914 Littlewood proved that $\text{li}(x)-\pi(x)$ takes on large positive and large negative values infinitely often (see the remark on Littlewood's result following Problem 17.1). The first value of x for which $\text{li}(x)-\pi(x)$ is negative has come to be known as the *Skewes number,* after S. Skewes, who in 1933 obtained a gigantic bound under which the first sign-change occurs. How gigantic? Well, it was considered a substantial improvement when R. S. Lehman, in 1966, proved that the Skewes number is less than 10^{1166}. Here too, detailed knowledge of the zeros of $\zeta(s)$ is relevant, and H. te Riele [tR3] has recently improved the bound to 10^{371}. Note the consequences of this phenomenon for those who would draw conclusions about the global behavior of a function solely from a finite amount of data. The Skewes number may be so large that centuries of computation of the difference $\text{li}(x)-\pi(x)$ would yield positive numbers, and steadily increasing ones at that. Thus a computation for x under, say, 10^{50} might not even scratch the surface of the true nature of this difference! For further discussion of the Skewes number, see [Boa].

The logarithmic integral of x, defined in Part One via

$$\text{li}(x)=\lim_{\epsilon\to 0}\left(\int_0^{1-\epsilon}\frac{1}{\log t}\,dt+\int_{1+\epsilon}^{x}\frac{1}{\log t}\,dt\right),$$

is easy to compute because, using Taylor series, one can show that (see [Rie, p. 55])

$$\mathrm{li}(x) = \gamma + \log\log x + \sum_{n=1}^{\infty} \frac{(\log x)^n}{n!n},$$

where γ is Euler's constant (see Section 24). It takes only about 400 terms of this series to get a value of $\mathrm{li}(10^{100})$ accurate to 12 significant digits.

The computation of $\pi(x)$ for large x is more difficult, since one would like to do it without counting all the primes less than x one by one. There are techniques for doing this, and the latest effort [LMO] is capable of computing the exact value of $\pi(x)$ for x near 10^{16} in a few hours. The exact value of $\pi(10^{100})$ is currently out of reach, however.

At the end of the last century F. Mertens conjectured an apparently stronger result than RH, namely that $|M(x)| < \sqrt{x}$ for all $x > 1$ (recall that $M(x)$ is the sum of $\mu(n)$ for all $n \leq x$; see Problem 17.2 for the property of $M(x)$ that is equivalent to RH). Although the Mertens conjecture is true for every value of x ever tried—M. Yorinaga in 1979 completed a check up to $x = 400{,}000{,}000$—the conjecture is now known to be false. In 1985 Odlyzko and te Riele [OtR] showed that, for infinitely many integers x, $|M(x)| > 1.06\sqrt{x}$. This area bears some similarity to the developments regarding the Skewes number in that the proof does not provide any specific example of an x for which the conjecture fails. The refutation shows only that some counterexample exists that is less than a gigantic number. Their methods have recently been refined by J. Pintz [Pin], who has shown that there is a counterexample less than $10^{10^{65}}$. But even if the smallest counterexample is near 10^{30}, we may never know a specific value of x for which $|M(x)| > \sqrt{x}$.

Various heuristic arguments based on probability can be used in attempts to explain why the Riemann hypothesis ought to be true. It can be argued that this type of argument, which occurs for several other problems in number theory (see Section 19, for example), may not be relevant to the truth of RH. But such heuristics can be helpful to an understanding of a conjecture by placing it in a larger context. See [Ios] for a discussion of probability arguments related to RH.

We conclude by mentioning that the Riemann hypothesis has many, many consequences in addition to those in Problems 17.1 and 17.2. For example, in Section 18 it is shown that RH is connected to the complexity of determining whether a number is prime. More precisely, if a more general assertion called the extended Riemann hypothesis is true, then primality can be determined by an efficient (i.e., polynomial time) algorithm.

Exercises

2. Verify identity (1).

3. How many terms of the alternating series on the right-hand side of (1) are necessary to obtain $\zeta(\frac{1}{2})$ to one decimal place? Write a computer program to verify that $\zeta(\frac{1}{2}) = 1.4\ldots$.

4. Show that the primes have zero density, i.e., limit $\lim_{x \to 0} \frac{\pi(x)}{x} = 0$ and conclude that they do not have positive upper density (definition in Section 16, Part Two).

18. PRIME FACTORIZATION

Problem 18

Is there a polynomial-time algorithm for obtaining a number's prime factorization?

The best known general-purpose factoring algorithms, that is, ones that work on all composite inputs rather than only on inputs of a special form, run in time $O(N^{0.25+e})$. There are faster algorithms that, like Miller's primality test, require various assumptions about prime distributions. Let's use the term heuristic complexity to refer to an algorithm's time complexity that is known to be valid only under certain, hopefully reasonable, assumptions about the distribution of primes. It seemed more than a coincidence that, up to early 1990, the best of the faster algorithms all had heuristic complexity functions of the form $O(e^{c\sqrt{\log N(\log\log N)}})$, where c is a modest constant (between 1 and 3). This function is far from polynomial in n, the length of the input N. Because the same function arose from several different algorithms, it seemed as if it might be the true complexity function for factoring. But then a 1988 idea of J. Pollard for factoring numbers of certain special forms was refined into a new general method by J. Buhler, H. Lenstra, and C. Pomerance; the method has heuristic complexity of the form $O(e^{c\sqrt[3]{\log N(\log\log N)^{2/3}}})$, which is a theoretical improvement over the previous complexity bound. These ideas have practical value as well, as A. Lenstra and M. Manasse used Pollard's ideas to find the complete factorization of the ninth Fermat number (see Section 16), $2^{512}+1$, a 155-digit number.

The best known general-purpose primality-testing algorithm (the Adelman–Rumely–Pomerance algorithm) runs in time $n^{c\log\log n}$, and for inputs of moderate size this algorithm works extremely well [CL]. A nice exposition of modern primality-testing appears in [Pom3]; see [Dix, Pom4, Wil] for a discussion of factoring algorithms. Riesel's book [Rie] gives a comprehensive treatment of factoring, primality testing, and other computational aspects of the theory of primes. To get an idea of the progress in factoring over the past few years note that the ten "most wanted" factorizations in the first edition (1983) of [BLSTW] were numbers having between 53 and 71 digits. Hundreds of new factorizations have been discovered since then, and the current list of ten most wanted factorizations consists of numbers having between 110 and 291 digits.

Note that the Miller–Rabin test described in Part One is a polynomial-time probabilistic test for compositeness: it is extremely likely to provide a definite proof of a number's compositeness, but if the number is prime no *proof* of that will result. An exciting recent breakthrough is the discovery by L. Adleman and M. Huang [AH] (building on work of S. Goldwasser and J. Kilian) of a probabilistic test for primality. Problems for which an efficient probabilistic algorithm exists are said to be in the class $\mathbb{RP}$ ("random polynomial"), which lies between the classes $\mathbb{P}$ and $\mathbb{NP}$. Thus both primes and composites are in $\mathbb{RP}$. For background on the various complexity classes, in particular the important class of nondeterministic polynomial-time problems ($\mathbb{NP}$), see [GJ].

One of the virtues of Miller's algorithm is its simplicity. It can even be implemented on a hand-held programmable calculator or a home computer. Then one can use only the values 2, 3, and 5 for b because it has been checked [PSW] that they decide primality correctly for all numbers less than 1.5×10^9, with four exceptions (see Exercise 6). For a clever way (due to A. K. Head) of forming $xy \bmod N$ (the main computational bottleneck in Miller's algorithm) on a 10-digit machine where x, y, and N are less than 1.5 billion, see [Ros, p. 100, Exercise 24]. Another point to remember is that most numbers are not prime; indeed, most odd numbers are divisible by some prime less than 20. So if one is testing random integers for primality one should precede the use of Miller's algorithm with a test for divisibility by some small primes.

The known approaches to primality testing fall short of providing a polynomial-time algorithm that depends on no unproved assumptions. At least we can be sure that the primes are in $\mathbb{NP}$, however; that is, there is a succinct (polynomial-length) certificate of a number's primality. Although the proof that composites lie in $\mathbb{NP}$ is obvious (a nontrivial divisor is the certificate of compositeness), the proof that primes also have such certificates, a theorem discovered in 1975 by V. Pratt, is more intricate ([Ros, Thm. 8.19]; see also [Dix, Wag6]).

There has been a lot of interest in the factoring problem recently because of applications to cryptography that assume there is no feasible way to factor large numbers. The significance of these applications is that two people, Alice and Bob, say, can arrange a secret code by which Bob can send secure messages to Alice over a public line, without Alice and Bob having had to share in advance some secret information regarding the code to be used. Assume that Bob is the sender and Alice the receiver of the messages. First Alice chooses distinct 100-digit primes, p and q, each of which is congruent to 3 $(\text{mod } 4)$. This can be done by generating random 100-digit numbers and testing them for primality using, say, the Miller–Rabin probabilistic technique, which has a negligible probability of error. Such a search won't take too long since if only numbers congruent 3 $(\text{mod } 4)$ are tested then approximately one out of every 230 numbers is prime. Then Alice forms $N = pq$ and makes N (but not p or q) available to the public. To send a message, Bob somehow codes it using a sequence of 0s and 1s (Bob can tell Alice publicly how the coding is done) and breaks the sequence up into blocks so that each block is the base-two representation of a number less than N. Then, for each block X, Bob computes $a \equiv X^2 \pmod{N}$ (where $0 < a < N$) and sends a instead of X. Now, it is easy for Alice, who knows p and q, to figure out the two square roots of a modulo p and the two square roots of a modulo q, and these numbers can be combined (Exercise 5) to yield the four square roots of a modulo N, of which presumably only one will correspond to a message in English. The assumption that no feasible method of factoring exists implies that no feasible method of taking square roots modulo N exists, so an eavesdropper learns nothing by listening in to the message as it is sent. The preceding technique is a variation, due to M. Rabin, of the Rivest–Shamir–Adleman (RSA) system.

When the original RSA method was proposed in 1977, the inventors used two primes having 64 and 65 digits to encode a secret message and offered $100 as a reward for its decryption. The product of the primes is a 129-digit number that has come to be known as the RSA number. The RSA number has not yet been factored, but the recent improvements in factoring methods may mean that their message will soon be unraveled. There are many other encryption schemes, each with certain advantages and disadvantages, all based on the idea of a function that's easy to compute (like forming the product pq) but hard to invert (discovering p or q from their product); see [Hel, Ros].

The security of coding schemes such as the one just discussed depends on more than just the nonexistence of a polynomial-time factoring algorithm. It actually requires that there is no fast algorithm that works any significant fraction of the time. For example, a message-sender would not be pleased if 10% of the messages were being read by eavesdroppers. The precise statement of the assumption under which prime-number-based codes are secure is rather complicated, and much stronger than just the nonexistence of a polynomial-time factoring algorithm.

Many apparently difficult problems have been proved to be $\mathbb{NP}$-complete, which, in practice, serves to certify their intractability. For an $\mathbb{NP}$-complete problem cannot be solved by a polynomial-time algorithm unless $\mathbb{P} = \mathbb{NP}$, which is widely disbelieved. However, this approach may not be possible for proving factoring intractable. To make this precise, note that the factoring problem is polynomially equivalent to the decision problem for the following set (Exercise 7):

$$A = \{(N, M) : N \text{ has a prime factor less than } M\}.$$

But A^c (the set of pairs not lying in A) lies in $\mathbb{NP}$: guess the prime factorization of N, using the fact that primes are in $\mathbb{NP}$ to (nondeterministically) verify that each prime factor is indeed prime, and check that these primes are all greater than M. Now, if A is $\mathbb{NP}$-complete, then any problem B in $\mathbb{NP}$ polynomially reduces to A, whence B^c polynomially reduces to A^c. But $A^c \in \mathbb{NP}$, so $B^c \in \mathbb{NP}$. In short, if

A is $\mathbb{NP}$-complete then $\mathbb{NP}$ is contained in co-$\mathbb{NP}$. This implies that co-$\mathbb{NP} \subseteq \mathbb{NP}$ ($A \in$ co-$\mathbb{NP} \Leftrightarrow A^c \in \mathbb{NP} \Rightarrow A^c \in$ co-$\mathbb{NP} \Leftrightarrow A \in \mathbb{NP}$), so if A is $\mathbb{NP}$-complete then $\mathbb{NP} =$ co-$\mathbb{NP}$, which, although it is seemingly weaker than $\mathbb{P} = \mathbb{NP}$, is also widely disbelieved.

The preceding discussion has focused on factoring arbitrary integers. If an integer has some special form, such as the numbers $2^r - 1$ for example, there may be special tests for factoring or determining primality that are much faster than general-purpose algorithms. Indeed, there is a polynomial-time algorithm for testing numbers to see if they are Mersenne primes (see Section 16). The largest[1] known prime, $2^{216,091} - 1$ (a number having more than 65,000 digits), was proved prime by such a special-purpose algorithm.

Exercises

4. Suppose $a \in [1, p]$ is a square mod p where p is prime and $p \equiv 3 \pmod 4$. Show that the $(\text{mod}\ p)$-square roots of a are given by $\pm a^{(p+1)/4} \pmod p$.

5. Suppose p and q are distinct odd primes, $N = pq$, $1 < a < N - 1$, $(\pm r)^2 \equiv a \pmod p$, and $(\pm s)^2 \equiv a \pmod q$. Whenever two integers x and y are relatively prime, there exist integers c and d such that $cx + dy = 1$, and these integers can be found quickly (see discussions of the Euclidean algorithm in [Knu] or [NZ]). Assuming c and d are such that $cp + dq = 1$, show that the four integers in $[1, N-1]$ congruent to $\pm r(1 - dq) \pm s(1 - cp) \pmod N$ are the square roots of $a \pmod N$.

6. Implement Miller's primality-testing algorithm on either a programmable calculator or home computer. Your program should work on all inputs less than 1.5 billion, assuming that the computing device has 10-digit accuracy. Only the values $b = 2$, 3, and 5 need be used, provided the input is not one of the four composite exceptions: 25,326,001; 161,304,001; 960,946,321; 1,157,839,381. See [Ros, p. 100, Exercise 24] for a fast way of forming $xy \pmod N$ for numbers near the overflow limit.

7. Suppose that an oracle that decides membership in the set $A = \{(N, M) : N$ has a prime factor less than $M\}$ is given. Using the oracle, describe a polynomial-time algorithm to factor a given integer N.

---

[1] No longer. On August 6, 1989, researchers at Amdahl found that $391{,}581 \times 2^{216{,}193} - 1$ is prime. This number has 37 more digits than $2^{216{,}091} - 1$, the largest known Mersenne prime. This is the first time since 1952 that the largest known prime has not been a Mersenne prime (see *Amer. Math. Monthly*, 97 (1990) p. 214.)

19. THE $3n+1$ PROBLEM

Problem 19

Is every positive integer eventually taken to the value 1 by the $3n+1$ function?

This problem is sometimes known as the Collatz conjecture, after L. Collatz of Germany who posed a similar problem in the 1930s. The problem involving the $3n+1$ function was posed in 1952 by B. Thwaites (see [Lag], which contains many results and a complete bibliography on this problem). The verification of the $3n+1$ conjecture for starting values up to three trillion was carried out by K. Ishihata and N. Yoneda of Tokyo.

It is convenient to introduce the function C, defined only on the odd integers, by $C(n) = (3n+1)/2^s$, where 2^s is the largest power of 2 dividing the even number $3n+1$. The function C merely combines each odd step in the trajectories of the original $3n+1$ function f with all the consecutive even steps that follow. Thus the $3n+1$ conjecture is true if and only if all C-trajectories of positive odd integers end up at 1 eventually, which is equivalent to the assertion that each odd number greater than 2 is eventually carried to a smaller integer by iterates of C. For the rest of this section the *stopping time* of a positive integer n will refer to the number of applications of C needed to transform n to a value smaller than n. Thus the stopping time of 27 is 37. For many integers, indeed, almost all integers, it is known that the stopping time is finite. For example, consider any $n > 1$ such that $n \equiv 1 \pmod 4$. Then $3n+1$ is divisible by 4 so $C(n) \leq \frac{1}{4}(3n+1)$ and the stopping time of n is 1. Similar computations show that, unless n is congruent modulo 256 to one of:

$$\begin{gathered} 27, 31, 47, 63, 71, 91, 103, 111, 127, 155, \\ 159, 167, 191, 207, 223, 231, 239, 251, 255, \end{gathered} \tag{1}$$

then the stopping time of n is less than 5 (Exercise 4). This yields that, asymptotically, at least 85% of the odd integers greater than 2 have a finite stopping time. This approach can be refined further, and one of the most striking results yet proved about this problem is the theorem of Terras and Everett (see [Lag, Thm. A]) that in fact, asymptotically, 100% of the odd integers beyond 2 have finite stopping time. More precisely, $\lim_{n\to\infty} \frac{F(n)}{n} = 1$ where $F(n)$ is the number of odd integers between 2 and n having finite stopping time. Note that this theorem does not imply even that infinitely many integers are transformed to 1 eventually (why not?). This is known, however; Crandall [Cra] has shown that for n sufficiently large, the trajectories of at least $n^{0.05}$ of the numbers between 1 and n reach 1 eventually.

It was mentioned in Part One that a counterexample can have two forms: a divergent trajectory or a nontrivial loop. At least it is known that there are no short nontrivial loops; Terras [Ter] and Garner [Gar] showed how a lower bound on a counterexample to the original conjecture could be used to obtain a lower bound on the size of a nontrivial loop. Their technique, combined with the fact that the conjecture is true for starting values up to 700 billion, yields that any nontrivial loop must have at least a half-million terms.

The list of congruence classes mod 256 in (1) can be used to speed up computation of the stopping time for large numbers of values. As one proceeds through the odd numbers between, say, 3 and 2,000,000,000 one need only examine the congruence classes in list (1). The stopping times for all the missed numbers are known in advance (Exercise 4), and can easily be included in the accumulation of the stopping times. Thus fewer than 15% of the stopping times need to be computed explicitly. Greater savings are possible if one is willing to work modulo a larger power of 2 (and put in more initial computation).

Another intriguing aspect of the $3n + 1$ problem is related to questions of computability. Let A denote the set of positive integers whose trajectory contains a 1 (and hence ends in the loop $(1, 2, 4)$). The $3n + 1$ conjecture asserts that A is the set of all positive integers. But it is not even known whether A is a computable set. In other words, no algorithm is known that:

(a) is guaranteed to halt for all n, and

(b) decides, for each input n, whether n lies in A.

Since no bound is known on the number of steps a trajectory might have (as a function of the starting value), just iterating the $3n + 1$ function is pointless: one wouldn't know when to give up and declare a trajectory unbounded. It would be remarkable if the set A, which is defined so simply, turns out to be noncomputable. It has been known since the 1930s that noncomputable sets exist, but to define one explicitly takes quite a bit of work (see any text on the theory of computation). John Conway has shown that problems similar to the $3n+1$ Problem can lead to stopping time questions whose solution set is noncomputable (see [Wag1, p. 76]).

We now discuss a heuristic attempt to understand the $3n + 1$ trajectories by studying a random process designed to imitate the iterations of C. Consider the question: How many iterations of C should it take, on average, to transform a starting value to a smaller number? If the odd and even steps of the $3n + 1$ function are as random as they appear to be, then this average should be well approximated by the average for a corresponding random process. Since, for odd n, it seems equally likely that $\frac{1}{2}(3n+1)$ is odd or even, and in the latter case, it seems equally likely that $\frac{1}{4}(3n + 1)$ is odd or even, and so on, we are led to the following model. Consider the random process that starts at x and, at each step, multiplies by $3/2^s$ where s is equal to 1 with probability $\frac{1}{2}$, 2 with probability $\frac{1}{4}$, 3 with probability $\frac{1}{8}$, and so on.

We are interested in the expected number of steps for this sequence to dip below x; since the x cancels out, this is the same as assuming a starting value of 1 and asking for the expected number of steps until a value less than 1 is reached. Theorem 19.1 computes this expected value.

Theorem 19.1. *For the random process just defined, but with starting value* 1, *the expected number of steps before a value below* 1 *is first obtained is* 3.49265. . . .

Sketch of proof. First note that the desired expected value is the same as the expected value of the number of $\frac{3}{2}$ steps before a dip under 1 for the random process that multiplies by $\frac{3}{2}$ and $\frac{1}{2}$ with equal probability. Let r_i be the number of sequences using terms equal to $\frac{3}{2}$ and $\frac{1}{2}$ and satisfying:

(a) there are i $\frac{3}{2}$s;
(b) the product of the entire sequence is less than 1;
(c) the product of any proper initial segment is greater than 1.

These conditions imply that the first term is $\frac{3}{2}$, the last term is $\frac{1}{2}$, and the number of $\frac{1}{2}$s is $j(i) = \lfloor Li \rfloor - i + 1$, where $L = \log_2 3$. The last follows from the fact that

$$\left[\frac{3}{2}\right]^i \left[\frac{1}{2}\right]^{j(i)} < 1 < \left[\frac{3}{2}\right]^i \left[\frac{1}{2}\right]^{j(i)-1}.$$

For example, $r_1 = 1$, $r_2 = 1$, and $r_3 = 2$ because the corresponding sequences satisfying conditions (a)–(c) must have, respectively, one, two, and two $\frac{1}{2}$s, and only the following sequences work:

$$\begin{aligned} i = 1: &\quad \frac{3}{2}, \frac{1}{2}; \\ i = 2: &\quad \frac{3}{2}, \frac{3}{2}, \frac{1}{2}, \frac{1}{2}; \\ i = 3: &\quad \frac{3}{2}, \frac{3}{2}, \frac{3}{2}, \frac{1}{2}, \frac{1}{2} \text{ and } \frac{3}{2}, \frac{3}{2}, \frac{1}{2}, \frac{3}{2}, \frac{1}{2}. \end{aligned}$$

As pointed out in Part One, the strong law of large numbers implies that, with probability 1, this random process has a finite stopping time (i.e., gets under 1 eventually). Therefore the desired expected value is the sum of the following infinite series:

$$\sum_{i=1}^{\infty} \frac{ir_i}{2^{i+j(i)-1}}, \tag{2}$$

since each summand is the product of i with the probability that the first dip under 1 occurs after i multiplications by $\frac{3}{2}$.

To calculate this partial sum we need to know several hundred values of the r_i. They can be computed by using the following recursion, which is not hard to establish (Exercise 6):

$$r_i = \binom{i + j(i) - 2}{i - 1} - \sum_{k=1}^{i-1} r_k \binom{i + j(i) - k - j(k) - 1}{i - k}.$$

Standard techniques (such as Stirling's formula to approximate factorials) imply that 600 terms of the series (2) suffice to get within 10^{-6} of the true sum. Using a computer to first calculate $r_1, \ldots, r_{600}$ and then sum the corresponding terms of the series, one gets a partial sum of 3.49265185.... Because the error is less than 10^{-6} this proves that the expected value is 3.49265.... □

With this theorem in hand, we can set up a computer experiment to compare the actual stopping times of the $3n + 1$ trajectories with 3.49265..., the expected stopping time in the probabilistic model. The result of such a computation is that the average value of the stopping times for all odd n between 2 and 2,000,000,000 is 3.4926497..., a very promising result. Indeed, this agreement is somewhat surprising because there are reasons for expecting the probabilistic model to do a poor job of approximating Cs stopping times. The probabilistic reasoning is based on an assumption of independence, which is definitely lacking in the one billion trials using C. For example, the stopping times for the odd numbers between, say, 1,000,000 and 1,001,000 are affected by the stopping times for values between 1,500,000 and 1,501,500. If the stopping times for the higher interval are larger than average, there will be a bias toward long stopping times in the lower interval as well, because trajectories starting in the lower interval must pass through the higher.

Since there is nothing random about the real $3n + 1$ sequences, this massive computation does not prove anything. Even if someone succeeded in proving that the average stopping times for integers in $[2, N]$ approach a finite limit as N approaches infinity, that would mean only that no divergent trajectory exists. This would be a major breakthrough, but it would still leave open the possibility of a nontrivial loop.

Finally, we mention another unsolved problem related to the numbers 3 and 2.

Problem 19.1

Consider the set of fractional parts of the numbers $(\frac{3}{2})^n$, $n = 1, 2, 3, \ldots$; that is, the set: $0.5, 0.25, 0.375, 0.0625, 0.59375, 0.390625, \ldots$. Is this set uniformly distributed in the interval $[0, 1]$?

Exercises

5 (a). Show that $16k+3$ has stopping time 2; $32k+11$ and $32k+23$ have stopping time 3; $128k + 7$, $128k + 15$, and $128k + 19$ have stopping time 4; and $256k + 39$, $256k + 79$, $256k + 95$, $256k + 123$, $256k + 175$, $256k + 199$, and $256k + 219$ have stopping time 5.

(b). Conclude that a program that computes stopping times only for those odd integers in list (1) can obtain the true average of stopping times by adding in 203 every time a multiple of 256 is passed.

6 (a). What are r_4 and r_5, the numbers used in the proof of Theorem 19.1?

(b). Verify the recursion for r_i used in the proof of Theorem 19.1. (Hint: This recursion is a formula describing how, for example, r_6 can be obtained from $r_1, \ldots, r_5$.)

20. DIOPHANTINE EQUATIONS AND COMPUTERS

Problem 20

> Is there an algorithm that decides, given a polynomial P with integer coefficients, whether there are rational numbers q_i such that $P(q_1, q_2, q_3, \ldots) = 0$?

The solution to Hilbert's tenth problem, a theorem that "is a remarkable achievement of 20th-century mathematics and deserves the close attention of anyone interested in the fundamental nature of the mathematical enterprise" (I. Kaplansky), was completed in 1970 when Yuri Matijasevič of the Soviet Union put the finishing touches on a body of work carried out by the American logicians Martin Davis, Hilary Putnam, and Julia Robinson. Roughly speaking the Americans showed that the problem would be solved if there was a sequence of integers that grew quickly (exponentially) but was defined simply (using polynomials and existential quantifiers). Matijasevič showed that the sequence consisting of every second Fibonacci number filled the bill. A complete proof can be found in [Dav]; some recent simplifications have been obtained by Jones and Matijasevič [JM1, 2].

Hilbert's problem asks whether the set A consisting of all Diophantine equations that do have solutions in positive integers is computable. Note that A, now known to be not computable, is listable (see [DH]). This means that there is an algorithm that lists all the members of A; the algorithm never completes the task, however, and would have to run infinitely long to complete the listing. Roughly speaking, the algorithm begins with a list of all polynomials and a list of all finite sequences of positive integers (both sets are countable); then it plugs the first sequence (if it has the right length) into the first polynomial, then the first sequence into the second polynomial; then the second sequence into the first polynomial, then the first sequence into the third polynomial, and so on. By this "dovetailing" technique, a solution to any solvable Diophantine equation is eventually discovered. This algorithm does not show that A is computable since, for a specific Diophantine equation having no solution, one would never learn in a finite amount of time that the equation will never appear on the list. Note, however, that a set is computable if and only if both it and its complement are listable. For if both listing programs exist, they can be run simultaneously; any input will, in finite time, appear on one or the other list, thus affirming or refuting the input's membership in the set. It follows that Problem 20 is equivalent to asking whether the set of polynomials for which there is no rational solution is listable.

We have already mentioned some of the surprising consequences of the solution to Hilbert's tenth problem, such as the existence of a prime-representing polynomial and an undecidable Diophantine equation. The result regarding primes is considerably more general. In fact, any listable subset of $\mathbb{N}$ (this includes all computable subsets of $\mathbb{N}$) has the same property, that is, it is the positive range of a polynomial evaluated over the integers (see [Dav, DH, DMR]). For arbitrary listable sets, such as the primes, the representing polynomial is usually quite complicated. An interesting example of a simple polynomial arises from the Fibonacci numbers: It is not too hard to show that the Fibonacci numbers are exactly those positive integers in the range of the polynomial $y(2-(y^2-yx-x^2)^2)$ (where x and y take on values in $\mathbb{Z}$; see Exercise 4).

To prove that an undecidable Diophantine equation exists, suppose the opposite; that is, assume that every Diophantine equation's solvability could be either proved or disproved (in some fixed axiom system). This would yield an algorithm for deciding whether Diophantine equations have solutions: given such an equation, examine all proofs (in order of length) until a proof that resolves the question one way or the other is found. The assumption implies that such a proof will eventually be found. But such an algorithm contradicts the negative solution to Hilbert's tenth problem.

Let's pause for a moment to clarify an important point. If a sentence S, such as Gödel's sentence or the unsolvability of a certain Diophantine equation, is unprovable, how can we know it is true? The point is that the sentence is constructed to be unprovable and unrefutable from a certain set of axioms, say A. Although A cannot prove either S or $\neg S$ ("not S"), stronger axiom systems might prove S. Typically, it happens that the system consisting of A together with $\mathrm{Con}(A)$ is strong enough to prove S, where $\mathrm{Con}(A)$ is the assertion that A is consistent. Why not then take $\mathrm{Con}(A)$ as an axiom, replacing A by the system of axioms $B = A \cup \{\mathrm{Con}(A)\}$? Then B would indeed be strong enough to prove S, but the method used to construct S from A could just as well be used to construct another sentence S' that is undecidable in the expanded system B.

The result about the existence of undecidable Diophantine equations alluded to in Part One is of great interest since it shows that, in theory, it is conceivable that some of the famous unsolved Diophantine problems such as Fermat's last theorem (which, as pointed out, is equivalent to a single Diophantine equation) have no solutions, but this fact is unprovable from the standard axioms. Indeed, there are several natural questions of arithmetic that are known to be unprovable from Peano's axioms, the usual axiom system for arithmetic. One such is Goodstein's theorem, described in Exercise 5. For more details, including a proof of Goodstein's theorem (from axioms stronger than Peano's), see [KP] or [Hen]. For more on undecidable Diophantine equations, see [DJS].

Hilbert's question puts no bound on the number of variables in the polynomial, or its degree. But even with certain such restrictions no algorithm exists. If solutions in $\mathbb{N}$ are sought, then there is no algorithm to solve Diophantine equations with 9 variables (see [Jon2]). When the number of variables is any of $8, 7, \ldots, 2$ the problem is open, but for polynomials in one variable there is an algorithm (Exercise 6). The situation with $\mathbb{Z}$ replacing $\mathbb{N}$ is even less well understood, with the dividing line at present being 21 variables rather than 9 (Y. Matijasevič and R. M. Robinson; see [FW]). Turning to a polynomial's degree rather than the number of its variables, it follows from a classical idea of T. Skolem that the solution to Hilbert's tenth problem means that there is no algorithm to decide the solvability of Diophantine equations of degree 4, in either $\mathbb{Z}$ or $\mathbb{N}$ (see [Dav]). The case of polynomials of degree 3 is unsolved. An algorithm exists for polynomials of degree 2 (see [Dav]), but this is not necessarily of practical value, since it does not run in polynomial time. Indeed, it is known that

$$\{a, b, c \in \mathbb{Z} : ax^2 + by = c \text{ is solvable in positive integers } x, y\}$$

is $\mathbb{NP}$-complete [MA].

Problem 20, Hilbert's tenth problem over the rationals, can be formulated in a way that refers only to equations where integer solutions are sought. A polynomial $P(x, y, z, \ldots)$ is called *homogeneous* if all terms have the same degree. For example, $3xy^2 + 2yz^2 - z^3$ is homogeneous of degree 3; $x^2 - y^2 - 1$ and $x^2 - y^3$ are not homogeneous. Note that a homogeneous polynomial P that is not just a constant has value zero when all the variables are set to zero (called a *trivial solution* to $P = 0$). Now, for every polynomial P there is a homogeneous polynomial Q such that $P = 0$ has a solution in rationals if and only if $Q = 0$ has a nontrivial solution in integers. See [Rob′] for the construction of Q; the converse is true as well (Exercise 7). So a $\mathbb{Q}$-algorithm exists if and only if there is an algorithm that decides, for a homogeneous polynomial P, whether $P = 0$ has a nontrivial solution in $\mathbb{Z}$.

Hilbert's tenth problem over the rationals is equivalent to a question about the existence of an algorithm that decides whether certain polyhedra can be imbedded in $\mathbb{R}^3$ so that all the vertices are at points with rational coordinates; see [Stu].

Note that Problem 20.1 asks for a definition that excludes $\neg$ ("not") and universal quantifiers ("for all x," $\forall x$). If these are permitted then, as was proved by Julia Robinson [Rob], $\mathbb{Z}$ is indeed definable in $\mathbb{Q}$. However, her definition is quite subtle, involving the Hasse–Minkowski theorem of number theory (an exposition appears in [FW]).

Exercises

4. Show that for any Fibonacci number F, there are integers x and y such that $F = y(2 - (y^2 - yx - x^2)^2)$. The converse is true as well (see [Jon1]); thus the positive range of this polynomial is precisely the set of Fibonacci numbers.

5. Consider the following algorithm for constructing a sequence of numbers, starting with some positive integer x_2. First write x_2 in super base 2, by which we mean writing x_2 as a sum of powers of 2, writing each of the exponents as a sum of powers of 2, writing each of the new exponents as a sum of power of 2, and so on. For example, the super base 2 representation of 44 ($= 10110$ in base 2) is

$$2^{2^2+1} + 2^{2+1} + 2^2.$$

Now, replace all the 2s in the super base 2 representation of x_2 by 3s; then subtract 1 from the resulting mass of 3s to get x_3, the next term of the sequence. To get x_4, write x_3 in super base 3, replace all the 3s by 4s, and subtract 1. Continuing yields a sequence of integers. To obtain x_{n+1} write x_n in super base n, replace all the ns by $(n+1)$s, and subtract 1. The sequence for starting value 44 begins with:

$$x_3 = 3^{3^3+1} + 3^{3+1} + 3^3 - 1 \quad (\approx 10^{13}),$$

$$x_4 = 4^{4^4+1} + 4^{4+1} + 2 \cdot 4 + 1 \quad (\approx 10^{155}),$$

$$x_5 \approx 10^{2185}.$$

The passage from x_3 to x_4 uses the fact that the base 3 representation of $3^3 - 1$ is $2 \cdot 3^2 + 2 \cdot 3 + 2$. Considering how quickly these sequences grow, the following result, known as Goodstein's theorem (see [KP] or [Hen, Chap. 10]), should seem quite surprising: For every starting value x_2 the sequence eventually strikes zero. If the starting value is 4, how large will n be when x_n first equals 0?

6. Construct an algorithm that accepts as input $P(x)$, a polynomial in one variable, and outputs "Yes" or "No" according as $P(x) = 0$ does or does not have an integer solution.

7. Show that if P is a homogeneous polynomial then there is another polynomial Q such that $P = 0$ has a nontrivial solution in $\mathbb{Z}$ if and only if Q has a solution in $\mathbb{Q}$.

REFERENCES

[AH] L. M. Adleman and M. A. Huang, Recognizing primes in random polynomial time, *Proceedings of the Nineteenth Annual ACM Symposium on Theory of Computing, New York City,* ACM Press, New York, 1987, pp. 462–469. [§18]

[Ahl] L. V. Ahlfors, *Complex Analysis,* second ed., McGraw Hill, New York, 1966. [§17]

[Alm] J. H. J. Almering, Rational quadrilaterals, *Indagationes Mathematicae,* 25 (1963) 192–199. [§14]

[Bac] E. Bach, Analytic Methods in the Analysis and Design of Number-Theoretic Algorithms, ACM Distinguished Dissertations, MIT Press, Cambridge, 1985. [§18]

[BBC] A. Beck, M. N. Bleicher, and D. W. Crowe, *Excursions into Mathematics,* Worth, New York, 1969. [§15]

[Ble] M. N. Bleicher, A new algorithm for the expansion of Egyptian fractions, *Journal of Number Theory,* 4 (1972) 342–382. [§15]

[Boa] R. P. Boas, The Skewes number, in *Mathematical Plums,* R. Honsberger, ed., Dolciani Mathematical Expositions, No. 4, Mathematical Association of America, 1979. [§17]

[BH] W. Borho and H. Hoffmann, Breeding amicable numbers in abundance, *Mathematics of Computation,* 46 (1986) 281–293. [§16]

[BC] R. P. Brent and G. L. Cohen, A new lower bound for odd perfect numbers, *Mathematics of Computation,* 53 (1989) 431–437. [§16]

[BCT] R. P. Brent, G. L. Cohen, and H. J. J. te Riele, An improved technique for lower bounds for odd perfect numbers, Australian National University Technical Report. [§16]

[BG] A. Bremner and R. K. Guy, A dozen difficult Diophantine dilemmas, *American Mathematical Monthly,* 95 (1988) 31–36. [§14]

[BS] R. Breusch, Solution to Problem E4512, *American Mathematical Monthly,* 61 (1954) 200–201. [§15]

[BLSTW] J. Brillhart, D. H. Lehmer, J. L. Selfridge, B. Tuckerman, and S. S. Wagstaff, Jr., Factorizations of $b^n - 1, b = 2, 3, 5, 6, 7, 10, 11, 12$ up to high powers, second ed., Contemporary Mathematics, vol. 22, American Mathematical Society, Providence, 1988. [§18]

[CL] H. Cohen and A. K. Lenstra, Implementation of a new primality test, *Mathematics of Computation,* 48 (1987) 103–121. [§18]

[Coh] G. L. Cohen, On the largest component of an odd perfect number, *Journal of the Australian Mathematical Society (Series A),* 42 (1987) 280–286. [§16]

[Cra] R. E. Crandall, On the "$3x + 1$" problem, *Mathematics of Computation,* 32 (1978) 1281–1292. [§19]

[Dav] M. Davis, Hilbert's tenth problem is unsolvable, *American Mathematical Monthly,* 80 (1973) 233–269. [§20]

[DH] M. Davis and R. Hersh, Hilbert's 10th problem, *Scientific American,* 229:5 (Nov., 1973) 84–91. [§20]

[DMR] M. Davis, Y. Matijasevič, and J. Robinson, Hilbert's tenth problem. Diophantine equations: Positive aspects of a negative solution, in *Mathematical Developments Arising From Hilbert Problems,* Proceedings of Symposia in Pure Mathematics, 28, Part 2, American Mathematical Society, Providence 1976, 323–378. [§20]

[Dic1] L. E. Dickson, *History of the Theory of Numbers,* vol. I, Chelsea, New York, 1971 (reprint of 1923 edition). [§16]

[Dic2] ——, *History of the Theory of Numbers,* vol. II, Chelsea, New York, 1971 (reprint of 1923 edition). [§14]

[Dix] J. Dixon, Factorization and primality tests, *American Mathematical Monthly,* 91 (1984) 333–352. [§18]

[DJS] V. H. Dyson, J. P. Jones, and J. C. Shepherdson, Some Diophantine forms of Gödel's theorem, *Archiv für Mathematische Logik und Grundlagenforschung,* 22 (1982) 51–60. [§20]

[Edw1] H. M. Edwards, *Riemann's Zeta Function,* Academic Press, New York, 1974. [§17]

[Edw2] ——, *Fermat's Last Theorem, A Genetic Introduction to Algebraic Number Theory,* Springer, New York, 1977. [§13]

[Edw3] ——, Fermat's last theorem, *Scientific American,* 239:4 (April, 1978) 104–122. [§13]

[Egg] H. G. Eggleston, Note 2347. Isosceles triangles with integral sides and two integral medians, *The Mathematical Gazette* 37 (1953) 208–209. [§14]

[Elk] N. Elkies, On $A^4 + B^4 + C^4 = D^4$, *Mathematics of Computation,* 51 (1988) 825–835. [§13]

[Erd] P. Erdős, The solution in whole numbers of the equation $\frac{1}{x_1} + \cdots \frac{1}{x_N} = \frac{a}{b}$, *Matematikai Lapok,* 1 (1950) 192–210. (Hungarian, with English summary; see *Math. Reviews* 13 (1952) 208.) [§15]

[EG] P. Erdős and R. Graham, *Old and New Problems and Results in Combinatorial Number Theory,* Monographie No. 28 de L'Enseignement Mathématique, Université de Genève, Geneva 1980. [§15]

[Fil] M. Filaseta, An application of Faltings' results to Fermat's last theorem, *Comptes Rendus/Mathematical Reports, Academy of Science, Canada,* 6 (1984) 31–32. [§13]

[FW] D. Flath and S. Wagon, How to pick out the integers in the rationals: an application of number theory to logic, *American Mathematical Monthly,* 98 (1991). [§20]

[GJ] M. R. Garey and D. S. Johnson, *Computers and Intractability: A Guide to the Theory of NP-Completeness,* San Francisco, W. H. Freeman, 1979. [§18]

[Gar] L. E. Garner, On the Collatz $3n + 1$ algorithm, *Proceedings of the American Mathematical Society,* 82 (1981) 19–22. [§19]

[Gra] R. L. Graham, On finite sums of unit fractions, *Proceedings of the London Mathematical Society,* 14 (1964) 193–207. [§15]

[GRS] R. L. Graham, B. L. Rothschild, and J. H. Spencer, *Ramsey Theory,* Wiley, New York, 1980. [§16]

[Gra′] A. Granville, The set of exponents, for which Fermat's last theorem is true, has density one, *Comptes Rendus/Mathematical Reports, Academy of Science, Canada,* 7 (1985) 55–60. [§13]

[Gün] R. Güntsche, Über rationale Tetraeder, *Archiv der Mathematik und Physik,* 3 (1907) 371. [§14]

[Guy] R. K. Guy, *Unsolved Problems in Number Theory,* Springer, New York, 1981. [§13]

[Hag1] P. Hagis, A lower bound for the set of odd perfect numbers, *Mathematics of Computation,* 27 (1973) 951–953. [§16]

[Hag2] ——, Sketch of a proof that an odd perfect number relatively prime to 3 has at least eleven prime factors, *Mathematics of Computation,* 40 (1983) 399–404. [§16]

[HW] G. H. Hardy and E. M. Wright, *An Introduction to the Theory of Numbers,* 4th ed., Oxford, London, 1960. [§20]

[Hea] T. L. Heath, *The Thirteen Books of Euclid's Elements,* vol. 2, Dover, New York, 1956. [§16]

[HB1] D. R. Heath-Brown, Fermat's last theorem for "almost all" exponents, *Bulletin of the London Mathematical Society,* 17 (1985) 15–16. [§13]

[HB2] ——, The first case of Fermat's last theorem, *The Mathematical Intelligencer,* 7 (1985) 40–47, 55. [§13]

[Hel] M. E. Hellman, The mathematics of public-key cryptography, *Scientific American,* 241:2 (Aug., 1979) 146–157. [§18]

[Hen] J. M. Henle, *An Outline of Set Theory,* Springer, New York, 1986. [§20]

[Ios] M. Iosifescu, On the random Riemann hypothesis, *Proceedings of the Seventh Conference on Probability Theory, Braşov, Romania,* 1982, Editura Academiei Republicii Socialiste România, Bucharest, 1984. [§17]

[IR] K. Ireland and M. Rosen, *A Classical Introduction to Modern Number Theory,* Springer, New York, 1982. [§13]

[Ivi] Aleksandar Ivić, *The Riemann Zeta-function,* Wiley, New York, 1985. [§17]

[Jon1] J. P. Jones, Diophantine representation of the Fibonacci numbers, *Fibonacci Quarterly,* 13 (1975) 84–88. [§20]

[Jon2] ——, Universal Diophantine equation, *Journal of Symbolic Logic,* 47 (1982) 549–571. [§20]

[JM1] J. P. Jones and Y. V. Matijasevič, Register machine proof of the theorem on exponential Diophantine representation of enumerable sets, *Journal of Symbolic Logic,* 49 (1984) 818–829. [§20]

[JM2] ——, Proof of recursive unsolvability of Hilbert's tenth problem, *American Mathematical Monthly,* 98 (1991) 689.

[JSWW] J. P. Jones, D. Sato, H. Wada, and D. Wiens, Diophantine representation of the set of prime numbers, *American Mathematical Monthly,* 83 (1976) 449–464. [§20]

[Knu] D. E. Knuth, *The Art of Computer Programming,* vol. 2, Addison-Wesley, Reading Mass., 1971. [§18]

[Kor1] I. Korec, Nonexistence of a small perfect rational cuboid, *Acta Mathematica Universitatis Comenianae,* 42–43 (1983) 73–86. [§14]

[Kor2] ——, Nonexistence of a small perfect rational cuboid, II, *Acta Mathematica Universitatis Comenianae,* 44–45 (1984) 39–48. [§14]

[KP] L. Kirby and J. Paris, Accessible independence results for Peano Arithmetic, *Bulletin of the London Mathematical Society,* 14 (1982) 285–293. [§20]

[Lag] J. C. Lagarias, The $3x+1$ problem and its generalizations, *American Mathematical Monthly,* 92 (1985) 3–23. [§19]

[LMO] J. Lagarias, V. S. Miller, and A. Odlyzko, Computing $\pi(x)$: The Meissel–Lehmer method, *Mathematics of Computation,* 44 (1985) 537–560. [§17]

[LM] E. J. Lee and J. S. Madachy, The history and discovery of amicable numbers I–III, *Journal of Recreational Mathematics,* 5 (1972) 77–93, 153–173, 231–249. [§16]

[Lee] J. Leech, The rational cuboid revisited, *American Mathematical Monthly,* 84 (1977) 518–533. Corrections, ibid. 85 (1978) 473. [§14]

[ME] N. MacKinnon and J. Eastmond, An attack on the Erdős conjecture, *The Mathematical Gazette,* 71 (1987) 14–19. [§16]

[MA] K. Manders and L. Adleman, NP-complete decision problems for binary quadratics, *Journal of Computer and System Sciences,* 16 (1978) 168–184. [§20]

[MW] H. B. Mann and W. A. Webb, A short proof of Fermat's theorem for $n = 3$, *The Mathematics Student,* 46 (1978) 103–104. [§13]

[McC] J. McCleary, How not to prove Fermat's last theorem, *American Mathematical Monthly,* 96, (1989) 410–420. [§13]

[Mel] R. A. Melter, Problem 6628, *American Malthematical Monthly,* 97 (1990) 350. Solutions by C. R. Maderer, J. H. Steelman, J. Buddenhagen, ibid. (1991) (to appear). [§14]

[NN] E. Nagel and J. R. Newman, *Gödel's Proof,* University Press, New York, 1958. [§20]

[NZ] I. Niven and H. S. Zuckerman, *An Introduction to the Theory of Numbers,* second ed., Wiley, New York, 1966. [§18]

[Odl] A. M. Odlyzko, On the distribution of spacings between zeros of the zeta function, *Mathematics of Computation,* 48 (1987) 273–308. [§17]

[OtR] A. M. Odlyzko and H. J. J. te Riele, Disproof of the Mertens conjecture, *Journal für die Reine und Angewandte Mathematik,* 357 (1985) 138–160. [§17]

[Pat] S. J. Patterson, *An Introduction to the Theory of the Riemann Zeta-Function,* Cambridge University Press, New York, 1988. [§17]

[Pin] J. Pintz, An effective disproof of the Mertens conjecture, *Astérisque,* No. 147–148 (1987) 325–333. [§17]

[Pom1] C. Pomerance, Multiply perfect numbers, Mersenne primes, and effective computability, *Mathematische Annalen,* 226 (1977) 195–206. [§16]

[Pom2] ——, Recent developments in primality testing, *The Mathematical Intelligencer,* 3 (1981) 97–105. [§18]

[Pom3] ——, The search for prime numbers, *Scientific American,* 247 6 (Dec. 1982) 136–147. [§18]

[Pom4] ——, Analysis and comparison of some integer factoring algorithms, in *Computational Methods in Number Theory,* Part I, H. W. Lenstra, R. Tijdeman, eds., Mathematical Centre Tracts 154, Amsterdam, 1982. [§18]

[PSW] C. Pomerance, J. L. Selfridge, and S. S. Wagstaff, Jr., The pseudoprimes to 25×10^9, *Mathematics of Computation,* 35 (1980) 1003–1026. [§18]

[Pri] P. A. Pritchard, Long arithmetic progressions of primes; some old, some new, *Mathematics of Computation,* 45 (1985) 263–267. [§16]

[RW] H. L. Resnikoff and R. O. Wells, Jr., *Mathematics in Civilization,* 2nd ed., Dover, New York, 1984. [§15]

[Rib1] P. Ribenboim, *13 Lectures on Fermat's Last Theorem,* Springer, New York, 1980. [§13]

[Rib2] ——, Recent results about Fermat's last theorem, *Expositiones Mathematicae,* 5 (1987) 75–90. [§13]

[Rib3] ——, *The Book of Prime Number Records,* Springer, New York, 1988. [§§15 and 17]

[Rie] H. Riesel, *Prime Numbers and Computer Methods for Factorization,* Birkhäuser, Boston, 1985. [§16]

[RS] G. Robins and C. S. Shute, *The Rhind Mathematical Papyrus,* Dover, New York, 1987. [§15]

[Rob] J. Robinson, Definability and decision problems in arithmetic, *Journal of Symbolic Logic,* 14 (1949) 98–114. [§20]

[Rob′] R. M. Robinson, Advanced problem 6415, *American Mathematical Monthly,* 91 (1984) 373–374. [§20]

[Ros] K. H. Rosen, *Elementary Number Theory and its Applications,* second ed., Addison-Wesley, Reading, Mass., 1987. [§§16 and 18]

[Ros′] M. I. Rosen, A proof of the Lucas–Lehmer test, *American Mathematical Monthly,*, 95 (1988) 855–856. [§16]

[Sch] M. R. Schroeder, *Number Theory in Science and Communication,* Springer, Berlin, 1984. [§16]

[Stu] B. Sturmfels, On the decidability of Diophantine problems in combinatorial geometry, *Bulletin of the American Mathematical Society,* 17 (1987)121–124. [§20]

[Syl1] J. J. Sylvester, Note on a proposed addition to the vocabulary of ordinary arithmetic, *Nature,* 37 (1887) 152–153 (correction, ibid. p. 179). [§16]

[Syl2] ——, On the divisors of the sum of a geometrical series whose first term is unity and common ratio any positive or negative integer, *Nature,* 37 (1888) 417–418. [§16]

[tR1] H. J. J. te Riele, On generating new amicable pairs from given amicable pairs, *Mathematics of Computation,* 42 (1984) 219–223. [§16]

[tR2] ——, Computation of all the amicable pairs below 10^{11}, *Mathematics of Computation,* 47 (1986) 361–368. [§16]

[tR3] ——, On the sign of the difference $\pi(x) - \mathrm{li}(x)$, *Mathematics of Computation,* 48 (1987) 323–328. [§17]

[Ter] R. Terras, A stopping time problem on the positive integers, *Acta Arithmetica,* 30 (1976) 241–252. [§19]

[TW1] J. W. Tanner and S. S. Wagstaff, Jr., New congruences for the Bernoulli numbers, *Mathematics of Computation,* 48 (1987) 341–350. [§13]

[TW2] ——, New bound for the first case of Fermat's last theorem, *Mathematics of Computation,* 53 (1989) 743–750. [§13]

[Vos] M. Vose, Egyptian fractions, *Bulletin of the London Mathematical Society,* 17 (1985) 21–24. [§15]

[Wag1] S. Wagon, The evidence: The Collatz problem, *The Mathematical Intelligencer,* 7:1 (1985) 72–76. [§19]

[Wag2] ——, The evidence: Odd perfect numbers, *The Mathematical Intelligencer,* 7:2 (1985) 66–68. [§16]

[Wag3] ——, The evidence: Fermat's last theorem, *The Mathematical Intelligencer,* 8:1 (1986) 59–61. [§13]

[Wag4] ——, The evidence: Primality testing, *The Mathematical Intelligencer,* 8:3 (1986) 58–61. [§18]

[Wag5] ——, The evidence: Where are the zeros of zeta of s?, *The Mathematical Intelligencer,* 8:4 (1986) 57–62. [§17]

[Wag6] ——, *Mathematica in Action,* W. H. Freeman, New York, 1991. [§§16 and 17]

[Wag′] S. S. Wagstaff, Jr., The irregular primes to 125000, *Mathematics of Computation,* 32 (1978) 583–591. [§13]

[Wei] A. Weil, *Number Theory: An Approach Through History; From Hammurapi to Legendre,* Birkhäuser, Boston, 1983. [§13]

[Wil] H. C. Williams, An overview of factoring, in *Advances in Cryptology,* D. Chaum, ed., Plenum, 1984. [§18]

[YB] J. Young and D. A. Buell, The twentieth Fermat number is composite, *Mathematics of Computation,* 50 (1988) 261–263. [§16]

[Zag] D. Zagier, The first 50 million prime numbers, *The Mathematical Intelligencer,* 0 (1977) 7–19. [§16]

CHAPTER 3

INTERESTING REAL NUMBERS

All numbers are not created equal. Although real numbers correspond to points on the number line and so, in one sense, they all look alike, there are tremendous differences among them. Some were created by the earliest humans $(1, 2, 3)$, others by the sophisticated Greeks $(\sqrt{2}, \pi)$, and still others by the discoverers of the calculus (e). In addition to the elementary distinction between integers and nonintegers, there are the classifications of real numbers into rational and irrational, or algebraic and transcendental. More modern properties shared by some, but not all, real numbers are normality and real-time computability. In this chapter we present a few unsolved problems about the properties of some famous numbers, and along the way we shall meet some other interesting, but less well known, numbers, such as Champernowne's number $(0.12345678910111213\ldots)$ and Liouville's number $(0.110001000000000000000001000\ldots)$.

Perhaps the most famous interesting real number is π. It has been known for 100 years that π is irrational (not a quotient of two integers) and transcendental (not the root of a polynomial with integer coefficients). But what about its decimal expansion, the famous $3.14159265\ldots$? Are there any unexpected patterns? Do all digits occur infinitely often? This is the essence of the question of π's *normality,* discussed in Section 21. The number e is also known to be irrational and transcendental, and is in some ways easier to deal with than π. But can it be combined with π in a simple way to yield a rational number? To be specific, Problem 21 asks whether there are integers m and n such that $m\pi = ne$. In other words, is e/π rational or irrational?

The square root of 2 holds a special position among irrational numbers, since it was the first. The ancient Greeks knew that the diagonal of a unit square had length

$\sqrt{2}$ and, what is much more remarkable, they were able to prove that this quantity is not a rational number. Much of modern mathematics is concerned with algorithms and speed of computation, and in this area too $\sqrt{2}$ is special since there are algorithms dating from Babylonian times that compute the digits very quickly. But the fastest known algorithms to compute the digits of $\sqrt{2}$ (or π or e) slow down as they get farther and farther out into the decimal expansion. Section 23 is concerned with the notion of real-time computability, a term used to describe algorithms that spew out their output without slowing down as they proceed. Is there a real-time algorithm for computing the digits of $\sqrt{2}$?

Of course, one can ask a myriad of related questions about other numbers that turn up in mathematics. In the final section we discuss one example that is interesting both because of its connection with one of the most beautiful formulas ever discovered (Euler's formula for the sum of the reciprocals of the even powers of the positive integers) and because of some recent progress that used only relatively simple mathematics.

21. PATTERNS IN PI

Problem 21

Are the digits in the decimal expansion of π devoid of any pattern?

No single number has attracted more attention than the number π, the ratio of the circumference of a circle to its diameter (and the ratio of the area of circle to the square of its radius). The number is involved in the famous and striking equation, $e^{i\pi} = -1$, and it shows up in an amazing variety of situations, both pure and applied, often in unexpected ways. For example,

$$\frac{1}{1} + \frac{1}{4} + \frac{1}{9} + \cdots + \frac{1}{n^2} + \cdots = \frac{\pi^2}{6}$$

(see Section 24); and: The probability that a one-inch stick, when tossed randomly onto a floor with parallel lines spaced one inch apart, will land so that it crosses one of the lines is $2/\pi$.

The properties of the number π have been studied intensely over the centuries (see [Bec, BB1]), in great part because of its connection with the famous circle-squaring problem (see Section 9). That problem was finally solved in 1882, when Lindemann proved that π is a transcendental number. It follows that π differs radically from the irrational number $\sqrt{2}$, for example, since the latter is a root of the polynomial equation $x^2 - 2 = 0$. Although π is well approximated by certain rationals, such as $\frac{22}{7}$ or $\frac{335}{113}$, its irrationality implies that it does not exactly equal any such ratio of integers. This is equivalent to saying that the decimal expansion of π does not ever terminate or repeat itself. So π cannot equal

$$3.14159265\ldots734734734734\ldots.$$

But this is not the same as saying the digits do not show any pattern. For instance, perhaps the digits of π look like

$$3.14159\ldots07007000700007\ldots.$$

This leads to the very difficult Problem 21, on which almost no progress has been made. In particular, it is not known whether all digits occur, on average, equally often. To make the preceding question more precise, consider the digit 7. The number of times it occurs in π's first 29 million digits (after the decimal point) is summarized in Table 21.1.

Number of digits:	100	1,000	10,000	100,000	1,000,000	10,000,000	29,360,000
Number of 7s:	8	95	970	10,025	99,800	1,000,207	2,934,083
Frequency of 7s:	8%	9.5%	9.7%	10.0%	9.98%	10.002%	9.99347%

TABLE 21.1
The number of 7s in the first 29 million digits of $\pi - 3$.

One would expect that each of the ten digits occurs, roughly, about 10% of the time. And indeed, the occurrence-ratios for the digit 7—0.08, 0.095, 0.097, 0.10025, 0.0998, 0.1000207, 0.0999347—seem to be converging to $\frac{1}{10}$. The conjecture is that these ratios (and the nine ratio-sequences for the other digits), if computed for the full infinite decimal representation of π, converge to the limit $\frac{1}{10}$. In particular, this would mean that no digit stops appearing after some point. The evidence for this conjecture is overwhelming—see Table 21.2, which illustrates these ratios (for the first ten million digits of π) for each of the ten digits—and no one has any doubt that the rest of the digits behave the same way as the first ten million. But no one has any idea how to go about proving it.

Anyone who thinks there might be a hidden, as yet unrecognized, pattern to π's digits might first consult Hans Eberstark. Eberstark is a talented individual who is extremely adept at performing feats of memorization and arithmetical computation, and he has memorized the first 11,000 digits of π. There were no patterns to help him, and he managed the extraordinary task by matching words or syllables

Digit:	0	1	2	3	4
Frequency:	999440	999333	1000306	999964	1001093
Digit:	5	6	7	8	9
Frequency:	1000466	999337	1000207	999814	1000040

TABLE 21.2
The frequencies of each of the ten digits in the first ten million digits of $\pi - 3$.

in various languages to sequences of digits (he knows 16 languages). See [Smi] for more details; for the current memorization record, which is at least 40,000 digits, consult the latest *Guinness Book of Records.* If you have trouble remembering even the first ten digits, you might try the following mnemonic phrase: "How I want a drink, alcoholic of course, after the heavy lectures involving quantum mechanics." The number of letters in the words corresponds to the digits of π!

The question as posed is not particularly important—even if π's digits are proved to be patternless, there are much more efficient ways of generating random numbers. But the techniques that have been developed to generate the millions of digits have had important ramifications for high-precision computations and they have led to very fast methods of computing, to a high degree of accuracy, functions such as $\sin x$ and $\log x$.

22. CONNECTIONS BETWEEN π AND e

Problem 22

Are π and e algebraically independent? Is their ratio rational?

Like π, the number e ($= 2.718281828459\ldots$) has been much studied, even though it was discovered over two thousand years later than π. The series representation of e^x can be used to quickly prove that e is irrational and, like π, e is known to be transcendental. Moreover, it is known that e^x is transcendental whenever x is a nonzero algebraic number. (Since, for example, $e^{\log 2} = 2$, this fails when $x = \log 2$; in particular, $\log 2$ must be transcendental.) But almost nothing is known about simple combinations[1] of the two famous numbers. For example, the second problem posed above is equivalent to asking whether the real number π/e is rational. Equally difficult unsolved problems ask whether any of the following are rational: $\pi + e$, $\pi \cdot e$, π^e, e^e, π^π. Such combinations are expected to be transcendental numbers, but proving transcendence of individual numbers is difficult. However, it is known that e^π is transcendental; this is related to the most famous theorem in this area: a^b is transcendental whenever a and b are both algebraic, a is neither 0 nor 1, and b is irrational (see Part Two).

A general conjecture that would imply the transcendence of many of the combinations is: The numbers π and e are *algebraically independent.* This means that for any polynomial $p(x, y)$ with integer coefficients, at least one of which is nonzero, $p(\pi, e) \neq 0$. This conjecture implies that the sum, product, and quotient of π and e are all transcendental, and much more.

[1] Here we are referring to operations involving real numbers; if complex numbers are allowed then there is the famous relation $e^{i\pi} = -1$.

23. COMPUTING ALGEBRAIC NUMBERS

Problem 23

If an irrational number is real-time computable, is it necessarily transcendental? Is $\sqrt{2}$ real-time computable?

From a computational point of view, rational numbers are simpler than irrational ones. This can be made precise by observing that the decimal digits of a rational can be computed in a way that requires a fixed amount of work per digit. If we define a *step* to be the writing of a digit from 0 to 9 (or an auxiliary symbol such as $*$ or $-$), then we can count how many steps are necessary when the usual long division technique is used to get the first n digits of, say, $17/108$. For each digit of the quotient we have to write at most 8 digits: a 0 to the right of the previous remainder, a digit of the quotient (obtained by consulting a multiplication table for 108), 3 digits (at most) corresponding to a multiple of 108, and 3 digits (at most) to get the remainder, which will be used to get the next digit of the quotient. If k denotes the amount of work needed to set up the multiplication table, then it follows that $8n + k$ is a bound on the number of steps needed to get the first n digits. So the average amount of work per digit (in this case, at most $8 + \frac{k}{n}$) is bounded. (Of course, the amount of work could be reduced since a rational's digits either terminate or eventually repeat, but this does not speed things up significantly—the number of steps is still, roughly, of the form cn.) This leads to the following definition.

Definition 23.1. A real number is *real-time computable* if there is an algorithm that computes its decimal digits and for which there is a constant c such that, for every n, the algorithm uses at most cn steps to compute the first n digits.

Roughly speaking, this definition says that on average the computation of each digit requires a fixed amount of work, independent of whether the algorithm is working on the tenth digit or the ten-millionth. Note, however, that the definition refers to base-ten digits. It is not known whether an irrational number can be real-time computable *in all bases.*

It is not hard to construct examples of irrational numbers that are real-time computable. For example, none of the following numbers (given in base 10 notation) is rational, but they are all real-time computable:

$$x_1 = 0.110001000000000000000001000000000000000\ldots$$

$$x_2 = 0.110100010000000100000000000000010000000\ldots$$

$$x_3 = 0.1001000010000001000000001000000000010000 \ldots$$

$$x_4 = 0.123456789101112131415161718192021222324 \ldots$$

$$x_5 = 0.1110100100001000000010000000000001000000 \ldots$$

$$x_6 = 0.110100110010110100101100110100110010110 \ldots .$$

The real number x_1, sometimes called Liouville's number, has a 1 in positions $1, 2, 6, 24, 120, \ldots, n!, \ldots$; x_2 has a 1 in positions corresponding to the powers of 2; x_3 has a 1 in positions corresponding to the squares; x_4 is Champernowne's number (see Section 21); and x_5 has a 1 in the positions corresponding to the Fibonacci numbers. The reader might try to figure out the pattern governing the placement of 1s in x_6 (a solution is in Part Two). Each of these numbers is irrational and real-time computable (Exercises 1, 2, 4).

Let's examine the Liouville number x_1 in more detail. The following method of generating its digits shows that it is real-time computable. After the digit 1 in the $(n-1)!$th position of x_1 is written we precede the writing of each subsequent 0 in x_1 by filling in, with an $*$, the next position in $n-1$ auxiliary rows underneath the segment consisting of x_1's first $(n-1)!$ digits. As soon as this array is filled in (some additional markers must be used to indicate when the last row of the array is completed), we put a 1 in our main string (see Figure 23.1).

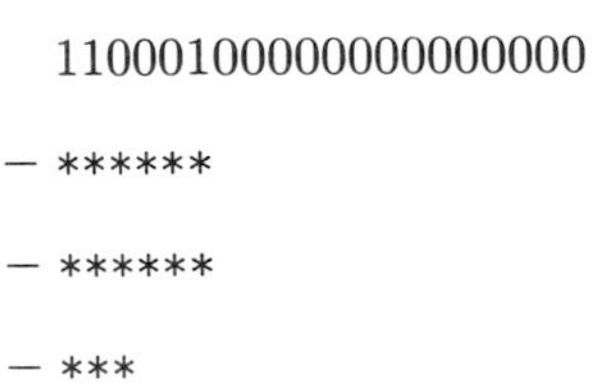

FIGURE 23.1
A typical stage in the generation of x_1's digits. After the 24th digit of x_1 (a "1") is generated, the array of $*$s is erased and another "—" is added to the column on the left.

Since $(n-1)! + (n-1)(n-1)! = n(n-1)! = n!$, the newly added 1 will be in the correct position. And the number of symbols written in all will be approximately $2 \cdot n!$ since one auxiliary $*$ is written for each digit of the main string. A few more symbols are necessary to mark the $n-1$ rows each time, but in all no more than $3m$

symbols are written to compute the first m digits of x_1. Therefore x_1 is real-time computable.

Liouville's number, x_1, is historically important since it was the first example of a transcendental number. In 1844 Liouville proved that numbers with very long runs of consecutive 0s, like x_1, are necessarily transcendental. Numbers such as x_3 have not yielded to his (or anyone elses) techniques, however, and that number's transcendence is unproved. Nevertheless, x_1 shows that at least some transcendental numbers are real-time computable. Since there are uncountably many transcendental numbers and only countably many computable numbers (because the number of n-line programs is finite for each positive integer n), most transcendental numbers are not real-time computable.

What about algebraic numbers? Essentially nothing is known; to make our extreme ignorance more precise, we don't know which of the following three possibilities holds:

(1) All algebraic irrational numbers are real-time computable.

(2) Some algebraic irrationals are real-time computable and some are not.

(3) No algebraic irrational is real-time computable.

In other words, there is not a single algebraic irrational number whose status regarding real-time computability is known!

The state of affairs will be clarified by examining another historically important number, $\sqrt{2}$, which was the first number to be proved irrational. The Pythagoreans of ancient Greece proved that $\sqrt{2}$ is irrational over two thousand years ago. Algorithms for obtaining the digits of $\sqrt{2}$ go back even further, and today we still use the Babylonian method: Let $a_0 = 2$ and define a sequence inductively by letting $a_{r+1} = \frac{1}{2}(a_r + 2/a_r)$. Then $a_r \to \sqrt{2}$ as $r \to \infty$. Moreover, the convergence is extremely rapid: the number of correct digits doubles with each iteration (see Table 23.2). So to get n digits of $\sqrt{2}$ only $\log_2 n$ terms of this sequence need to be computed. However, the work becomes more complicated as more digits are computed. In order to compute n digits of $\sqrt{2}$ one must, when computing the final a_r, do an n-digit division. Using the long division method takes at least n^2 steps, so at least n^2 steps are used overall. Since n^2 is not bounded by cn for any constant c, this algorithm is not a real-time algorithm.

The argument of the preceding paragraph does not settle the question of whether $\sqrt{2}$ is real-time computable, however. The Babylonian algorithm is not fast enough, but perhaps there is an as-yet-undiscovered algorithm that does compute n digits of $\sqrt{2}$ in cn steps. If so, $\sqrt{2}$ would be an irrational algebraic number that is real-time computable, ruling out possibility (3) above; if not, then possibility (1) is ruled out. Despite a lot of progress recently in the construction of ultra-fast algorithms, it seems unlikely that the digits of $\sqrt{2}$ are real-time computable. Unfortunately there has been almost no progress on techniques to prove that something

$$a_0 = 2$$
$$a_1 = 1.5$$
$$a_2 = 1.41666666\ldots$$
$$a_3 = 1.414215686274509804\ldots$$
$$a_4 = 1.414213562374689911\ldots$$
$$a_5 = 1.414213562373095049\ldots$$

TABLE 23.2
The first few iterations of the Babylonian method of computing $\sqrt{2} = 1.4142135623730950488\ldots$.

cannot be computed in a certain amount of time, so if $\sqrt{2}$ is not real-time computable then this fact may be very difficult to prove.

It is generally expected that no algebraic irrational number is real-time computable. This is the content of the Hartmanis–Stearns conjecture [HS], and it stems from the belief that the digits of an algebraic irrational ought to be quite random. However, the notion of randomness can be interpreted many ways. Even if the Hartmanis–Stearns conjecture is true, from the point of view of applications to cryptography requiring random digits, the digits of an algebraic irrational number are definitely *not* sufficiently random. In other words, if someone reveals the digits of an algebraic irrational number one at a time, the receiver can, after learning a relatively small number of digits, predict the entire sequence. This striking theorem was proved recently by R. Kannan, A. K. Lenstra, and L. Lovász [KLL].

Exercises

1. Why is each of x_1, x_2, x_3, x_4, x_5 irrational?
2. Show that each of x_2, x_3, x_4, x_5 is real-time computable.

24. SUMMING RECIPROCALS OF POWERS

Problem 24

Is $1 + \frac{1}{2^5} + \frac{1}{3^5} + \frac{1}{4^5} + \cdots$ irrational?

It is impossible to add up the reciprocals of all positive integers. This is because the harmonic series, $1+\frac{1}{2}+\frac{1}{3}+\frac{1}{4}+\frac{1}{5}+\cdots$, diverges to infinity. But if enough of the integers are omitted then the series may well converge; for example, the geometric series $1+\frac{1}{2}+\frac{1}{4}+\frac{1}{8}+\cdots = 2$. Most calculus texts contain a proof that the reciprocals of the squares form a convergent series (the easiest way is to apply the integral test). It is easy enough, with the help of a calculator or computer, to see that this sum is approximately 1.6449, but it is harder to compute the sum exactly. Euler was the first to realize that

$$\sum_{k=1}^{\infty} \frac{1}{k^2} = \frac{\pi^2}{6},$$

and he gave an essentially correct proof.

Because of the comparison test for convergence, it follows from the convergence of $\sum \frac{1}{k^2}$ that the reciprocals of the cubes, the fourth powers, and so on, all lead to convergent series. To simplify notation, let $\zeta(n)$ be the sum of the reciprocals of all nth powers; thus $\zeta(1)$ is undefined and $\zeta(2) = 1.6449\ldots = \pi^2/6$. Table 24.1 gives approximate values of the sums corresponding to higher powers.

n:	2	3	4	5	6	7	8
$\zeta(n)$:	1.644934	1.202057	1.0823232	1.0369278	1.0173431	1.00834928	1.00407736

TABLE 24.1
Values of the first seven reciprocal-power series.

The values are approaching 1, as they must (Exercise 2), and the distance from 1 is roughly halved with each higher power. But can Euler's neat formula in the case of squares, $\pi^2/6$, be extended to the higher powers? In fact, Euler himself discovered such a formula, but it applies only to the case of even powers. As in the case of squares, the corresponding power of π is involved:

$$\zeta(4) = \frac{\pi^4}{90}, \qquad \zeta(6) = \frac{\pi^6}{945}, \qquad \zeta(8) = \frac{\pi^8}{9450},$$
$$\zeta(10) = \frac{\pi^{10}}{93555}, \qquad \zeta(12) = \frac{691\pi^{12}}{638512875};$$

in general $\zeta(2n)$ is a rational multiple of π^{2n}.

But very little is known about odd powers. In fact, prior to 1978, nothing was known about the nature of the numbers $\zeta(3)$, $\zeta(5)$, $\zeta(7), \ldots$. Then, in a tour de force of elementary though difficult reasoning based on a curious expression for $\zeta(3)$ in terms of binomial coefficients, R. Apéry of France proved that $\zeta(3)$ is irrational. His proof was quite controversial and has been described [VdP] as "a mixture of miracles and mysteries." Apéry's presentation of his result met with a mixed reception; the following description is from [Beu]: "... in June 1978 R. Apéry confronted his audience with a miraculous proof for the irrationality of $\zeta(3) = 1^{-3} + 2^{-3} + 3^{-3} + \cdots$. The proof was elementary but the complexity and the unexpected nature of Apéry's formulas divided the audience into believers and disbelievers. Everything turned out to be correct, however." Presumably $\zeta(5)$, $\zeta(7)$, $\zeta(9)$, and so on are all irrational, as the even values are (Exercise 3), but this is not known.

Another natural question is whether the odd values follow any pattern similar to the even values. In particular, is $\zeta(3)$ the product of π^3 with a rational number? Although this is unknown, there are some subtle differences between the even and odd cases and it is not generally believed that such a pattern exists for $\zeta(n)$, n odd. Computations of the decimal expansion of $\zeta(3)/\pi^3$ have shown no periodicity.

Although the harmonic series diverges, there is one especially famous question concerning its rate of divergence. If $H(r)$ denotes the sum of the first r terms of the harmonic series, $H(r) = 1 + \frac{1}{2} + \frac{1}{3} + \cdots + \frac{1}{r}$, then $H(r)$ is very close to $\log r$. The reason for this closeness is that the partial sum is well approximated by the integral $\int_1^{r+1} \frac{1}{t}\,dt = \log(r+1)$, which for large r is only slightly greater than $\log r$. It is easy to see that the integral is smaller than the partial sum (sketch the integrand), so that $H(r) > \log r$. It is an important fact of analysis that this difference converges to a definite constant as r gets larger; this constant is called Euler's constant, denoted by γ. In other words, $\lim_{r\to\infty} H(r) - \log r = \gamma$ (for a proof that this limit exists, see [Bur] or [Knu1, pp. 74, 108–111]). The value of γ is $0.5772156649\ldots$, and so this yields reasonably precise information about how fast the harmonic series diverges: The sum of the reciprocals of the first million integers, $H(1{,}000{,}000)$, is approximately $\gamma + \log 1{,}000{,}000 = 14.39272622\ldots$ (the real value of this sum is $14.39272672\ldots$, computed using formula (3) on page 74 of [Knu1]). Now, like π and e, γ is a fundamental constant of mathematics. But γ has proved much more troublesome than the other two. The transcendence of π and e was proved over a century ago, but we still do not know the answer to the following question.

Problem 24.1

Is γ an irrational number?

Exercises

1. Show that the sum of the first k terms of the series defining $\zeta(n)$ approximates the value of $\zeta(n)$ with error at most

$$\frac{1}{k^{n-1}(n-1)}.$$

(Hint: Use the integral test.)

2. (a) Show $\lim_{n\to\infty} \zeta(n) = 1$. (Hint: Use Exercise 1.)
   (b) Show $\lim_{n\to\infty} (\zeta(n+1)-1) / (\zeta(n)-1) = \frac{1}{2}$.
3. Why does Euler's formula for $\zeta(2n)$ imply that $\zeta(2n)$ is irrational?

INTERESTING REAL NUMBERS: Part 2

21. PATTERNS IN PI

Problem 21

Are the digits in the decimal expansion of π devoid of any pattern?

The question regarding π is usually phrased more generally: Is π *normal*? A normal number is one for which any finite pattern of digits, for example, 123456789, occurs with the expected limiting frequency. This means that on average the preceding pattern occurs once in every billion (10^9) digits. Moreover, a normal number must satisfy this condition when written in any base. It has been known for some time that there are lots of normal numbers (almost all numbers are normal; see [Niv3, Chap. 8]), but explicit examples are hard to find. D. Champernowne, while an undergraduate at Cambridge University, proved that 0.123456789101112131415... is normal in base 10 ([Cha]; see [Niv3, Chap. 8]), and it has even been proven [Cha, CE] that the numbers 0.23571113171923... and 0.46891012141516..., obtained by writing down all the primes and composites, respectively, are normal in base 10. But an explicit example of a number that is normal in all bases is still lacking. Indeed, it is unknown whether or not Champernowne's number, when written in base 2 (where it becomes 0.000111111100110101101...), is normal in base 2. It is known, however, that a number can be normal in some bases but not in others [Cas, Sch]. And although it is not known that each of the ten digits keeps occurring in π, it has been proved that some integer multiple of π (indeed, of any irrational number) has this property [Mah1]. Another positive result about digit-frequencies was discovered by J. Franel in 1917 (see [TW]); he proved that for any $a \in (0, 1]$, for any place p, and for any digit d, the frequency of occurrence of d in the pth place of the real numbers in the sequence $1^a, 2^a, 3^a, 4^a, 5^a, \ldots, n^a$ approaches the limit $\frac{1}{10}$ as $n \to \infty$. And Stoneham [Sto] proved that, for example, all blocks of 1230 digits occur somewhere in the (periodic) decimal expansion of $1/17^{1000}$. In that same paper Stoneham also constructs numbers that are normal in several bases, and transcendental.

The proof that π is transcendental is quite difficult, but π's irrationality can be proved using only a few intricate maneuvers involving the integral of $p(x) \sin x$ for a certain polynomial $p(x)$. See [Niv1] for the proof. It is also possible, using only elementary means, to prove the stronger result that π^4 is irrational [Han]. A proof of π's transcendence can be found in Niven's book [Niv3] or in [HW, §11.14] or in [BB1, §11.2].

For several centuries the fastest way to compute π was to use Gauss's formula,

$$\pi = 48 \arctan \frac{1}{18} + 32 \arctan \frac{1}{57} - 20 \arctan \frac{1}{239}$$

or one of many similar ones (these can be proved using trigonometric identities, although it is more elegant to use complex numbers; see [Rib, pp. 201–204; Rie, p. 297]), using Taylor series to evaluate the three arctangents. Each of the three series alternate in sign, and so the error in a partial sum is no greater than the first omitted term. Thus one could proceed to evaluate 1,000,000 digits of π as follows. Add up terms of the three series until the last term added in is (in absolute value) less than $10^{-1,000,002}$, multiply by the coefficients 48, 32, and -20, and add up the three numbers. The first series will be the slowest to converge; obtaining the desired accuracy will take approximately 400,000 terms. The other two will converge much faster, but still it will take about 900,000 terms in all (Exercise 2) and therefore over two-and-a-half million operations ($+$, $-$, $\times$, $\div$) on 1,000,002-digit numbers to get the desired million digits of π.

In 1976 R. P. Brent and E. Salamin independently realized that a very rapidly converging sequence discovered by Gauss almost 200 years ago could be used to compute π. Their method [Sal] is a dramatic improvement over the method of the preceding paragraph—the number of full-precision operations for a million digits is reduced to a mere 155. In order to state their remarkable formula for π we need to introduce the arithmetic-geometric mean of two numbers. The reader is probably aware that, given two positive numbers a and b, their *arithmetic mean* is the average, $\frac{1}{2}(a+b)$, and their *geometric mean* is $\sqrt{ab}$. Moreover, the arithmetic mean is always greater than the geometric mean provided $a \neq b$ (can you prove this?). Now we may, following Gauss, iterate this procedure, taking the two means of the two means of a and b. More precisely, define two sequences $\{a_n\}$, $\{b_n\}$ by setting $a_0 = a$ and $b_0 = b$ and inductively defining a_{n+1} to be the arithmetic mean, and b_{n+1} to be the geometric mean of a_n and b_n. These two sequences converge to a common limit, called the *arithmetic-geometric mean* of a and b. More important (see Exercise 4), they converge very quickly—the number of correct digits doubles at each step, and so it takes only about 19 terms for million-digit accuracy.

We can now state the formula for π. Let $a_0 = 1$ and $b_0 = 1/\sqrt{2}$, let a_n, b_n denote the two sequences of means as just defined, let a be their common limit, and let $d_n = a_n^2 - b_n^2$. Then the Brent–Salamin formula states that

$$\pi = \frac{4a^2}{1 - \sum_{j=1}^{\infty} 2^{j+1} d_j}.$$

If we define π_n to be

$$\frac{4a_{n+1}^2}{1 - \sum_{j=1}^{n} 2^{j+1} d_j},$$

thenπ_n will converge very quickly to π. Indeed, π_{19} will be accurate to over a million digits and π_{26} to almost 200,000,000 digits. Because each iteration requires eight arithmetic operations, where a square root is considered one operation since techniques exist to compute square roots very quickly, 155 multiple-precision operations are needed to compute a million digits of π this way (Exercise 5). The data presented in Tables 21.1 and 21.2 are based on computations done in Japan [KTYU] using the Brent–Salamin formula. That computation is not the record, however. David Bailey of the NASA Ames Research Center, working on a CRAY 2, computed just over 29,000,000 digits in early 1986 [Bai2]. His computation, which took 28 hours of processing time, utilized a variation on the arithmetic-geometric mean formula discovered by the Borwein brothers. Their variation is quartically convergent: each iteration of the formula multiplies the number of correct digits by four. Not to be outdone, Y. Kanada recaptured the record by computing 33 million digits using 5 hours and 36 minutes of processing time on July 3, 1986. And at the beginning of 1987 he completed a computation of the first 2^{27}—more than 134 million—digits. And recently the record crossed the Pacific Ocean yet again as D. Chudnovsky and G. Chudnovsky of Columbia University completed a computation of the first one billion digits of π.

For more on the history of π-calculations see [Bec, BB1, Gar, Wag]. A detailed disussion of the use of the arithmetic-geometric mean in high-precision computations of π and the elementary functions can be found in [BB1] (see also [AB]).

Exercises

1. How many times do the digits 0 and 1 occur in the first thousand digits of Champernowne's number?

2. Look up the formula for the arctangent series and confirm that for $x = \frac{1}{18}$, the 400,000th term is less than $10^{-1,000,002}$. How many terms will the other two series take before falling under this bound?

3. If we apply the arctangent series to the equation $\arctan 1 = \pi/4$, we obtain the series

$$\frac{\pi}{4} = 1 - \frac{1}{3} + \frac{1}{5} - \frac{1}{7} + \frac{1}{9} - \cdots.$$

How many terms of this series must be added to obtain 10 digits of π?

4. If $\{a_n\}$, $\{b_n\}$ are the two sequences used to define the arithmetic-geometric mean of a and b (where $b < a$), show that $c_n = c_{n-1}^2/8a_{n+1}$, where $c_n = a_n - b_n$. This implies that $c_n < c_{n-1}^2/8b$. It follows that once the terms of the sequence are within 1 of each other, the convergence becomes very rapid since the error (c_n) gets squared at each step. This means that the number of correct digits roughly doubles at each iteration. Try it on some sample values of a and b to see this speed of convergence.

5. Show that π_{19} can be computed using 155 multi-precision operations (i.e., $+, -, \times, \div, \sqrt{\ }$).

6. Try programming the two techniques (arctangent and arithmetic-geometric mean) for getting digits of π. It would be best to work on a computer that has available a package for dealing with storage and arithmetic on multi-precision numbers, say 30 or 50 digits. The trickiest part will be getting the square root of a multi-precision number, and for this Newton's method (see Section 23) should be used: To get $\sqrt{a}$ to, say, 50 digits, let $b_0 = \frac{1}{2}(1 + a)$ and compute b_n sequentially where $b_n = \frac{1}{2}(b_{n-1} + a/b_{n-1})$ and all arithmetic is done using 50 digits. When the b-sequence stops changing, the value of b_n will be the first 50 digits of $\sqrt{a}$. Although the arithmetic-geometric mean method is faster for computing large numbers of digits, for 50- or 100-digit accuracy the arctangent method is faster, as you will discover.

22. CONNECTIONS BETWEEN π AND e

Problem 22

Are π and e algebraically independent? Is their ratio rational?

The irrationality of e is easy to prove. The series for e^x implies that

$$e = 1 + \frac{1}{2!} + \frac{1}{3!} + \frac{1}{4!} + \cdots.$$

If e were rational, then, if n is larger than e's denominator, $n!e$ would be an integer. But this would mean that

$$\frac{1}{n+1} + \frac{1}{(n+1)(n+2)} + \frac{1}{(n+1)(n+2)(n+3)} + \cdots$$

is a positive integer. This is a contradiction since this series is less than $\frac{1}{2}+\frac{1}{4}+\frac{1}{8}+\cdots$, which equals 1.

The transcendence of both e and π are special cases of Lindemann's 1882 theorem (see [Niv3, Chap. 9] or [Bak, Chap. 1]; the latter contains concise proofs of the transcendence of e and π). He proved that e^b is transcendental whenever b is a nonzero algebraic number. Moreover, b is not restricted to being real, but may also be a complex number. Letting $b = 1$ yields e's transcendence (proved nine years earlier by Hermite). But π's transcendence follows as well. For if π were algebraic then, since i is algebraic (satisfying $x^2+1=0$), $i\pi$ would be algebraic. Lindemann's theorem would then imply that $e^{i\pi}$ is transcendental, contradicting $e^{i\pi} = -1$.

The seventh in Hilbert's celebrated list of problems asked whether the e in Lindemann's theorem could be replaced by any algebraic number (see [Tij]). Since an algebraic raised to a rational power is algebraic (Exercise 3), the case $b = m/n$, m, n integers, must be excluded. This problem was solved in 1934 by A. O. Gelfond (Russia) and T. Schneider (Germany), independently. More precisely, they proved the following theorem.

Theorem 22.1. *If a and b are algebraic complex numbers with $a \neq 0$, $a \neq 1$, and $b \neq \frac{m}{n}$ $(m, n$ integers$)$, then a^b is transcendental.*

One application is that $2^{\sqrt{2}}$ is transcendental. Another example illustrating the power of complex numbers: the equation $e^{i\pi} = -1$ implies that $(-1)^{-i} = (e^{i\pi})^{-i} = e^{\pi}$; it follows from Theorem 22.1 that e^{π} is transcendental.

See [Niv3, Chap. 10] for a proof of Theorem 22.1, which uses both complex analysis and modern algebra (field extensions). Note that a complex number $u+iv$, where u and v are real, is algebraic if and only if both u and v are algebraic reals (Exercise 4). Hence Theorem 22.1 asserts that at least one of a^b's real and imaginary parts is transcendental. A proof that, in fact, *both* the real and imaginary parts are transcendental, provided b is neither purely real nor purely imaginary, appears in [Rob].

C. L. Siegel, who had a hand in proving the transcendence of $2^{\sqrt{2}}$ before the discovery of the more general Theorem 22.1, has pointed out [Rei, p. 164] that Hilbert felt that problems such as the transcendence of $2^{\sqrt{2}}$, like so many other problems in number theory, seemed to be incredibly difficult. In 1919 Hilbert asserted that the Riemann hypothesis would probably be proved before he died and that Fermat's last theorem might be proved by the end of the century, but that $2^{\sqrt{2}}$'s transcendence would turn out to be more troublesome than either of the others. Fifteen years later the transcendence of that number and the stronger Gelfond–Schneider theorem had been established. The other two problems are still unsolved. "This shows that one cannot guess the real difficulties of a problem before having solved it" [Sie, p. 84].

Recently R. Kannan and L. McGeoch [KM] have shown, using the Lovász basis reduction algorithm, that if $e + \pi$ is the root of a polynomial with integer coefficients, then the Euclidean length of the vector of coefficients of the polynomial is at least 500,000,000. David Bailey [Bai1] used an algorithm due to Ferguson and Forcade, and a couple of hours on a CRAY 2 supercomputer, to show that if $e + \pi$ is the solution of a degree-eight polynomial, then the height of the polynomial is at least 1018, with similar results for π/e and some other combinations of π, e, and γ (Euler's constant γ is discussed in Section 24).

Exercises

1. (a) Prove that at least one of $\pi + e$ and $\pi - e$ is irrational.
   (b) Prove that at least one of $\pi + e$ and $\pi \cdot e$ is irrational.
   (c) Prove that at least one of $\pi \cdot e$ and π/e is irrational.

2. (a) Show that a consequence of the conjecture that e and π are algebraically independent is that the decimal expansion of π does not contain the entire decimal expansion of e; that is, π's digits do not look like 3.1415926...2718281828459... .

   (b) Why is it impossible (regardless of the truth of the conjecture mentioned in Exercise 2(a)) for both the decimal expansion of π to contain e and the decimal expansion of e to contain π?

3. Prove that a rational power of an algebraic number is algebraic.

4. Let a and b be real numbers. Prove that $a + bi$ is algebraic if and only if a and b are each algebraic. (Hint: For the forward direction, show that $a - bi$ is algebraic if $a + bi$ is.)

5. Prove that $\log_m n$ is transcendental if n and m are positive integers greater than 1 and n is not a rational power of m.

6. Prove that $e^{\pi\sqrt{163}}$ is transcendental. Nevertheless, this number is remarkably close to an integer: it equals 262537412640768743.99999999999925007. . . .

23. COMPUTING ALGEBRAIC NUMBERS

Problem 23

If an irrational number is real-time computable, is it necessarily transcendental? Is $\sqrt{2}$ real-time computable?

Liouville's proof of x_1's transcendence is not too difficult and, as with so many other results in this chapter, we refer the reader to Niven's book [Niv3] for a clear discussion of the details (see also [Niv2; HW, p. 161]). The key point is that if a number x is algebraic then, for some positive integer d, there are only finitely many rationals $\frac{m}{n}$ satisfying

$$\left| x - \frac{m}{n} \right| < \frac{1}{n^d}.$$

But in the case of Liouville's number there are, for any d, infinitely many rationals satisfying the inequality (Exercise 3). This proof leans heavily on the very long strings of zeros in x_1, and does not apply to numbers such as $x_2, \ldots, x_6$.

The discovery by G. Cantor in 1873 that the reals are uncountable while the algebraic numbers are countable yielded a shorter, if less concrete, proof that transcendental numbers exist, and the same technique implies that some numbers are not real-time computable. We now know that several important constants such as π and e are not algebraic; it is suspected that they are not real-time computable either, but this is not known. Striking partial results toward the Hartmanis–Stearns conjecture have been obtained by Loxton and van der Poorten [LV1, 2], building on earlier work by Cobham [Cob] and Mahler. Their work yields that certain real-time computable numbers—those computable on a finite automaton (a special sort of machine that has less power than the full power of a computer or a human)—must be transcendental if they are irrational. This yields the transcendence of the numbers x_2, x_5, and x_6, although the transcendence of these examples was already known by more classical techniques (see [Mah2, refs. 4, 7, 8, 35, and 36]). The new aspect of the work of Loxton, van der Poorten, and Cobham is the derivation of transcendence from a number's computability properties. By the way, x_6 was constructed as follows. A "1" is placed in the nth position if and only if the base-2 representation of n has an odd number of 1s. This number is irrational and real-time computable (Exercise 4).

The history of the transcendence proof for x_4 is interesting. In 1937 Mahler received morphine injections while undergoing bone surgery. Wanting to convince himself that the drug had not damaged his brain, he turned to the problem of whether 0.123456789101112... is transcendental, which he settled affirmatively. Mahler's proof for x_4 is a special case of his result that $0.P(1)P(2)P(3)\ldots$ is transcendental whenever P is a nonconstant polynomial with integer coefficients such that $P(i) \in \mathbb{N}$ whenever i is a positive integer.

There are other transcendence results for sequences that are not necessarily real-time computable. The transcendence of sequences of 0s and 1s generated by repeatedly folding a strip of paper was proved by Mendés-France and van der Poorten [DMV, MV]).

Our definition of real-time computable is an informal one since the nature of the machine carrying out the computation was not made precise. This is well illustrated by x_6. Under our definition a step is the writing of a single symbol; movement left or right on the paper, to read a digit halfway between the current digit and the start of the string, for example, is not counted. But in the real world such movements take time and should be counted when measuring complexity of an algorithm. In the algorithm for x_6 (solution to Exercise 4) the amount of left and right movement increases as each digit is computed, so the number may not be real-time computable on a real-world machine. Although greater precision is possible (using either the notion of a Turing machine or random-access machine), in the problem at hand it seems unimportant. All real-time computable (irrational) numbers whose transcendence is decided are transcendental, regardless of the sort of machine used to define real-time computable. So, as Cobham has observed, we may as well consider the conjecture to state that an irrational number that is real-time computable on any sort of machine is not algebraic. For a rigorous summary of some of the different ways of looking at the complexity of a single number, see [BB2]. The Babylonian technique for evaluating $\sqrt{2}$ is faster than it looks because of modern developments in the speed of multiplication and division. Let's assume that two n-bit numbers are given (i.e., two integers having n digits when written in base 2; a bit is a position that can take on one of the values 0 or 1). Let $M(n)$ be the number of bit operations ($x + y$, $x \cdot y$, $1 - x$ all computed modulo 2, i.e., $1 + 1 = 0$) needed in the fastest algorithm to multiply the two given numbers. The elementary school algorithm for multiplication can be used (the move to base 2 causes no siginificant change) but it takes roughly n^2 steps (by which we now mean bit operations). There are faster ways (see [Knu2, Chap. 4] or [AHU, §§2.6, 7.5]); the best result known is that $M(n) \leq cn(\log n)(\log\log n)$ for some constant c. Since each of the $2n$ input bits must be examined, multiplication takes at least $2n$ steps; this means that, ignoring the leading constants, $n \leq M(n) \leq n(\log n)(\log\log n)$. Moreover, it is known that division can be done in $cM(n)$ steps and this can be combined with Newton's method (as the Babylonian square-root algorithm is usually called) to compute square roots in $cM(n)$ steps. In summary, the first n digits of $\sqrt{2}$ can be computed in $cM(n)$ $(= cn(\log n)(\log\log n))$ steps. This is quite fast, but still not a real-time computation since the number of steps has not been reduced to a constant multiple of n. The arithmetic-geometric mean algorithm for digits of π (Section 21) shows that n digits of π can be computed in $cM(n)\log n$ steps.

Newton's method applies to more than just square roots. It can be used to get a sequence of approximations to a root of any equation $f(x) = 0$, and if the starting value is sufficiently close to a root and f is a sufficiently nice function (in particular, any polynomial whose derivative at the root is nonzero), the convergence is very rapid. For details see a numerical analysis text or almost any calculus text. The equation governing Newton's method is:

$$x_{n+1} = x_n - \frac{f(x_n)}{f'(x_n)},$$

which has a natural geometric interpretation (Exercise 5). Newton's method and modern improvements have led to very fast algorithms for computing roots of polynomials. For example, it is known (A. Schönhage) that for any algebraic number x, the first n digits of x can be computed in $cM(n)$ steps. This means that, except for the possibility of a larger constant c, an arbitrary algebraic number can be computed as fast as $\sqrt{2}$ can.

Exercises

3. Show that for any d, the equation $|x_1 - m/n| < n^{-d}$ has infinitely many solutions in integers m, n. (Hint: Let m/n be the sum of the first s terms of the series defining x_1, where $s > d$.)

4. Show that x_6 is irrational and real-time computable.

5. (a) Show that Newton's method for finding the roots of $x^2 - 2 = 0$ yields the same iteration formula that the Babylonians used.

(b) Draw the graph of $f(x) = x^2 - 2$, and consider the following method of finding a root. Let $x_0 = 2$, draw the tangent to the graph at $(2, f(2))$, let x_1 be the x-intercept of this line, draw the tangent at $(x_1, f(x_1))$, let x_2 be the x-intercept of this line, and so on. Show that this method is exactly the same as Newton's method (and hence, by Exercise 5(a), the same as the Babylonian method).

24. SUMMING RECIPROCALS OF POWERS

Problem 24

Is $1 + \frac{1}{2^5} + \frac{1}{3^5} + \frac{1}{4^5} + \cdots$ irrational?

The ζ-function, extended to the domain of complex numbers, is one of the most intriguing functions of mathematics, primarily because of the unsolved Riemann hypothesis. See Section 17 for more on this problem and its connections with conjectures involving the distribution of prime numbers.

There are a variety of ways to prove that $\zeta(2) = \pi^2/6$. One elegant approach starts from the infinite product expansion discovered by Euler:

$$\frac{\sin x}{x} = \prod_{k=1}^{\infty}\left(1 - \frac{x^2}{k^2\pi^2}\right).$$

As motivation for this formula note that it is clearly correct when $x = k\pi$, k an integer; for a proof of its validity for all x see [Kob, §2.1]. Now, the coefficient of x^2 in the Maclaurin series expansion of $\frac{\sin x}{x}$ is $-\frac{1}{6}$ (Exercise 4). And the right-hand side can be expanded into a power series by the usual method of multiplying out (for justification see [Tit, p. 34]), which yields

$$-\sum_{k=1}^{\infty}\frac{1}{k^2\pi^2}$$

as the coefficient of x^2. Since these coefficients must be equal (each is the second derivative at $x = 0$, divided by 2), it follows that $\zeta(2) = \pi^2/6$. A more elementary proof using only ideas of trigonometry (first discovered by A. M. and I. M. Yaglom in 1953) is outlined in Exercise 5. For details of how Euler originally discovered that $\zeta(2) = \pi^2/6$, see [Ayo] or [Wei, Chap. III, §§XVII–XIX]; one of Euler's proofs is relatively elementary, using only manipulations with Taylor series and integrals (see [Ayo, p. 1079]). Further history related to this fascinating formula can be found in [Sta].

One consequence of the fact that $\zeta(2) = \pi^2/6$ is the following argument showing that whenever two positive integers are chosen randomly, the probability that they are relatively prime (i.e., have no common divisor greater than 1) is $6/\pi^2$. An informal proof of this[1] can be given as follows (see [Knu2, §4.5.2, exercise 10] for a

[1] More precisely, it is being shown that if the integers are chosen randomly from the interval $[1, N]$ then the probability that they are relatively prime converges to $6/\pi^2$ as $N \to \infty$.

more rigorous treatment). First observe that the probability that 2 does not divide both of the chosen numbers is the probability that at least one of the numbers is odd, which is $1 - \frac{1}{4}$. Similarly, the probability that 3 does not divide both of them is $1 - \frac{1}{9}$, and likewise for 5, 7, 11, 13, and so on for all primes. We need not consider nonprime possibilities since if, for example, 2 does not divide both of the numbers, then neither does any multiple of 2. Now, the probability that the two numbers have no common divisor is the probability that all these events occur simultaneously, which is the product of all the probabilities. So the desired probability is the infinite product

$$\left(1 - \frac{1}{4}\right)\left(1 - \frac{1}{9}\right)\left(1 - \frac{1}{25}\right)\left(1 - \frac{1}{49}\right)\cdots.$$

By using the formula for the sum of a geometric series, one can see that the infinite product coincides with

$$\frac{1}{\left(1 + \frac{1}{4} + \frac{1}{16} + \cdots\right)} \frac{1}{\left(1 + \frac{1}{9} + \frac{1}{81} + \cdots\right)} \frac{1}{\left(1 + \frac{1}{25} + \frac{1}{100} + \cdots\right)} \cdots.$$

If we multiply the denominator out in the usual way we obtain the reciprocal of the square of each integer exactly once. For example, $\frac{1}{100}$ is the term obtained by choosing $\frac{1}{4}$ from the first series, 1 from the second series, $\frac{1}{25}$ from the third series, and 1 from the rest. This implies that the desired probability is the reciprocal of $\sum_{k=1}^{\infty} \frac{1}{k^2}$, as claimed. You can easily verify this by experimenting: Ask each of two friends to pick, say, a hundred integers at random, and check the pairs consisting of their first, second, etc., choices to see if they have a common divisor. Approximately 61% of the pairs should have no common divisor greater than 1.

Euler's general formula for $\zeta(n)$, where n is a positive *even* integer, is

$$\frac{2^{n-1}\pi^n |B_n|}{n!},$$

where B_n is the nth Bernoulli number. The Bernoulli numbers, an important sequence of rational numbers that also plays a role in the analysis of Fermat's last theorem (Section 13), are defined as follows. The Bernoulli number B_n is the coefficient of $x^n/n!$ in the Maclaurin series expansion of

$$f(x) = \frac{x}{e^x - 1}$$

(where $f(0)$ is defined to be 1). Therefore $B_n = f^{(n)}(0)$, the nth derivative of f at 0. Except for B_1, the odd-indexed B_n are all zero. For more on the Bernoulli numbers, see [IR, Chap. 15]. For proofs of Euler's formula for $\zeta(n)$, n even, which has been called "one of the most beautiful results of elementary analysis" [Ber1], see [Apo1; Ber1; IR, Chap. 15; Kno, §7.3]. There do exist formulas that relate $\zeta(n)$, n odd, to the corresponding power of π, but they are much more complicated than Euler's formula for n even (see [Ber2]).

Apéry's proof of the irrationality of $\zeta(3)$ uses the curious formula

$$\sum_{k=1}^{\infty} \frac{1}{k^3} = \frac{5}{2} \sum_{k=1}^{\infty} \frac{(-1)^k}{k^3 \binom{2k}{k}} \tag{$*$}$$

There are related formulas for squares and fourth powers:

$$\begin{aligned} \sum_{k=1}^{\infty} \frac{1}{k^2} &= 3 \sum_{k=1}^{\infty} \frac{1}{k^2 \binom{2k}{k}} \\ \sum_{k=1}^{\infty} \frac{1}{k^4} &= \frac{36}{17} \sum_{k=1}^{\infty} \frac{1}{k^4 \binom{2k}{k}}. \end{aligned} \tag{$**$}$$

Perhaps such formulas exist for larger powers as well—if so, they might be useful in a proof of $\zeta(n)$'s irrationality—but so far no relationship between $\zeta(n)$ and either

$$\sum_{k=1}^{\infty} \frac{1}{k^n \binom{2k}{k}} \qquad \text{or} \qquad \sum_{k=1}^{\infty} \frac{(-1)^k}{k^n \binom{2k}{k}}$$

is known for any $n \geq 5$. Formula $(**)$ leads to another proof that $\zeta(2) = \pi^2/6$. Since

$$2\left(\arcsin \frac{x}{2}\right)^2 = \sum_{k=1}^{\infty} \frac{x^{2k}}{k^2 \binom{2k}{k}},$$

$$\zeta(2) = 3 \cdot 2 \left(\arcsin \frac{1}{2}\right)^2 = 6 \left(\frac{\pi}{6}\right)^2.$$

See [Mel, §3.1] for a proof of $(**)$. A proof of the arcsin formula appears in [Leh], which also contains a discussion of the formula above for $\zeta(3)$ and other remarkable formulas relating π to series involving binomial coefficients.

Some of the mystery has been taken out of the proof of $\zeta(3)$'s irrationality by F. Beukers [Beu] who, guided by Apéry's ideas, found a proof based on double and triple integrals (see also [BB2, Thm. 11.5]). These integrals can be used to give yet another proof that $\zeta(2) = \pi^2/6$ [Apo2].

Finally, we mention that recent work of H. Ferguson has produced an efficient algorithm to search for algebraic relations for individual numbers. Concrete applications include:

(1) if $\zeta(3)/\pi^3$ satisifies a polynomial of degree 5 or less with integer coefficients then at least one of these coefficients has more than 50 digits;

(2) if Euler's constant γ satisfies a polynomial of degree one or two with integer coefficients then one of the coefficients has at least 100 digits.

In particular, neither of these numbers is rational unless one of the denominator is very large. Additional applications of Ferguson's algorithm to gather evidence for the transcendence of γ and other famous constants can be found in [Bai1].

Exercises

4. Evaluate the first three terms of the Maclaurin series expansion of

$$\frac{\sin x}{x}.$$

5. (a) Prove that $\cot^2 x < \frac{1}{x^2} < 1 + \cot^2 x$ when $0 < x < \pi^2/6$.

(b) Assuming the identity

$$\sum_{k=1}^{m} \cot^2 \frac{k\pi}{2m+1} = \frac{m(2m-1)}{3}$$

(which can be proved using de Moivre's formula: $(\cos x + i \sin x)^n = \cos nx + i \sin nx$; see [Pap, YY]), show how part (a) can be used to prove

$$\frac{m(2m-1)}{3} < \frac{(2m+1)^2}{\pi^2} \sum_{k=1}^{m} \frac{1}{k^2} < m + \frac{m(2m-1)}{3}.$$

(c) Show that the result of (b) implies that $\zeta(2) = \pi^2/6$.

6. Compute the first few Bernoulli numbers. Instead of repeatedly taking derivatives of $\frac{x}{e^x-1}$, it is easier to proceed as follows. The definition implies that

$$x = \left(B_0 + B_1 x + \frac{B_2 x^2}{2!} + \cdots\right)\left(x + \frac{x^2}{2!} + \frac{x^3}{3!} + \cdots\right)$$

(using the series expansion of e^x). Multiplying out and setting all coefficients to 0 (except the coefficient of x, which equals 1) yields values for the B_n.

7. Compute the first five digits of $\zeta(3)$ two ways, first by adding up terms of the series used to define $\zeta(3)$, and second, by the right-hand side of formula $(*)$. Which method is faster?

REFERENCES

[AB] G. Almkvist and B. Berndt, Gauss, Landen, Ramanujan, the arithmetic-geometric mean, ellipses, π, and the *Ladies Diary, American Mathematical Monthly,* 95 (1988) 585–608. [§21]

[AHU] A. V. Aho, J. E. Hopcroft and J. D. Ullman, *The Design and Analysis of Computer Algorithms,* Addison-Wesley, Reading, Mass., 1974. [§23]

[Apo1] T. M. Apostol, Another elementary proof of Euler's formula for $\zeta(2n)$, *American Mathematical Monthly,* 80 (1973) 425–431. [§24]

[Apo2] T. M. Apostol, A proof that Euler missed: Evaluating $\zeta(2)$ the easy way, *The Mathematical Intelligencer,* 5 (1983) 59–60. [§24]

[Ayo] R. Ayoub, Euler and the zeta function, *American Mathematical Monthly,* 81 (1974) 1067–86. [§24]

[Bai1] D. Bailey, Numerical results on the transcendence of constants involving π, e, and Euler's constant, *Mathematics of Computation,* 50, (1988) 275–281. [§22]

[Bai2] D. Bailey, The computation of π to 29,360,000 decimal digits using Borweins' quartically convergent algorithm, *Mathematics of Computation,* 50 (1988) 283–296. [§21]

[Bak] A. Baker, *Transcendental Number Theory,* Cambridge University Press, London, 1975. [§22]

[Bec] P. Beckmann, *A History of π,* fifth ed., Golem Press, Boulder, 1982. [§21]

[Ber1] B. Berndt, Elementary evaluation of $\zeta(2n)$, *Mathematics Magazine,* 48 (1975) 148–154. [§24]

[Ber2] ——, Modular transformations and generalizations of several formulas of Ramanujan, *Rocky Mountain Journal of Mathematics,* 7 (1977) 147–189. [§24]

[Beu] F. Beukers, A note on the irrationality of $\zeta(2)$ and $\zeta(3)$, *Bulletin of the London Mathematical Society,* 11 (1979) 268–272. [§24]

[BB1] J. M. Borwein and P. B. Borwein, On the complexity of familiar functions and numbers, *SIAM Review,* 30 (1988) 589–601. [§21]

[BB2] ——, *Pi and the AGM,* Wiley, New York, 1987. [§§23 and 24]

[Bur] F. Burk, Euler's constant, *The College Mathematics Journal,* 16 (1985) 279. [§24]

[Cas] J. W. S. Cassels, On a problem of Steinhaus about normal numbers, *Colloquium Mathematicae,* 7 (1959) 95–101. [§21]

[Cha] D. Champernowne, The construction of decimals normal in the scale of ten, *Journal of the London Mathematical Society,* 8 (1933) 254–260. [§21]

[Cob] A. Cobham, Uniform tag sequences, *Mathematical Systems Theory,* 6 (1972) 164–192. [§23]

[CE] A. Copeland and P. Erdős, Note on normal numbers, *Bulletin of the American Mathematical Society,* 52 (1946) 857–860. [§21]

[DMV] M. Dekking, M. Mendès-France, and A. van der Poorten, Folds!, *The Mathematical Intelligencer,* 4 (1983) 130–138, 173–181, 190–195. [§23]

[Gar] M. Gardner, Slicing π into millions, *Discover,* 6 January, 1985, 50–52. [§21]

[Han] J. Hančl, A simple proof of the irrationality of π^4, *American Mathematical Monthly,* 93 (1986) 374–75. [§21]

[HW] G. H. Hardy and E. M. Wright, *An Introduction to the Theory of Numbers,* fourth ed., Oxford, London, 1960. [§§21 and 22]

[HS] J. Hartmanis and R. Stearns, On the computational complexity of algorithms, *Transactions of the American Mathematical Society,* 117 (1965) 285–306. [§23]

[IR] K. Ireland and M. Rosen, *A Classical Introduction to Modern Number Theory,* Springer, New York, 1982. [§24]

[KTYU] Y. Kanada, Y. Tamura, S. Yoshino, and Y. Ushiro, Calculation of π to 10,013,395 decimal places based on the Gauss–Legendre algorithm and Gauss arctangent relations, Computer Centre, University of Tokyo, 1983. [§21]

[KLL] R. Kannan, A. K. Lenstra, and L. Lovász, Polynomial factorization and nonrandomness of bits of algebraic and some transcendental numbers, *Mathematics of Computation,* 50 (1988) 235–250. [§23]

[KM] R. Kannan and L. A. McGeoch, Basis reduction and evidence for transcendence of certain numbers, in *Sixth Conference on Foundations of Software Technology and Theoretical Computer Science Conference,* Lecture Notes in Computer Science, No. 241, Springer, Berlin 1986 263–269. [§22]

[Kno] K. Knopp, *Infinite Sequences and Series,* Dover, New York, 1956. [§24]

[Knu1] D. E. Knuth, *The Art of Computer Programming,* vol. 1, Addison-Wesley, Reading, Mass., 1968. [§24]

[Knu2] ——, *The Art of Computer Programming,* vol. 2, Addison-Wesley, Reading, Mass., 1971. [§§23 and 24]

[Kob] N. Koblitz, *p-adic Numbers, p-adic Analysis, and Zeta-Functions,* 2nd ed., Springer, New York, 1984. [§24]

[Leh] D. H. Lehmer, Interesting series involving the central binomial coefficient, *American Mathematical Monthly,* 92 (1985) 449–457. [§24]

[LV1] J. H. Loxton and A. J. van der Poorten, Arithmetic properties of the solutions of a class of functional equations, *Journal für die reine und angewandte Mathematik,* 330 (1982) 159–172. [§23]

[LV2] ——, Arithmetic properties of the solutions of a class of functional equations. II, forthcoming. [§23]

[Mah1] K. Mahler, Arithmetical properties of the digits of the multiples of an irrational number, *Bulletin of the Australian Mathematical Society,* 8 (1973) 191–203. [§21]

[Mah2] ——, Fifty years as a mathematician, *Journal of Number Theory,* 14 (1982) 121–155. [§23]

[Mel] Z. A. Melzak, *Companion to Concrete Mathematics,* Wiley, New York, 1973. [§24]

[MV] M. Mendés-France and A. J. van der Poorten, Arithmetic and analytic properties of paperfolding sequences, *Bulletin of the Australian Mathematical Society,* 24 (1981) 123–131. [§23]

[Niv1] I. Niven, A simple proof that π is irrational, *Bulletin of the American Mathematical Society,* 53 (1947) 509. [§21]

[Niv2] ——, *Numbers: Rational and Irrational,* New Mathematical Library, vol. 1, Random House, New York, 1961. [§23]

[Niv3] ——, *Irrational Numbers,* Carus Mathematical Monographs, No. 11, The Mathematical Association of America. Distributed by Wiley, New York, 1967. [§§21, 22, and 23]

[Pap] I. Papadimitriou, A simple proof of the formula $\sum_{k=1}^{\infty} k^{-2} = \pi^2/6$, *American Mathematical Monthly,* 80 (1973) 424–425. [§24]

[Rei] C. Reid, *Hilbert,* Springer, New York, 1970. [§22]

[Rib] P. Ribenboim, Consecutive powers, *Expositiones Mathematicae,* 2 (1984) 193–221. [§21]

[Rie] H. Riesel, *Prime Numbers and Computer Methods for Factorization,* Birkhäuser, Boston, 1985. [§21]

[Rob] M. L. Robinson, On certain transcendental numbers, *Michigan Mathematical Journal,* 31 (1984) 95–98. [§22]

[Sal] E. Salamin, Computation of π using arithmetic-geometric mean, *Mathematics of Computation,* 30 (1976) 565–570. [§21]

[Sch] W. Schmidt, On normal numbers, *Pacific Journal of Mathematics,* 10 (1960) 661–672. [§21]

[Sie] C. L. Siegel, *Transcendental Numbers,* Annals of Mathematics Studies, No. 16, Princeton University Press, Princeton, 1949. [§22]

[Smi] S. B. Smith, *The Great Mental Calculators,* Columbia University Press, New York, 1983. [§21]

[Sta] E. L. Stark, The series $\sum_{k=1}^{\infty} k^{-s}$, $s = 2, 3, 4, \ldots$, once more, *Mathematics Magazine,* 47 (1974) 197–202. [§24]

[Sto] R. G. Stoneham, On the uniform ϵ-distribution of residues within the periods of rational fractions with applications to normal numbers, *Acta Arithmetica,* 22 (1973) 371–389. [§21]

[TW] E. Thorp and R. Whitley, Poincaré's conjecture and the distribution of digits in tables, *Compositio Mathematica,* 23 (1971) 233–250. [§21]

[Tij] R. Tijdeman, Hilbert's seventh problem: on the Gelfond Baker method and its applications, in *Mathematical Developments Arising From Hilbert Problems,* Proceedings of Symposia in Pure Mathematics, 28, Part 1, American Mathematical Society, Providence, 1976. [§22]

[Tit] E. C. Titchmarsh, *The Theory of Functions,* 2nd ed., Oxford University Press, London, 1939. [§24]

[VdP] A. van der Poorten, A proof that Euler missed, *The Mathematical Intelligencer,* 1 (1979) 195–203. [§24]

[Wag] S. Wagon, The evidence: Is π normal?, *The Mathematical Intelligencer,* 7:3 (1985) 65–67. [§21]

[Wei] A. Weil, *Number Theory, An approach through history from Hammurapi to Legendre,* Birkhäuser, Boston, 1984. [§24]

[YY] A. M. Yaglom and I. M. Yaglom, *Challenging Mathematical Problems with Elementary Solutions,* Vol. II, Holden-Day, San Francisco, 1967. [§24]

HINTS AND SOLUTIONS: TWO-DIMENSIONAL GEOMETRY

1. ILLUMINATING A POLYGON

1.1. Let H and L denote the edges that are incident to the corner. Note that if a ray R' is parallel to R, very close to R, and on the H side of R, then R' strikes H, travels a short distance to strike L, and then leaves the corner along a path that is parallel to R''s initial segment. The situation is similar when R' is on the L side of R.

1.2. Measure angles counterclockwise from the x-axis, as usual. As measurements of direction of travel, two angles are equivalent if and only if their difference is divisible by 2π. As measurements of inclination of undirected lines, two angles are equivalent if and only if their difference is divisible by π. If a light ray traveling at an angle θ is reflected from a line whose angle is ϕ, the ray's next segment travels at an angle $2\phi - \theta$. In this problem, it may be assumed without loss of generality that the angles of A's sides are 0 and α, and then the initial direction of travel of both R_H and R_L is $\frac{1}{2}\alpha - \pi$. Immediately after striking H, R_H's direction of travel is $\pi - \frac{1}{2}\alpha$, and then, after striking L, its direction β of travel is given by

$$\beta = 2\alpha - (\pi - \frac{1}{2}\alpha)) = \frac{5}{2}\alpha - \pi.$$

Immediately after striking L, R_L's direction of travel is $2\alpha - (\frac{1}{2}\alpha - \pi)$, and then, after striking H, its direction γ of travel is $-\frac{3}{2}\alpha + \pi$. For $2\pi/5 < \alpha < 2\pi/3$ (as is assumed here), the angles β and γ are both between 0 and α, so they represent the final directions of travel of R_H and R_L in A. And the difference $\beta - \gamma$ is divisible by 2π if and only if $\alpha = \pi/2$.

1.3. Let x denote the point at which the ray crosses pq, and y the point at which the ray strikes the ellipse. By the basic property of foci, the segments py and qy make equal angles with the tangent to the ellipse at the point y. From the fact that the segment xy lies inside the angle pyq, it follows that the reflected segment also lies in this angle and hence crosses pq.

1.4. Assume without loss of generality that the circumference of the circle C is 1. Use arc-length as a parameter, so that each point of C is represented by a decimal $0.a_1a_2\ldots$, where each a_i is a digit between 0 and 9. Now consider an arbitrary sequence of lines $L_1, L_2, \ldots$. Each L_i intersects C in at most two points, so all these points of intersection can be arranged in a sequence $p_1, p_2, \ldots$. Let q be the point of C that corresponds to the decimal $0.b_1b_2\ldots$, where b_i is chosen to be 0 unless p_i's ith digit is 0, and in that case set $b_i = 1$.

1.5. Consider an arbitrary triangle W with interior angles α, β, and γ. Let x, y, and z denote W's vertices, and let p denote the point of intersection of W's angle bisectors. Construct lines L_x, L_y, and L_z through x, y, and z, respectively, and perpendicular to the segments from p to these points. These lines form a triangle T circumscribed to W, and the angles of T opposite the angles α, β, and γ are, respectively,

$$A = \frac{1}{2}(\beta + \gamma), \qquad B = \frac{1}{2}(\gamma + \alpha), \qquad C = \frac{1}{2}(\alpha + \beta).$$

Since each of A, B, and C is less than the sum of the other two, the triangle T is acute. Clearly W is a billiard path of period 3 in T.

The above system of equations is equivalent to the system,

$$\alpha = B + C - A, \qquad \beta = C + A - B, \qquad \gamma = A + B - C,$$

which yields positive α, β, and γ whenever A, B, and C are the angles of an acute triangle T. Hence in that case we may start with α, β, and γ as the angles of a triangle W, and construct a triangle similar to T that has W as a billiard path of period 3.

1.6. Place the ball at a boundary point p_0, and aim very close to (but not directly at) the center. (For each i, let A_i denote the part of the table between the ray segments $p_{i-1}p_i$ and p_ip_{i+1}. Aim in such a way that A_1 is too narrow to contain any disk of radius ϵ. Since the A_is are congruent and the table is the union of the A_is, some ray segment crosses the opponent's disk.)

1.7. Choose the coordinate system so that the point v_i is the origin, the tangent L_i is the x-axis, and the smooth convex table lies in the upper half-plane. There is a differentiable function f such that in a neighborhood of the origin, the boundary curve coincides with the graph of f; then $f(0) = 0$ and $f'(0) = 0$. Let

$$g(t) - (\text{dist. from } v_i \quad 1 \text{ to } (t, f(t)) + (\text{dist. from } v_i + 1 \text{ to } (t, f(t)).$$

Then $g'(0) = 0$ because g is differentiable and the perimeter-maximizing choice of the vertices v_j implies that g attains a local maximum at 0. With

$$v_{i-1} = (\xi, \eta) \qquad \text{and} \qquad v_{i+1} = (\rho, \sigma),$$

we have

$$g(t) = \left((\xi - t)^2 + (\eta - f(t))^2\right)^{1/2} + \left((\rho - t)^2 + (\sigma - f(t))^2\right)^{1/2}$$

Calculation shows that

$$g'(0) = \frac{\rho}{(\rho^2 + \sigma^2)^{1/2}} - \frac{\xi}{(\xi^2 + \eta^2)^{1/2}}.$$

Since $g'(0) = 0$, the angles with the line L_i made by the segments $v_{i-1}v_i$ and $v_{i+1}v_i$ have equal cosines, and hence the angles are equal.

1.8. By combining triangulations of Q and R, we obtain the desired triangulation of P. For the coloring of P, start with colorings of Q and R and then permute the colors for R so that these colorings agree at the ends of the diagonal D.

1.10. For the first part, note that if the angles are commensurate, then their sum ($(n-2)\pi$) is a rational multiple of β_i and hence β_i is a rational multiple of π. For the second part, recall (Exercise 1.2) that if a ray travels in direction θ to strike an edge whose angle of inclination is ϕ, then the reflected direction of travel is $2\phi - \theta$. From this it follows by induction that after the ray has struck edges whose angles of inclination are successively $\alpha_1, \ldots, \alpha_k$, then the ray's current direction of travel is

$$2(\alpha_k - \alpha_{k-1} + \alpha_{k-2} - \alpha_{k-3} + \cdots + \alpha_1) - \theta \qquad \text{if } k \text{ is odd,}$$
$$2(\alpha_k - \alpha_{k-1} + \alpha_{k-2} - \alpha_{k-3} + \cdots - \alpha_1) + \theta \qquad \text{if } k \text{ is even.}$$

When there are only finitely many choices for the α_is, and when all the choices are commensurate with π, there are only finitely many possibilities (mod 2π) for the direction of travel of a ray that starts in a given direction θ.

1.11. See pages 378–381 of Hardy and Wright [HW].

1.12. Choose an equilateral triangle xyz of edge-length δ, and a positive number ϵ. Let $R^+(x,y)$ denote the half-line from x through y, and $R^-(x,y)$ the opposite half-line issuing from x. With x as center, construct a minor circular arc of radius $\delta+\epsilon$ from $R^+(x,y)$ to $R^+(x,z)$ and a minor circular arc of radius ϵ from $R^-(x,y)$ to $R^-(x,z)$. Proceed similarly with y and z as centers. The smooth convex figure C bounded by the six circular arcs has the property that whenever a line is perpendicular to C's boundary at one of its points of intersection with the boundary, it is also perpendicular at the other point of intersection. This *double normal* property actually characterizes the so-called convex bodies of *constant width,* which appear also in Sections 4 and 11. (A plane convex body is said to be of constant width w if each supporting line of the body is at distance w from the parallel supporting line.)

1.13. When C is an ellipse, T is a larger ellipse having the same foci. When C is an equilateral triangle, T is a smooth curve bounded by six elliptical arcs, each having two vertices of C as foci.

1.14. Since the boundary J is a Jordan curve, there is a continuous mapping ζ of the unit interval $[0,1]$ onto J such that $\zeta(\sigma) = \zeta(\tau)$ if and only if $\sigma = \tau$ or $\{\sigma,\tau\} = \{0,1\}$. Assume without loss of generality that the point $\zeta(0)$ is not an inner point of any maximal segment in J. Then for each such maximal segment S there are numbers α_S and β_S such that $0 \leq \alpha_S < \beta_S \leq 1$ and the segment is the image under ζ of the interval $[\alpha_S,\beta_S]$. Since the maximal segments have at most endpoints in common, the same is true of the intervals $[\alpha_S,\beta_S]$ in $[0,1]$. This implies that for each positive integer n there are at most n segments S such that $\beta_S-\alpha_S > 1/n$. But then by 1.4 there is a direction that is not attained by any of the segments S. (Note that convexity is not actually used here. The conclusion holds for an arbitrary Jordan curve J. However, in $\mathbb{R}^3$ there exists a topological 2-sphere S (not the boundary of a convex body) such that every line in $\mathbb{R}^3$ is parallel to some segment in S.)

1.15. For each point $\sigma = (\sigma_1,...,\sigma_d) \in \mathbb{R}^d$ having $|\sigma_i| = 1$ for all i, let $R(\sigma)$ denote the half-line consisting of all positive multiples of σ. For each point $p = (p_1,\ldots,p_d) \in B$ there exists i such that $|p_i| = 1$. If $p_i = 1$, then the half-line $p+R(1,\ldots,1)$ misses C, and if $p_i = -1$ the half-line $p+R(-1,\ldots,-1)$ misses C.

Hence each point of B is $R(1,\ldots,1)$-lighted or $R(-1,\ldots,-1)$-lighted. Also, if σ is such that $\sigma_i = p_i$ for all i such that $|p_i| = 1$, then p is strongly $R(\sigma)$-lighted. Note, finally, that if $R_1,\ldots,R_k$ are rays issuing from the origin, and $k < 2^d$, then there exists a $\sigma \in B$ whose sign-pattern does not agree with that of any of the (generators of the) R_is, and consequently σ is not strongly lighted by any R_i.

2. EQUICHORDAL POINTS

2.1. With $a = f(0)$ and $b = f(\pi)$, $\mu < a + b$ is the condition for the equichordal construction and

$$\frac{1}{\mu} < \frac{1}{a} + \frac{1}{b}$$

for the equireciprocal construction. No condition is needed for the equiproduct construction.

2.3. If an ellipse E of eccentricity e has one focus at the origin $(0,0)$ and the other at the point $(2c,0)$ (in rectangular coordinates), then E's equation in polar coordinates may be written as

$$\frac{1}{r} = \frac{1}{k}(1 - e\sin\theta)$$

where k is an appropriate constant.

2.4. For notational simplicity, let the interior point be the origin. Write the circle's equation in polar coordinates (r,θ), and find a quadratic equation that determines two values of r for the circle's intersection with the line $\tan\theta = \mu$. Note that although each of the values involves μ, their product does not.

2.5. If c is the curvature function for the boundary of $R(\gamma + f)$, then c's denominator is positive and c's numerator is equal to $(\gamma + f)(\gamma + f - f'') + 2f'^2$. Since f'' is continuous, there exists a positive real β such that $f''(\theta) > -\beta$ for all θ. For $\gamma \geq \beta$, the curvature c is everywhere positive.

2.7. To show that $\phi(x+y) \leq \phi(x) + \phi(y) + 2\epsilon$ for each $\epsilon > 0$, choose α and β so that $\phi(x) < \alpha < \phi(x) + \epsilon$ and $\phi(y) < \beta < \phi(y) + \epsilon$. Let $u = x/\alpha$ and $v = y/\beta$. Then u and v both belong to the convex set of points for which $\phi \leq 1$. Hence with

$\gamma = \alpha + \beta$, the convex combination

$$\frac{1}{\gamma}(x+y) = \frac{\alpha}{\gamma}u + \frac{\beta}{\gamma}v$$

also belongs to that set. But then $\phi(\frac{1}{\gamma}(x+y)) \leq 1$, whence

$$\phi(x+y) \leq \gamma \leq \alpha + \beta < (\phi(x)+\epsilon) + (\phi(y)+\epsilon).$$

3. PUSHING DISKS TOGETHER

3.1. With xy denoting $\mathrm{dist}(x, y)$, assume that

$$\frac{p_1p_2}{q_1q_2} \leq \min\left(\frac{p_2p_3}{q_2q_3}, \frac{p_3p_1}{q_3q_1}\right).$$

The desired continuous shrinking of triangle $p_1p_2p_3$ onto $q_1q_2q_3$ can be accomplished in five stages, as follows:

(i) Translate $p_1p_2p_3$ so as to make p_1 coincide with q_1.

(ii) Rotate $p_1p_2p_3$ about the new point p_1 $(= q_1)$ so as to bring the point p_2 onto the ray from q_1 through q_2. Note that because the triples $p_1p_2p_3$ and $q_1q_2q_3$ had the same orientation, the new point p_3 is on the same side of the line through q_1 and q_2 as q_3.

(iii) If $p_1p_2 = q_1q_2$, this step is omitted. Otherwise, $p_1p_2 > q_1q_2$ and the triangle $p_1p_2p_3$ can be shrunk uniformly toward the point p_1 until p_2 coincides with q_2. We now have $p_1 = q_1$ and $p_2 = q_2$. For the new points p_1, p_2, and p_3, it follows from the minimizing choice of the original p_1 and p_2 that $p_1p_3 \geq q_1q_3$ and $p_2p_3 \geq q_2q_3$.

(iv) If $p_1p_3 = q_1q_3$, this step is omitted. If $p_1p_3 > q_1q_3$, rotate the segment p_2p_3 about the point p_2 and toward the point p_1. This steadily decreases the distance p_1p_3, and the rotation can be continued until $p_1p_3 = q_1q_3$.

(v) For the current versions of p_1, p_2, and p_3 we have $p_1 = q_1$, $p_2 = q_2$, and $p_1p_3 = q_1q_3$. If, in addition, $p_2p_3 = q_2q_3$, then $p_3 = q_3$. If $p_2p_3 > q_2q_3$, then we rotate the segment p_1p_3 about the point p_1 and toward the point p_2, thus decreasing the distance p_2p_3 until it becomes equal to q_2q_3.

3.2. Let p_1, p_2, and p_3 be vertices of an equilateral triangle, and let p_4 be the third vertex of the other equilateral triangle that has p_1 and p_2 as vertices. Set $q_1 = p_1$, $q_2 = p_2$, $q_4 = q_3 = p_3$. In any continuous shrinking of the p_i into the positions q_i, the distances $p_1p_2, p_1p_3, p_2p_3, p_1p_4$, and p_2p_4 cannot vary. Hence for any given position of (points corresponding to) p_1 and p_2 during the shrinking, there are only two possible positions for p_3 and p_4, both on the perpendicular bisector L of the segment p_1p_2. But the point p, confined to L, cannot move continuously from its initial position to its final position (which coincides with the final position of p_4) without reducing its distance to p_1 and p_2. The validity of the argument is unchanged when the configuration of the p_is is replaced by its reflection in a line. By a slight dilatation of the p_i configuration, an example is obtained in which the interpoint distances for the p_is are actually greater than those for the q_is.

3.3. (a) Use Theorem 3.3 and Exercise 1.

(b) (M. Kneser and W. Habicht [Kne2].) Let C_ρ denote the circle of radius ρ centered at the origin, and for $0 < \rho < \sigma$ let $A(\rho, \sigma)$ denote the closed annulus between C_σ and C_ρ. The construction involves a radius ρ and a distance δ that are later specified in more detail. It is required always that $\rho \geq 3$ and $\delta < 1$, and it will become clear that the desired example is attained whenever ρ is sufficiently large and δ sufficiently small.

For $1 \leq i \leq k$, the point p_i will lie on C_ρ and we set $q_i = p_i$. For $k < i \leq k+m$, the point p_i will lie on $C_{\rho-1+\delta}$ and we set $q_{i+m} = q_i = p_i$. We set $n = k + 2m$, and for $k + m < i \leq n$ let p_i denote the point at which $C_{\rho+1-\delta}$ is intersected by the ray from the origin through p_{i-m}. The numbers ρ, δ, k, and m are still to be specified, but it is clear that for any specification the interpoint distances of the q_is do not exceeed those of the corresponding p_is.

Let the points $p_i, \ldots, p_k$ of C_ρ be chosen so that their mutual distances are all at least 2, and so that, subject to this condition, k is as large as possible. It can be verified that $k/\rho \to \pi$ as $\rho \to \infty$. Let $U_1 = \bigcup_{i=1}^{k} B(p_i, 1)$. Then the perimeter of U_1 is $2\pi k$, the part of U_1's perimeter that lies outside C_ρ exceeds πk, and for δ sufficiently small the part of U_1's perimeter that lies outside $C_{\rho+\delta}$ still exceeds πk. Let the points $p_{k+1}, \ldots, p_{k+m}$ of $C_{\rho-1+\delta}$ be so densely distributed that the following conditions are satisfied:

(a) the union $U_2 = \bigcup_{i=k+1}^{k+m} B(p_i, 1)$ covers the annulus $A(\rho - 1 + \delta, \rho)$;

(b) the union $U_3 = \bigcup_{i=k+m+1}^{k+2m} B(p_i, 1)$ covers the annulus $A(\rho, \rho + 1 - \delta)$;

(c) the part of U_2's perimeter that lies inside $C_{\rho-1+\delta}$ differs by less than δ from the perimeter $2\pi(r - 2 + \delta)$ of $C_{\rho-2+\delta}$;

(d) the part of U_3's perimeter that lies outside $C_{\rho+1}$ differs by less than δ from the perimeter $2\pi(\rho+2-\delta)$ of $C_{\rho+2-\delta}$.

These choices may require large m, but the fact that they can be made follows from the Heine–Borel theorem and some elementary geometric estimates.

Now set

$$U_p = \bigcup_1^n B(p_i, 1) = U_1 \cup U_2 \cup U_3 \quad \text{and} \quad U_q = \bigcup_1^n B(q_i, 1) = U_1 \cup U_3.$$

Then the perimeter of U_p is at most $2\pi(\rho-2+\delta)+\pi(2\rho+4+2\delta)+2\delta$, while the perimeter of U_q is at least $2\pi(\rho-2+\delta)+\pi k-\delta$. Since $k/\rho \to \pi$ as $\rho \to \infty$, it is clear that $\pi(2\rho+4+2\delta)+2\delta < \pi k - \delta$ for all sufficiently large ρ and sufficiently small δ.

3.6. (J. Pach) Let the points p_i, the points q_i, and the convex hulls P and Q be as in Theorem 3.4. For each $\rho > 0$, let P_ρ denote the union of all disks of radius ρ centered in P (this is called the *parallel body* of P with radius ρ). It is known that

$$\mu(P_\rho) = \mu(P) + \rho \operatorname{per}(P) + \rho^2\pi.$$

An easy calculation shows that

$$\mu\left(\bigcup_i B(q_i,\rho)\right) \le \mu(Q_\rho) \le \mu\left(\bigcup_i B(q_i,\rho)\right) + \phi(\rho)$$

with $\phi(\rho) \to 0$ as $\rho \to \infty$. A similar statement applies to P_ρ and $\mu\left(\bigcup_i B(p_i,\rho)\right)$. But then it follows that

$$\rho(\operatorname{per}(P) - \operatorname{per}(Q)) \ge (\mu(P_\rho) - \mu(P)) - (\mu(Q_\rho) - \mu(Q)).$$

and that the expression on the right side of this inequality is greater than or equal to

$$\mu\left(\bigcup_i B(p_i,\rho)\right) - \mu\left(\bigcup_i B(q_i,\rho)\right) - \phi(\rho) - \mu(P) + \mu(Q).$$

If the answer to Problem 3 is known to be negative, then this last expression is bounded below as $\rho \to \infty$, whence it follows that $\operatorname{per}(P) \ge \operatorname{per}(Q)$.

3.7. The intersection of parallelepipeds

$$R(a_1^1,\ldots,a_d^1;b_1^1,\ldots,b_d^1),\ldots,R(a_1^n,\ldots,a_d^n;b_1^n,\ldots,ba-d^n)$$

is equal to $R(a_1, \ldots, a_d; b_1, \ldots, b_d)$ where for each i, $a_i = \max\{a_i^1, \ldots, a_i^n\}$ and $b_i = \min\{b_i^1, \ldots, b_i^n\}$.

3.8. (a) Suppose that $p_1, \ldots, p_n$ and $q_1, \ldots, q_n$ are points of the line $\mathbb{R}$, with

$$|p_i - p_j| \geq |q_i - q_j|.$$

For $0 \leq t \leq 1$, let $x_i(t) = (1-t)p_i + tq_i$. Then

$$|x_i(t) - x_j(t)| = |(1-t)(p_i - p_j) + t(q_i - q_j)|.$$

If it is true for all i and j that $|p_i - p_j| \geq |q_i - q_j|$ and the differences $p_i - p_j$ and $q_i - q_j$ are both nonnegative or both nonpositive, then

$$|x_i(t) - x_j(t)| = (1-t)|p_i - p_j| + t|q_i - q_j|$$

and each distance $|x_i(t) - x_j(t)|$ is monotone in t. In this case, the x_is describe a continuous shrinking of the p_is onto the q_is.

Now suppose, on the other hand, that for some i and j, one of the numbers $p_i - p_j$ and $q_i - q_j$ is positive and the other is negative. Then the pair (p_i, p_j) cannot be shrunk continuously onto the pair (q_i, q_j), because in any such motion the order of the points would eventually be reversed and hence by continuity there would be a time at which the image of p_i coincides with that of p_j.

(b) To see that $m_d \geq 2d+1$ when $d \geq 2$, consider the case in which $p_1, \ldots, p_d$ are the standard basis vectors of $\mathbb{R}^d$, $p_{d+j} = -p_j$ for $1 \leq j \leq d$, and p_{2d+1} is a large positive multiple of p_1. Let $q_i = p_i$ for $1 \leq i \leq 2d$, and $q_{2d+1} = -p_i$. Then each set of $2d$ of the p_is can be shrunk continuously onto the corresponding q_is, but it is not possible to shrink the entire set of $2d+1$ p_is continuously onto the corresponding q_is.

3.9. The proof due to Radon [Rad] is based on the following observation, which is known as Radon's theorem: *If $n \geq d+2$ and $p_1, \ldots, p_n$ are points of $\mathbb{R}^d$, then the set of indices $\{1, \ldots, n\}$ can be partitioned into two sets J and K such that the convex hull of $\{p_j : j \in J\}$ intersects the convex hull of $\{p_k : k \in K\}$.* To prove Radon's theorem, embed each of the points p_i in $\mathbb{R}^{d+1}$ by appending a $(d+1)$st coordinate equal to 1. The resulting set of more than $d+1$ points in $\mathbb{R}^{d+1}$ is linearly dependent and hence there are real numbers α_i, not all zero, such that $\sum_{i=1}^n \alpha_i p_i = 0$ and $\sum_{i=1}^n \alpha_i = 0$. Let

$$J = \{i : \alpha_i < 0\}, \qquad K = \{i : \alpha_i < 0\},$$

$$\sigma = -\sum_{i \in J} \alpha_i = \sum_{i \in K} \alpha_i > 0,$$

and

$$q = \frac{1}{\sigma} \sum_{j \in J} \alpha_j p_j = \frac{1}{\sigma} \sum_{k \in K} \alpha_k p_k.$$

Then q belongs to the convex hull of the p_js and also to the convex hull of the p_ks.

To prove Helly's theorem from Radon's theorem, use induction on the number of sets. Suppose that $C_1, \ldots, C_n$ are convex subsets of $\mathbb{R}^d$, with $n \geq d+2$, and that p_i belongs to each of these sets with the possible exception of C_i. If the sets J and K and the point q are as in Radon's theorem, then q belongs to all the sets C_i.

3.10. To prove Theorem 3.4 it suffices, in view of Helly's theorem, to deal with the case in which $n = d+1$. If $\bigcap_{i=1}^{d+1} B(p_i, \rho_i) \neq \emptyset$, then for each $\epsilon > 0$ the set $\bigcap_{i=1}^{d+1} B(p_i, \rho_i + \epsilon)$ contains a ball of radius ϵ and hence by Theorem 3.2 the volume of $\bigcap_{i=1}^{d+1} B(q_i, \rho_i + \epsilon)$ is positive. But then, of course, $\bigcap_{i=1}^{d+1} B(q_i, \rho_i) \neq \emptyset$.

3.11. (J. Pach) Let $\mathbb{R}^d$ be regarded in the usual way as a subspace of $\mathbb{R}^{n-1}$. For each $x \in \mathbb{R}^d$ and $\rho > 0$, let $C(c, \rho)$ denote the $(n-1)$-ball that is centered at x and has radius ρ. Under the hypotheses of Theorem 3.4, the volume ($(n-1)$-measure) of $\bigcap_{i=1}^{n} C(p_i, \rho_i + \epsilon)$ is positive for each $\epsilon > 0$, whence by Theorem 3.2 the same is true of $\bigcap_{i=1}^{n} C(q_i, \rho_i + \epsilon)$. But then $\bigcap_{i=1}^{n} C(q_i, \rho_i) \neq \emptyset$, and from this it follows readily that $\bigcap_{i=1}^{n} B(q_i, \rho_i) \neq \emptyset$.

3.12. The intersection $\bigcap_i B(x_i, \rho_i)$ contains a ball of radius ρ if and only if the intersection $\bigcap_i B(x_i, \rho_i - \rho)$ is nonempty. The set $\{x_1, \ldots, x_n\}$ is contained in a ball of radius ρ if and only if the intersection $\bigcap_i B(x_i, \rho)$ is nonempty.

3.13. Let $\rho_i = \|p - x_i\|$, whence $p \in \bigcap_i B(x_i, \rho_i)$. Then apply Theorem 3.5.

3.14. For the case in which $m = d$, repeated application of the result of Exercise 13 serves to extend the given f to a contraction that is defined on a dense subset of $\mathbb{R}^d$. Then appeal to uniform continuity to extend f to a contraction that carries all of $\mathbb{R}^d$ into $\mathbb{R}^d$. To treat the case in which $d < m$, assume without loss of generality that $\mathbb{R}^d$ is a subspace of $\mathbb{R}^m$, and then directly apply the result for $d = m$. For the case in which $d > m$, assume without loss of generality that $\mathbb{R}^m$ is a subspace of $\mathbb{R}^d$. Apply the result for $d = m$ to extend f to a contraction $g: \mathbb{R}^d \to \mathbb{R}^d$. The composition $h = p \circ g: \mathbb{R}^d \to \mathbb{R}^m$ is the desired extension of f, where p is the orthogonal projection of $\mathbb{R}^d$ onto $\mathbb{R}^m$.

4. UNIVERSAL COVERS

4.2. (a) The circle-sectors with center v and arcs A_v and B_v have central angle $\theta = \pi/n$ and radii $\delta + \rho$ and ρ, respectively. Each A_v sector contributes $\theta(\delta + \rho)$ to the perimeter of the body, and $\theta(\delta + \rho)^2/2$ to the area. Each B_v sector contributes $\theta\rho$ to the perimeter and $\theta\rho^2/2$ to the area. Hence the total perimeter is π and the total area is

$$n\frac{\theta\rho^2}{2} + n\left(\frac{\theta(\delta+\rho)^2}{2} - \beta\right) + \gamma$$

where γ is the area of the n-gon and β is the area of an isosceles triangle whose equal sides of length δ enclose an angle of θ.

(b) Let the vertices of the underlying n-gon be numbered $v_0, \ldots, v_{n-1}$ in order of clockwise traversal of the boundary. For each vertex v, let v' denote the vertex at distance δ from v that is encountered first in traversing the boundary from v, and let v'' denote the other vertex at distance δ from v. (Thus $v_i' = v_j$ and $v_i'' = v_{j+1}$, where $j \equiv i + (n-1)/2 \pmod n$.) Note that if $w = v''$ then $v = w'$. Define $w_0 = v_0$, $w_1 = v_0''$, $w_2 = w_1'', \ldots$. For each vertex v, let $S(v)$ and $T(v)$ denote the unit segments through v that lie in $R(n, \delta)$ and pass through v' and v'', respectively. Now starting with the segment $S(w_0)$, rotate it about w_0 to the position $T(w_0) = S(w_1)$. Then rotate it about w_1 to the position $T(w_1) = S(w_2)$. In the final rotation, the segment rotates about w_{n-1} from the position $S(w_{n-1})$ to the position $T(w_{n-1}) = S(w_0)$. Now it is back in its original position, but its ends are reversed.

4.3. (a) Let p and q be points of C whose distance is equal to C's diameter, and let L and M be lines that pass through p and q, respectively, and are perpendicular to the segment pq. From the fact that each of p and q is a point of C farthest from the other one, it follows that C lies in the strip bounded by L and M. But then, since C is of constant width 1, the distance between p and q is 1.

4.5. Assuming that H is a universal cover, reason as follows [Pál1]. Since the parallel cuts are at unit distance, each point that is interior to one of the corner triangles is at distance greater than 1 from each point in the opposite corner triangle. Thus whenever a set of unit diameter is placed in H, it misses at least three of the corner triangles. Now note that in each set of three vertices of H, there are two vertices that are neither adjacent nor opposite.

4.6. Each of the three congruent parts is a pentagon that has one of its vertices at the center of H.

4.7. For each positive integer k, the unit d-cube $[0,1]^d$ can be cut into k^d smaller cubes by dividing the underlying interval $[0,1]$ into k subintervals of length $1/k$. The diameter of each of the smaller cubes is equal to $\sqrt{d}/k$. Now take $k = \lfloor\sqrt{d}\rfloor + 1$.

5. FORMING A CONVEX POLYGON

5.4. For a set W of 5 points in general position in the plane, consider the convex 4-gon of smallest area among those with all four vertices in W.

5.5. Consider the function $\phi(i,j) = a_j - a_i$, defined whenever $1 \leq i < j \leq s+1$.

5.6. An especially nice solution appears on p. 502 of Lovász [Lov].

5.7. Use Ramsey's theorem, letting the d-simplices with vertices in X play the role of the triangles in the proof for $\mathbb{R}^2$. Divide the vertex-sets of the d-simplices into d classes, according to the residue class (mod d) of the number of other points of X in the simplex.

5.8. Divide the 3-sets $\{a,b,c\}$ in the plane into $2m$ classes, according to the values of the residues (mod 2) and (mod m) of the number of points of W that are interior to the triangle abc. Choose $p > n$ such that $p \equiv 2 \pmod{m}$, and let q denote the Ramsey number $N(3,p,2m)$. For W in general position with $|W| > q$, there is a p-set S in W all of whose 3-sets belong to the same class. As was noted in Part One, S is the vertex-set of a convex p-gon P. The polygon P can be dissected into $p-2$ triangles, and they all have (mod m) the same number of points of W. Since m divides $p-2$, the total number of points of W interior to P is divisible by m.

5.9. Use the result of Exercise 5.1.

5.10. As extended to allowable sequences, the conjecture that $g(6)$ is finite says that there exists an n for which the following condition is satisfied by all allowable n-sequences in which only substrings of length 2 are reversed.

There is a set S of six indices in $\{1,\ldots,n\}$ such that:

(i) for each $j \in S$ there is a permutation in the sequence in which j precedes all five indices in $S\backslash\{j\}$;

(ii) for each index $i \notin S$, there is no permutation in the sequence in which i precedes all members of S.

6. POINTS ON LINES

6.2. For the seven points of S, take the three vertices of a triangle, the centroid of the triangle, and the midpoints of the three edges.

6.3. With $n = |S|$, suppose that p lies on each ordinary line for S. Let U (respectively, V) denote the set of all points of S that lie on ordinary lines for S (respectively, $S \cup \{p\}$). Then $|U| \geq 2n/3$, for S admits at least $n/3$ ordinary lines and no point of S belongs to two such lines. Also $|V| \geq (n+1)/3$, for $S \cap \{p\}$ admits at least $(n+1)/3$ ordinary lines and no point of S belongs to two such lines. It is clear also that $U \cap V = \emptyset$, whence $|U \cup V| > n$, and this is a contradiction completing the proof. (This argument is due to Elliott [Ell].)

6.4. We use the inner-product notation $\langle\, , \rangle$ for vectors in $\mathbb{R}^d$, and we speak of spheres and hyperplanes rather than circles and lines. In this form, the result is valid for all d. Let us assume without loss of generality that C is the unit sphere $\{x\colon \langle x, x\rangle = 1\}$ and p is the origin. Then with $x' = \phi(x)$, we have $x = \langle x', x'\rangle^{-1}x'$. Attention is restricted here to the case in which the origin does not belong to the sphere or hyperplane that is to be inverted, for the remaining cases are similar but simpler.

With the aid of straightforward linear algebra, the following three statements about a subset S of $\mathbb{R}^d$ are seen to be equivalent:

(i) S is a sphere in $\mathbb{R}^d \setminus \{0\}$;

(ii) for some point q and some $r > 0$ with $r^2 \neq \langle q, q\rangle$,

$$S = \{x\colon \langle x - q, x - q\rangle = r^2\};$$

(iii) for some point v and some real α with $0 \neq \alpha < \frac{1}{4}\langle v, v\rangle$,

$$S = \{x\colon \langle x, x\rangle + \langle v, x\rangle + \alpha = 0\}.$$

Thus if (i) holds, then for appropriate α and v, the following statements are equivalent: $x' \in \phi S$;

$$\frac{\langle x', x'\rangle}{\langle x', x'\rangle^2} + \frac{\langle v', x'\rangle}{\langle x', x'\rangle} + \alpha = 0.$$
$$\langle c', x'\rangle + \langle \alpha^{-1} v, x'\rangle + \alpha^{-1} = 0.$$

Since

$$0 \neq \alpha^{-1} < \frac{1}{4}\langle v, v\rangle/\alpha^2 = \frac{1}{4}\langle \alpha^{-1} v, \alpha^{-1} v\rangle,$$

it follows that ϕS is a sphere that misses the origin.

Also, the following two statements are equivalent:

(i) H is a hyperplane that misses the origin;
(ii) for some point $v \neq 0$ and some $\alpha > 0$, $H = \{x\colon \langle x, v\rangle = \alpha\}$.

Thus for appropriate α and v, the following statements are equivalent:

$$x' \in \phi H;$$
$$\left\langle \frac{x'}{\langle x', x'\rangle}, v \right\rangle = \alpha;$$
$$\langle x', x'\rangle - \langle x', v\rangle/\alpha = 0;$$
$$\left\langle x' - \frac{1}{2\alpha} v, x' - \frac{1}{2\alpha} v \right\rangle = \frac{1}{4\alpha^2}\langle v, v\rangle.$$

The last equation is that of a sphere with center $\frac{1}{2\alpha} v$ and radius $\|\frac{1}{2\alpha} v\|$.

6.5. Choose $p \in S$, let C be a circle centered at p, and let ϕ denote the inversion in C. Since the set S is not collinear or concyclic, the set $\phi(S \setminus \{p\})$ is not collinear and hence, by Exercise 3, $\phi(S \setminus \{p\})$ admits an ordinary line L that misses p. But then $\phi(L) \cup \{p\}$ is a circle that contains p and exactly two other points of S. (This use of inversion is due to Motzkin [Mot].)

6.6. The theorem and the following proof are due to de Bruijn and Erdős [DE1], who also provide a complete description of the situations in which $m = n$. For simplicity, call the sets A_i *lines* and the members $x_1, \ldots, x_n$ of X *points.* Denote by p_i the number of points on the line A_i, and by l_j the number of lines on the point x_j. Observe that $p_i \leq l_j$ whenever $x_j \notin A_i$, because for each of the p_i points y of A_i there is a line $A(y)$ that contains $\{x_j, y\}$, and $y \neq y'$ implies $A(y) \neq A(y')$.

Now assume without loss of generality that $p_i \geq 2$ for all i, that $l_j \geq l_n$ for all j, and that $A_1 \ldots, A_t$ are the lines on x_n (thus $t = l_n$). Counting incidences in two ways shows that

$$\sum_{i=1}^{m} p_i = \sum_{j=1}^{n} l_j.$$

It is shown in the next paragraph that after some permissible shuffling of notation,

$$\sum_{i=1}^{t} p_i \leq \sum_{j=1}^{t} l_j,$$

whence, of course,

$$\sum_{i=t+1}^{m} p_i \geq \sum_{j=t+1}^{n} l_j.$$

But for $t < i \leq m$, it follows from the observation in the preceding paragraph that $p_i \leq l_n$, and hence (by the minimizing choice of l_n) $p_i \leq l_t$. From there, the desired inequality, $m \geq n$, follows from the fact that

$$\sum_{i=t+1}^{m} p_i \geq \sum_{j=t+1}^{n} l_j.$$

For $1 \leq i \leq t$, choose $x_{k(i)} \in A_i \setminus \{x_n\}$. When h and i are distinct indices between 1 and t, $x_{k(h)} \neq x_{k(i)}$, because equality here would imply $\{x_{k(h)}, x_n\} \subseteq A_h \cap A_i$. And, since $x_{k(h)} \neq x_{k(i)}$, $x_{k(h)} \notin A_i$, because otherwise $\{x_{k(h)}, x_{k(i)}\} \subseteq A_h \cap A_i$. Thus the notation may be chosen so that $k(i) = i$ for $1 \leq i \leq t$. Now with $x_1 \notin A_2$, we have $p_2 \leq l_1$. Similarly, $p_3 \leq l_2, \ldots, p_t \leq l_{t-1}, p_1 \leq l_t$. But then

$$\sum_{i=1}^{t} p_i \leq \sum_{j=1}^{t} l_j,$$

and the proof is complete.

(The inequality, $m \geq n$, can be extended to the situation in which there is a $\lambda > 1$ such that each pair of points appears in precisely λ of the sets. See the theory of incomplete balanced block designs, and especially Fisher's inequality on p. 99 of [Rys].)

6.7. For the desired set of five points, take the center and four vertices of a square. Now consider a noncollinear set $S = \{p_1, \ldots, p_5\}$, and suppose that S is neither a near-pencil nor in general position. Then there is a line L that contains precisely three points of S, and we may assume they are p_1, p_2, and p_3. Since at most one of these three points is collinear with p_4 and p_5, there are distinct indices i and j in $\{1, 2, 3\}$ such that the set $\{p_i, p_j, p_4, p_5\}$ is in general position.

6.8. Clearly $\lambda(4) = 6$, and$\lambda(5) = 6$ by Exercise 7. Now for each set S, let $\gamma(S)$ denote the number of connecting lines determined by S. With $n \geq 6$, consider a set S of n points that does not form a near-pencil, and let p and q be the points of an ordinary line L for S. If the set $S \setminus \{p\}$ is not a near-pencil, then $\gamma(S) \geq \lambda(n-1)+1$ because L is not the same as any line determined by two points of $S \setminus \{p\}$. Suppose, then, that $S \setminus \{p\}$ is a near-pencil, whence there is a line M that contains $n-2$ points of $S \setminus \{p\}$ and omits only a single point p' of $S \setminus \{p\}$. Note that $n - 2 \geq 4$. If $q \in M$, then $p \notin M$ and $S \setminus \{q\}$ is not a near-pencil because the line M contains at least three points of $S \setminus \{q\}$ but does not contain either p or p'; hence, again, $\gamma(S) \geq \lambda(n-1)+1$.

If $q \notin M$, then $\gamma(S) \geq 2(n-2)+2$ because in addition to L and M there are $n-2$ lines joining p to points of M and $n-2$ lines joining q to points of M. We have proved, therefore, that

$$\lambda(n) \geq \min\{\lambda(n-1)+1, 2n-4\}$$

for each $n \geq 6$. Since $\lambda(4) = \lambda(5) = 6$, it follows that $\lambda(n) > n$ for all $n \geq 4$.

6.11. The assertion is obvious when $n = d+1$. Suppose that the assertion fails for some choice of $d \geq 2$ and $n \geq d+2$. Consider the least such d and, with respect to this choice of d, the least such n. And consider an arbitrary finite set S of n points in $\mathbb{R}^d$ such that no hyperplane contains S. Let H be an ordinary hyperplane for S, so that there is a $(d-2)$-flat G in H and a point p of $H \setminus G$ such that $(S \cap H) \setminus \{p\} \subset G$. If the set $S \setminus \{p\}$ does not lie in a hyperplane, then the inductive hypothesis as applied to $(d, n-1)$ implies that $S \setminus \{p\}$ determines at least $n-1$ ordinary hyperplanes, none of which is H. If $S \setminus \{p\}$ does lie in a hyperplane J, then the inductive hypothesis as applied to $(d-1, n-1)$ implies that $S \setminus \{p\}$ determines at least $n-1$ ordinary $(d-2)$-flats in J. When extended to include p, each of these is an ordinary hyperplane for S in $\mathbb{R}^d$, and each is different from H.

6.13. Let u and v be the points of $S \cap J$ and $S \cap K$, respectively, that are closest to q, and consider the set of lines consisting of J, K, each line through u and a point of $S \cap K$, and each line through v and a point of $S \cap J$. The number of lines in this set is $k + l + 1$. Since every line in the set intersects every other line in the set, no two determine the same direction.

6.14. It follows from Exercise 6.8 that $p_i + p_j + 1 \leq d$ for all i, j. There are $r(r-1)/2$ pairs of indices (i, j) with $1 \leq i < j \leq r$, and each p_i appears in $r-1$

such pairs. Summing over all such pairs yields the inequality

$$(r-1)\sum_{i=1}^{r} p_i \le \frac{1}{2} r(r-1)(d-1).$$

But $\sum_{i=1}^{r} p_i = n-1$, whence

$$d \ge \frac{2(n-1)}{r} + 1 \ge \frac{2(n-1)}{d} + 1, \qquad d^2 - d - 2(n-1) \ge 0,$$

and

$$d \ge \frac{1}{2}(1+\sqrt{8n-7}) > \frac{\sqrt{n}}{2}.$$

7. TILING THE PLANE

7.1. The sum of the n interior angles is $(n-2)\pi$. If the angles are all of measure α then $(n-2)\pi = n\alpha$, whence $2\pi/\alpha = 2n/(n-2)$. Thus n must be 3, 4, or 6 if $2\pi/\alpha$ is to be an integer. Now suppose that the plane is tiled by congruent regular n-gons, and consider a vertex v of one of the tiles. If v is a vertex of each of the k tiles surrounding v, then $2\pi = k\alpha$, whence k is 3, 4, or 6 and n is, respectively, 6, 4, or 3. If, on the other hand, v belongs to a tile T of which v is not a vertex, and there are m other tiles incident to v, then v is an inner point of an edge of T and it follows that $\pi = m\alpha$. But $\alpha = \pi(n-2)/n$ and hence $n = m(n-2)$, implying that (n, m) is $(4, 2)$ or $(3, 3)$.

7.4. Think of the underlying rectangles as arranged in horizontal strips $S_j = \bigcup_{-\infty<i<\infty} R_{ij}$. Since $\alpha \ne \beta$, it is true that no matter how the diagonals are placed to produce the tiling T by triangles, every symmetry of T carries each strip onto a strip. Now think of marking each underlying rectangle with $+$ when the added diagonal has positive slope, $-$ when it has negative slope. When all marks are $+$, T is isohedral. When successive strips are marked

$$\ldots, \text{ all } +, \text{ all } -, \text{ all } -, \text{ all } +, \text{ all } -, \text{ all } -, \ldots$$

(repeating the pattern of a $+$ strip followed by two $-$ strips), T is not isohedral but it is periodic because its symmetry group includes both the horizontal translation by distance α and the vertical translation by distance 3β. When, for each j, the marking is

$$\cdots - - - - + + + - - + - - + + + - - - - \cdots,$$

(with $+$ for R_{0j}, $-$ for each of $R_{(-2)j}$, $R_{(-1)j}$, R_{1j}, R_{2j}, then three $+$s at each end, then four $-$s at each end, etc.), the symmetry group consists of vertical translations and reflections in horizontal lines. This can be modified to eliminate the translations. To obtain a trivial symmetry group, mark exactly one strip $\cdots + + + \cdots$ (all $+$s), one $\cdots + - + - + - \cdots$ (alternating $+$ and $-$), one $\cdots + + - - + + - - + + - - \cdots$ (alternating $++$ and $--$), etc.

7.5. See Figure 3.1.6 on page 119 of Grünbaum and Shephard [GS4]. By adjusting the sizes of the convex heptagons, the idea behind this tiling can be used to produce both $\mathcal{T}$ and $\mathcal{T}'$.

7.6. For a tiling of the requested sort, see Figure 3.4.1 on page 136 of [GS4]. (As reported by Grünbaum, experimental evidence in favor of the conjecture stated here consists of the fact that no tiling known to him contradicts the conjecture, and he and others have tried without success to produce such a tiling.)

7.7. (a) For a nice p-hexagon P that is really a pentagon, indexing can be chosen so that the successive vertices and edges are v_0, S_1, v_1, S_2, v_2, S_3, v_3, S_4, v_4, S_5, v_5 $(= v_0)$, with S_1 and S_4 parallel and of equal length. Now rotate the edge S_5 slightly about v_4, moving v_5 away from v_0, and fill the gap between (the new) v_5 and v_0 by a new edge S_6 joining v_5 to v_0. The resulting p-hexagon has six edges, and can be made arbitrarily close to P. The other cases can be handled similarly.

(b) Start with a regular pentagon P, whose interior angles are all $3\pi/5$. Truncate P at one vertex, introducing a new edge that is perpendicular to the angle bisector at that vertex. The result is a convex hexagon H that has four interior angles of $3\pi/5$ and two of $4\pi/5$. Since no combination of the terms $4\pi/5$, $4\pi/5$, $3\pi/5$, $3\pi/5$, $3\pi/5$, $3\pi/5$, π adds up to 2π, no point can be surrounded by nonoverlapping copies of H.

7.8. Let v_1, v_2, v_3, v_4 be the successive vertices of the quadrilateral Q, indexed so that the diagonal v_2v_4 is interior to Q. If Q is convex, the diagonal v_1v_3 is also interior to Q. Note that Q can be 2-dissected only by using v_2v_4 or, when Q is convex, using v_1v_3. Hence Q is 2-equidissectable if and only if either

(i) the vertices v_1 and v_3 are equidistant from the line determined by v_2v_4, or

(ii) Q is convex and the vertices v_2 and v_4 are equidistant from the line determined by $v_1 v_3$.

The cases of dissectability into two congruent triangles or two similar triangles are easy to analyze. Laczkovich [Lac2] for deeper results about similar triangles.

With the aid of the determinant formula used in the proof of Theorem 7.6, it follows that when an invertible linear transformation of the plane is represented in the usual way by an invertible matrix A, then for each triangle T it is true that the area of T's image is equal to the area of T times the absolute value of A's determinant. Hence the property of k-equidissectability is invariant for each k. However, it is easy to produce a convex quadrilateral Q and an invertible linear transformation L such that Q can be dissected into two congruent triangles but the image $L(Q)$ cannot be dissected into two similar triangles.

7.9. Suppose that, in a dissection of a convex n-gon P into q quadrilaterals, precisely k of the quadrilaterals are nonconvex. Each of those k quadrilaterals Q has exactly one vertex v at which Q's interior angle exceeds π. Since P is convex, in the dissection each such v is interior to P and acute interior angles of other quadrilaterals fit together to complete the task of surrounding v. Since the sum of P's interior angles is $(n - 1\pi$, and the sum of the interior angles of all the q quadrilaterals is $2\pi q$, it follows that $2\pi k + \pi(n-2) \leq 2\pi q$ and hence $2(q-k) \geq n-2 > 0$. (We are indebted to D. Gale for this argument.) It is also worth mentioning that when a polygonal region P is to be dissected into *convex* quadrilaterals, but P itself is permitted to have holes (and hence, of course, to be nonconvex), then it may be very difficult to decide whether any dissection of the desired sort exists (see Lubiw [Lub]).

7.11. The assertion is obvious for $n \leq 4$. With $n \geq 5$, suppose the assertion proved for all smaller values of n and consider a dissection of a convex n-gon P into k ($\leq n-2$) triangles. Let $\mathcal{I}$ denote the set of all pairs (T, L) such that T is a triangle in the dissection and L is a line that contains a side of T; then $|\mathcal{I}| = 3k$. Each of the lines that contains an edge of P is associated with at least one triangle, and each of the m additional lines that contains an edge of a triangle is associated with at least two triangles (one on either side of the line). Hence

$$n + 2m \leq |\mathcal{I}| = 3k \leq 3n - 6,$$

and it follows that $m \leq n - 3$. Now we claim that, in fact, $m = n - 3$ and the m lines arise from segments added in the stated manner. To prove this it suffices to establish the following.

(∗) At least one of the triangles has an edge vw that connects two nonadjacent vertices of P.

Suppose for the moment that (∗) has been established. Then the n-gon P is dissected by the segment vw into two convex polygons P' and P''. If these polygons have n' and n'' vertices, respectively, and are dissected (by the dissection of P) into k' and k'' triangles, respectively, then $n' + n'' = n + 2$ and $k' + k'' = k$. By the inductive hypothesis, $k' \geq n' - 2$ and $k'' \geq n'' - 2$, whence

$$n + 2 = n' + n'' \leq k' + k'' + 4 = k + 4 \leq n + 2.$$

It follows that $k' = n' - 2$, $k'' = n'' - 2$, and by the inductive hypothesis each of P' and P'' is dissected in the indicated manner. But then the same is true of P.

It remains only to prove (∗). If (∗) fails and each vertex of P lies on one or more of the m additional lines, it can be verified that no two of P's vertices lie in the same triangle of the dissection. But then $k \geq n$, contradicting the hypotheses. Hence there's a vertex y of P that is not on any of the m additional lines. If x and z are the vertices on either side of y, then one of the triangles of the dissection has a vertex at y, a second vertex on the segment yx, and a third vertex on the segment yz. If the second and third vertices are at the ends of their respective segments, then (∗) holds. Otherwise, discard the triangle xyz and denote by Q the part of P that remains. Then Q is an n-gon or an $(n+1)$-gon and we have a $(k-1)$-dissection of it, contradicting the inductive hypothesis.

7.12. When $k < n - 2$, no k-dissection is possible. When $k = n - 2$, try all ways of placing $n - 3$ segments in the indicated manner, thus obtaining an $(n-2)$-dissection. By using the determinant formula from the proof of Theorem 7.6, each of the $(n-2)$-dissections can be tested to see whether it is an equidissection. For a more sophisticated approach, see Dantzig, Hoffman, and Hu [DHH]. Considering the dissections of a convex n-gon P into $n - 2$ triangles, they assume that each admissible triangle (one whose vertices are vertices of P) has an associated cost, and they develop an algorithm for finding a dissection of minimum total cost. Hence we may proceed as follows:

(i) Produce any $(n-2)$-dissection, and use it to compute P's area.
(ii) Define the cost of any triangle T to be

$$\left|(\text{area of } T) - \frac{1}{n-2}(\text{area of } P)\right|.$$

(iii) Note that the $(n-2)$-equidissections of P are precisely the $(n-2)$-dissections whose total cost is zero.

7.13. For the first assertion, note that the entire arrangement is symmetric with respect to the sphere's center, and that for $k \geq 1$, no region includes two antipodal points. When a great circle is added to the arrangement, it cuts each region that it intersects into two parts, and because of the symmetry it intersects an even number of regions. For the second assertion, let v, e, and n denote, respectively, the numbers of vertices, edges, and regions in a subdivision of the sphere formed by k great circles in general position. Each circle C contains $2(k-1)$ points of intersection with the other circles. Hence C contains $2(k-1)$ edges, and $e = 2k(k-1)$. Each pair of circles produces two vertices, so $v = k(k-1)$.

Now use the fact that by Euler's theorem, $v - e + n = 2$.

8. PAINTING THE PLANE

8.1. First rule out the possibility that each point of X has precisely three friends in X. Then consider the case in which x_1 has at most two friends. Finally, show that if x_1 has four friends then they can be painted 1-chromatically in two colors, leaving the third color for x_1.

8.2. If x_1 has at most two friends, paint the set $\{x_2, \dots, x_6\}$ with three colors (possible by Exercise 1) and use for x_1 the color different from those used for x_1's friends. Next consider separately the case in which x_1 has precisely five friends and the case in which x_1 has precisely four friends. Finally, discuss the case in which each of the points x_i has precisely three friends.

8.4. This construction shows that $s_2 \leq 6$. Woodall [Woo] obtained the same conclusion by a different construction in which each of the six sets is closed.

8.5. If p/q and r/s are rationals in their lowest terms such that $(p/q)^2+(r/s)^2 = 1$, then exactly one of p and r is even, and the other, along with q and s, is odd. (This uses the characterization of Pythagorean triples stated in Exercise 13.2.) Now let us say that two points (a, b) and (c, d) of $\mathbb{Q}^2$ are *equivalent* if and only if the differences $a - c$ and $b - d$ both have odd denominators when written in lowest terms. Then any two points of $\mathbb{Q}^2$ whose distance is 1 are equivalent, and the various equivalence classes are translates of each other. Hence it suffices to color a single equivalence class in the required way. We do this for the class containing the origin by assigning one color to points of the form (odd/odd, odd/odd) or (even/odd, even/odd) and the other color to points of the form (odd/odd, even/odd) or (even/odd, odd/odd).

8.6. It suffices to do this for any particular value of δ. In constructing the configuration of $2d+3$ points, the choice of $\delta = \sqrt{2}$ is convenient because then d of P's points may be taken as the points $p_1, \ldots, p_d$ on the coordinate axes at distance 1 from the origin. Define two other points q_+ and q_- of P by the condition that

$$q_{\pm} = \frac{1 \pm \sqrt{d+1}}{d}(1, 1, \ldots, 1),$$

so that each of q_+ and q_- is at distance $\sqrt{2}$ from each of the points $p_1, \ldots, p_d$. To obtain the remaining $d+1$ points of P, "rotate" this configuration of $d+2$ points about q_+ so that the distance between q_- and its image is equal to $\sqrt{2}$.

8.7. The proof is a straightforward extension of the proof of Theorem 8.1, using the configuration of $2d+3$ points mentioned in Exercise 8.6 (see Woodall [Woo] for a different proof that $s_d \geq d+2$).

8.9. See p. 76 of [HDK] for the construction showing that no value of δ other than $\sqrt{3}$ has the stated property. To see that $\sqrt{3}$ has the property, assume without loss of generality that all three closed sets are nonempty and let p be a point that belongs to two of the sets. Let q and r be the other two vertices of an equilateral triangle that is inscribed in the circle and has p as one of its vertices. Then at least one of the three sets contains at least one of the pairs $\{p, q\}$, $\{q, r\}$, and $\{r, p\}$.

8.10. Let S denote the square of edge-length 3, and T the square of edge-length 1 that is obtained by contracting S toward its center. Let the sets covering S be $X_1, \ldots, X_6$, and for each $\rho > 0$ define the sets Y_i and their components as in the proof of Theorem 8.2. If each component is of diameter less than 1 then some point p of D belongs to three of the Y_is—say to $Y_1 \cap Y_2 \cap Y_3$. Let C denote the circle of radius 1 centered at p, and note that C is contained in S. If at least one of Y_1, Y_2, and Y_3 intersects C, then that Y_i includes two points at distance 1. If not, then C is covered by the remaining three Y_is, and the preceding exercise yields two points at distance $\sqrt{3}$ in one of these Y_is.

8.11. Once the lemmas are available, the proof that $u_d \geq 4d-2$ is a straightforward extension of the proof of Theorem 8.2.

8.12. Choose a basis $b_1, \ldots, b_d$ for M, and let each point x of M be represented by the sequence $(x_1, \ldots, x_d)$ of its coordinates with respect to this basis. For each $\tau > 0$, let Q_τ denote the parallelotope $\{x : 0 \leq x_i \leq \tau \text{ for all } i\}$. For $1 \leq i \leq d$, define the opposite facets F_i^- and F_i^+ of Q_τ in the natural way, and let η_τ denote the minimum distance that is realized between any pair of opposite facets. Now choose τ large enough to insure that $\eta_\tau > 1$, and then imitate the proof for Exercise 8.11 to show that $\overline{\chi}(M) \geq d + 1 + \overline{\chi}(S, 1)$. Having done that, complete the discussion by noting that when $d = 2$, it follows from the connectedness of S that $\overline{\chi}(S, 1) \geq 3$, and when $d \geq 3$ a slightly more complicated argument shows that $\overline{\chi}(S, 1) \geq 3$.

8.13. For each δ_i $(1 \leq i \leq k)$, the plane can be covered by a collection C_i of 7 sets, none of which includes two points at distance δ_i. For the desired covering by 7^k or fewer sets, take all sets of the form

$$X_1 \cap X_2 \cap \cdots \cap X_k \qquad \text{with } X_i \in C_i.$$

8.14. If the origin can be joined by paths of the specified sort to points x and y of $\mathbb{Q}^d$, then x can be joined also to $x + y$ and hence the origin can be joined to $x + y$. Thus it suffices to show that the origin can be joined to each point of $\mathbb{Q}^d$ in which all coordinates are zero except for a single coordinate $1/n$ in the ith position, where n is a nonzero integer. Since distances from the origin are unchanged by a permutation of coordinates, it suffices to consider the case $i = 1$. By Lagrange's four squares theorem, there are integers α, β, γ, and δ such that

$$4n^2 - 1 = \alpha^2 + \beta^2 + \gamma^2 + \delta^2$$

and hence

$$1 = 1\left(\frac{1}{2n}\right)^2 + \left(\frac{\alpha}{2n}\right)^2 + \left(\frac{\beta}{2n}\right)^2 + \left(\frac{\gamma}{2n}\right)^2 + \left(\frac{\delta}{2n}\right)^2.$$

This implies that the points

$$\left(\frac{1}{n}, 0, 0, 0, 0, 0, 0, \ldots, 0\right) \qquad \text{and} \qquad \left(\frac{1}{2n}, \frac{\alpha}{2n}, \frac{\beta}{2n}, \frac{\gamma}{2n}, \frac{\delta}{2n}, 0, 0, \ldots, 0\right).$$

are at unit distance from each other and the latter is at unit distance from the origin.

9. SQUARING THE CIRCLE

9.1. (a) The following diagram is a complete solution, since any triangle can be placed so that the altitude drawn from the top vertex lies entirely within the triangle.

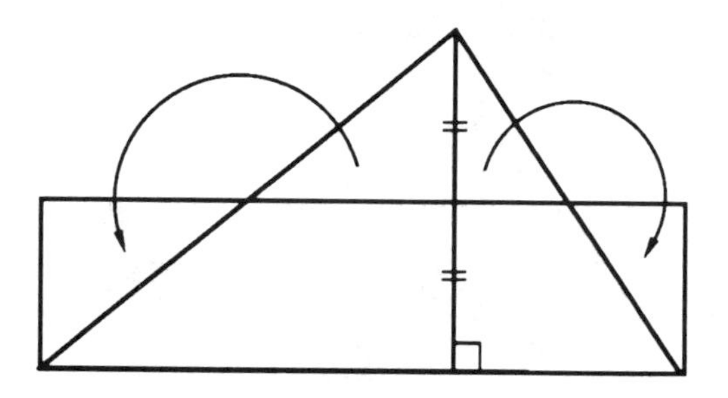

FIGURE 9.3
The transformation of a triangle into a rectangle.

9.2. You should end up with a five-piece decomposition as illustrated in Figure 9.4.

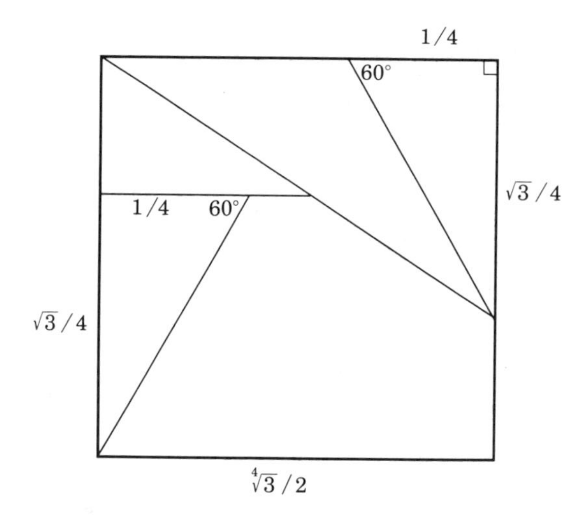

FIGURE 9.4
A dissection of a square into pieces that can be moved to form an equilateral triangle.

9.3. If T' is superimposed on T as shown in Figurc 9.5, then one need only cut along the two dotted lines.

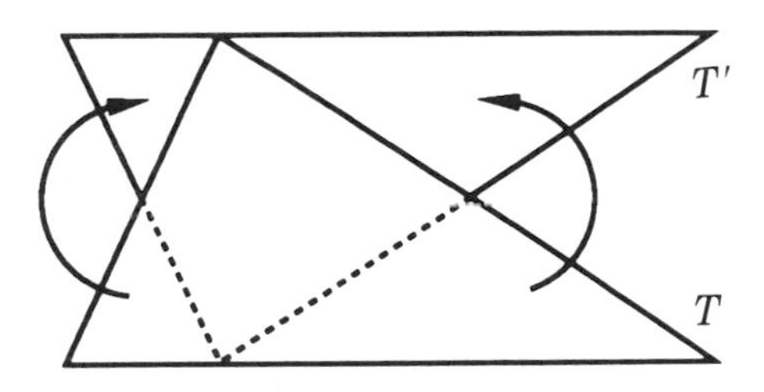

FIGURE 9.5
The transformation of T into its mirror image.

9.4. (a) Let D denote the disk, S the given segment outside D, and r the counterclockwise rotation through 1 radian about D's center. Then let A be a part of a radius of D that is congruent to S (it is important that A not contain D's center, which is easily arranged since S is shorter than D's radius); let $B = \{\rho^n(A) : n = 1, 2, 3, \ldots\}$; let $C = D \backslash (A \cup B)$. Then the sets A, B, and C partition D, while the sets S, $A \cup B$, and C partition $D \cup S$. Moreover, ρ maps $A \cup B$ onto B and a translation exists mapping S onto A. Using the identity on C completes the proof of equidecomposability of $D \cup S$ with D.

(b) First break up the line segments into finitely many pieces each of which is shorter than the disk's radius. Then use (a) repeatedly until all segments are absorbed into D.

9.5. (a) It suffices to show that the representation of each number in E as a polynomial in e^i is unique. But if some number had two distinct representations, subtraction would yield a polynomial relation satisfied by e^i. This would contradict the transcendence of e^i, which follows from Lindemann's theorem that e^b is transcendental whenever b is algebraic.

(b) Let ρ denote a one-radian counterclockwise rotation; ρ corresponds to multiplication of a complex number by e^i. Using the uniqueness of representations mentioned in the solution to part (a), it follows that $\rho(E) = A$. Similarly, if τ denotes the one-unit rightward translation, then $\tau(E) = B$.

10. APPROXIMATION BY RATIONAL SETS

10.4. If r and s are positive numbers and

$$a = r(s^2 - \cos^2\theta + 1), \qquad b = 2rs, \qquad c = r((s + \cos\theta)^2 - 1),$$

then a, b, and c are positive and $a^2 = b^2 + c^2 - 2bc\cos\theta$, whence a, b, and c are the side-lengths of a triangle with θ as one angle. Now given any triangle with side-lengths a, b, and c, solve the above system of equations to obtain positive real values for r and s in terms of a, b, and c. Approximate those closely by rational values r' and s', and then use the above equations to find rational a', b', and c' that are close to a, b, and c.

10.5. For positive real numbers w, x, y, and z, it is easy to see that $2(w^2 + x^2) = y^2 + z^2$ if and only if there exists a parallelogram whose sides are of lengths w and x and diagonals are of lengths y and z. Moreover, if $r > 1$, $s > 0$, $t > 1 + \sqrt{2}$, and

$$w = \frac{2rs}{r^2+1}, \qquad x = \frac{s(r^2-1)}{r^2+1}, \qquad y = \frac{s(t^2+2t-1)}{t^2+1}, \qquad z = \frac{s(t^2-2t-1)}{t^2+1}$$

then $w^2 + x^2 = s^2$ and $y^2 + z^2 = 2s^2$, so there is a parallelogram with sides w and x and diagonals y and z. Since the displayed system of equations can be solved for r, s, and t in terms of w, x, y, and z, the strategy here is similar to the one used in Exercise 4.

10.8. With $\mathbb{R}^d$ embedded as usual in $\mathbb{R}^{d+1}$, let S denote the sphere in $\mathbb{R}^d$ that is centered at the origin and has radius ρ. Let q (respectively, c) be the point of $\mathbb{R}^{d+1}$ whose first d coordinates are 0 and whose last coordinate is $\rho x/y$ (respectively, $-\rho(y^2 - x^2)/(2xy)$). Then the distance from q to any point of S is $\rho z/y$. And the set $S \cup \{q\}$ lies on the sphere in $\mathbb{R}^{d+1}$ that is centered at c and whose radius is equal to ρ times the quantity

$$\frac{x}{y} + \frac{y^2 - x^2}{2xy} = \frac{z^2}{2xy} = \left(1 + \left(\frac{y^2 - x^2}{2xy}\right)^2\right)^{1/2}$$

For the second part of the exercise, note that each positive rational number can be expressed in the form $\rho z^2/(2xy)$ for some choice of rational ρ and of integers x, y, and z as described.

11. INSCRIBED SQUARES

11.1. (a) Let Q be a concentric (and homothetic) square of edge-length a. Remove from K all the points interior to Q, and then add the two edges of Q that are not parallel to edges of K.

(b) use the fact that in polar coordinates, the equation of J_p is $r^p(|\cos\theta|^p + |\sin\theta|^p) = 1$, and (since $|\cos(\theta+\pi/2)| = |\sin\theta|$ and $|\sin(\theta+\pi/2)| = |\cos\theta|$) the curve J_p is invariant under rotation through the angle $\pi/2$.

11.2. If T is the linear transformation that carries (x, y) onto $(x/a, y/b)$, then T carries E onto the unit circle C and carries parallel lines onto parallel lines. Hence the parallelograms inscribed in E are carried onto parallelograms inscribed in C. The latter are in fact rectangles concentric with C, so it follows (by inverting T) that each parallelogram inscribed in E is concentric with E. Now consider the two diagonals of such a parallelogram and suppose that $(a\cos\sigma, b\sin\sigma)$ and $(a\cos\tau, b\sin\tau)$are the ends of these diagonals that lie in the upper half-plane. The condition for perpendicularity of the diagonals is

$$a^2\cos\sigma\cos\tau + b^2\sin\sigma\sin\tau = 0,$$

and with $\xi = \tan\sigma$ and $\eta = \tan\tau$ this yields $\xi\eta = -a^2/b^2$. The condition for equal length of the diagonals is

$$a^2\cos^2\sigma + b^2\sin^2\sigma = a^2\cos^2\tau + b^2\sin^2\tau,$$

which with the aid of standard trigonometric identities yields $\xi = -\eta$. Hence $\eta = a/b$, $\tau = \arctan a/b$, and the square's edge- length is $2a\cos\tau = 2ab/\sqrt{a^2+b^2}$.

11.3. For each point v of the circular arc qr, the body C lies in the strip between the lines L and M that are orthogonal to the segment pv and pass through p and v, respectively. This strip is of width 1. Verify that C also lies in another strip of width 1 whose bounding lines are orthogonal to L and M.

11.4. If J admits an inscribed regular n-gon P with $n \geq 7$, one of J's three circular arcs—say the arc qr—includes three vertices u, v, and w of P that are consecutive in the natural cyclic order on P's boundary. The perpendicular bisectors of the segments uv and vw intersect at p. This implies that the center of P's circumcircle is at p, and from this a contradiction can be derived. A special argument is needed for the case $n = 5$. (Note that C is of *constant width* in the sense discussed in Part Two of Section 4. Eggleston [Egg1] shows how to modify C to obtain a convex body of constant width whose boundary does not have an inscribed regular n-gon for any $n \geq 5$.)

11.5. Let the successive vertices of the quadrilateral Q be a, b, c, d, with $90°$ angles at a and c, an angle of $150° + \epsilon$ at b, and an angle of $30° - \epsilon$ at d. If all four vertices of Q lie on the boundary of an equilateral triangle E, then some side S of Q is contained in a side G of E. If S is dc, an immediate contradiction results, because if a is to lie on E's boundary, then B must be interior to E. Similarly, S cannot be the side da, and we may assume without loss of generality that S is cb. On the ray from c through b, we encounter first a vertex v of E and then the intersection w of this ray with the ray from d through a. The triangle awv has an interior angle of $60°$ at w and an exterior angle of $60°$ (an angle of E) at v. However, the exterior angle is the sum of the two nonadjacent interior angles, so a contradiction results.

11.7. For notational simplicity, let $b = 2$. Let S_1 denote the square

$$\{(\xi, \eta) \colon \max(|\xi|, |\eta|) = 1\},$$

of edge-length 2 and centered at the origin c_0. Let ϵ and ρ be real numbers with $0 < \epsilon < \rho < 1$. For each $n > 1$, let $S_n = c_n + \rho^n S_1$, where the center c_n of S_n is given by

$$c_n = (c_{n-1} + \rho^{n-1} + \epsilon^n + \rho^n, 0).$$

Then $S_1, S_2, \ldots$ is a sequence of pairwise disjoint homothets of S_1, marching steadily to the right and converging to the point

$$q = (1 + \sum_{n=1}^{\infty} \epsilon^n + 2\sum_{n=1}^{\infty} \rho^n, 0).$$

To produce a Jordan curve $J(\rho, \epsilon)$, add the limit point q and connect each S_n to S_{n+1} by an isthmus of width ϵ^n surrounding part of the x-axis. For each fixed ρ with $1/\sqrt{2} < \rho < 1$, it is true for all sufficiently small ϵ that $J(\rho, \epsilon)$ contains an inscribed square of each length between 0 and 2.

11.8. Let $v_1, \ldots, v_4$ be the vertices of a parallelogram inscribed in $C \cap P$. Then $v_1, \ldots, v_4, -v_1, \ldots, -v_4$ are the vertices of a parallelotope inscribed in C.

11.9. Use the fact that $f(x + \epsilon) = f(x) - \epsilon \sin^2(\pi/\epsilon)$.

11.11. Assertion (a) is Theorem 11.10. For (b), define a continuous function g on $[0, 1-\epsilon]$ by setting

$$g(x) = f(x+\epsilon) - f(x).$$

Then $g(0) \geq 0$ and $g(1-\epsilon) \leq 0$, so g attains the value 0. For (c), use the argument of Meyerson [Mey4], changing the scale (for notational simplicity) by replacing $[0, 1]$, δ, and ϵ by $[0, L]$, λ, and 1, respectively. With $\rho = 1 - \lambda$, we want to show that for each $f \in F_{[0,L]}$ there exists $x \in [0, L]$ such that

$$f(x-\lambda) = f(x+\rho).$$

Suppose the contrary. Then by continuity we may assume

$$f(x-\lambda) < f(x+\rho) \qquad \text{for all } x \in [0, L].$$

(The case $>$ is similar.) Since $f(x-\lambda) = 0$ for all $x \leq \lambda$, it follows that $f(x+\rho) > 0$ for all $x \in [0, \lambda]$, whence

$$f(x+1+\rho) > 0 \qquad \text{for all } x \in [0, \lambda].$$

By induction, $f(x+n+\rho) > 0$ whenever $x \in [0, \lambda]$ and $x+n+\rho \leq L$. Hence in each unit length subinterval of $[0, L]$, the set of points with $f(x) > 0$ has measure greater than λ. A similar argument shows that in each such subinterval, the set of points with $f(x) < 0$ has measure greater than ρ. Since $\lambda + \rho = 1$, that is a contradiction completing the proof.

12. FIXED POINTS

12.1. Suppose that f is a continuous mapping of $X \cup Y$ into itself, with $f(p) \in X$. Define the retraction $r : X \cup Y \to X$ by setting $r(x) = x$ for all $x \in X$, $r(y) = p$ for all $y \in Y$. Then consider the mapping $rf : X \to X$. Since X has the fixed-point property, there exists $q \in X$ such that $r(f(q)) = q$. It must be the case that $f(q) \in X$, for otherwise $f(q) \in Y$ and $rf(q) = p$, whence $p = q \in X$ and $f(q) = f(p) \in X$, a contradiction. From the fact that $f(q) \in X$ it follows that $r(f(q)) = f(q)$ and hence $f(q) = q$.

12.4. If there were a retraction r of U onto B, the composition of r with a rotation of B would be a fixed-point-free continuous mapping of U into itself. Conversely, if f were a fixed-point-free continuous mapping of U into itself, then a retraction r

of U onto B could be obtained as follows: for each point $p \in U$, form the ray from p through $f(p)$ and let $r(p)$ be the point at which this ray intersects B.

12.5. Let e denote the number of edges of the graph, and for each i let v_i denote the number of i-valent nodes. Then $2e$ is the number of ordered pairs consisting of an edge and an incident node. Count these pairs in a different way to see that

$$2e = v_1 + 2v_2 + 3v_3 + \cdots + 2kv_{2k} + (2k+1)v_{2k+1} + \cdots.$$

This is an even number and the sum

$$2(v_2 + v_3) + \cdots + 2k(v_{2k} + v_{2k+1}) + \cdots$$

is even. Hence the same is true of their difference

$$v_1 + v_3 + \cdots + v_{2k+1} + \cdots,$$

which is just the number of nodes of odd degree.

12.7. If a point y of triangle T belongs to the side whose ends are labeled i and j, then $\lambda_i(x) + \lambda_j(x) = 1$. Since

$$\lambda_i(f(y)) + \lambda_j(f(y)) + \lambda_k(f(y) = 1$$

and each of these numbers is nonnegative, it must be true that

$$\lambda_i(f(y)) \leq \lambda_i(y) \quad \text{or} \quad \lambda_j(f(y)) \leq \lambda_j(y).$$

12.8. Because of the symmetry of the construction, it suffices to consider the subtriangle with vertex-set $V(u, v, w)$, where u, v, and w are the vertices of T. The edges of this triangle join

$$u \text{ to } \frac{1}{2}u + \frac{1}{2}v, \quad u \text{ to } \frac{1}{3}u + \frac{1}{3}v + \frac{1}{3}w, \quad \text{and} \quad \frac{1}{2}u + \frac{1}{2}v \text{ to } \frac{1}{3}u + \frac{1}{3}v + \frac{1}{3}w,$$

and thus their lengths are, respectively,

$$\frac{1}{2}\|v - u\|, \quad \frac{1}{3}\|(w - u) + (v - u)\|, \quad \text{and} \quad \frac{1}{6}\|(w - u) + (w - v)\|.$$

12.9. For continuous $f : A \to A$ and for each $a \in A$, let $\xi(a)$ denote the x-coordinate of a and let $\delta(a) = xi(a) - \xi(f(a))$. The function δ is continuous, is nonnegative at the point p and nonpositive on the segment S. If δ vanishes at a point of $A \setminus S$, that is a fixed point of f. Otherwise δ is nonnegative on all of $A \setminus S$, hence also on S, and thus vanishes everywhere on S. But then f maps S into S and hence has a fixed point in S.

For each point p of C, let $\lambda(p)$ be the length of the arc from u to p. (Thus $\lambda(u) = 0$, $\lambda(v) = 2$, etc.) For a continuous mapping f of C into itself, let $\delta(p) = \lambda(p) - \lambda(f(p))$. Then $\delta(u) \leq 0$. If there is a point of C for which $\delta \geq 0$, then the function δ attains the value 0 and that yields a fixed point. If δ is everywhere negative, then f maps S into S and hence f has a fixed point in S.

12.10. Let K denote either X or its closure. Consider a continuous mapping f of K into K, and suppose that $f(0) \in S_i$. Let r denote the retraction of K onto S_i that takes all of $K \setminus S_i$ into 0. Then there exists $p \in S_i$ such that $rf(p) = p$. If $f(p) \in S_i$, then $rf(p) = f(p)$ and p is a fixed point of f. If $f(p) \notin S_i$ then $rf(p) = 0$, whence $p = 0$ and in fact $f(p) \in S_i$. A separate argument is needed when K is the closure of X and $f(0)$ belongs to the segment that is the limit of the S_is.

Comment. The space $X \times X$ lacks the fixed-point property [Kle2].

12.11. Arcs and the $\sin \frac{1}{x}$ arc are snakelike, hence also treelike. In addition, the continuum of Figure 12.1a is treelike.

HINTS AND SOLUTIONS: NUMBER THEORY

13. FERMAT'S LAST THEOREM

13.2. (b) If x and y are both even then 2 divides each of x, y, and z. If x and y are both odd, then z is even so 4 divides z^2. But

$$x^2 + y^2 = (2i+1)^2 + (2j+1)^2 = 4k + 2,$$

which is not divisible by 4.

(c) First show that

$$\frac{1}{2}(z+x) \qquad \text{and} \qquad \frac{1}{2}(z-x)$$

are relatively prime and that their product is a square. Then let r and s be such that

$$r^2 = \frac{1}{2}(z+x) \qquad \text{and} \qquad s^2 = \frac{1}{2}(z-x).$$

13.3. The hint yields that $x^n > (z-y)ny^{n-1}$. Therefore,

$$z - y < \frac{x^n}{ny^{n-1}} < \frac{x}{n},$$

whence

$$y + 1 \leq z < y + \frac{x}{n}.$$

This yields that $n < x$.

13.4. (a) Suppose not. Then there are primitive solutions (x_i, y_i, z_i), $i = 0, 1, 2, \ldots$, to Fermat's equation with exponent pm_i for some increasing sequence $\{m_i\}$ satisfying $m_{i+1} > m_i z_i$. But then $(x_i^{m_i}, y_i^{m_i}, z_i^{m_i})$ are primitive solutions to the equation with exponent p. Moreover, the condition on m_{i+1} implies that

$$z_{i+1}^{m_{i+1}} > z_{i+1}^{z_i m_i} > z_i^{m_i},$$

so the solutions are distinct, contradicting Faltings's theorem.

(b) It follows from (a) that there are integers $a, b, c, d, \ldots$ such that FLT is true for the exponents $3^a, 5^b, 7^c, 11^d, \ldots$.

13.5. The formula for the sum of a geometric series yields that

$$\left(1 - \frac{1}{p_i}\right) \leq \prod_{i=1}^{n} \left(\frac{1}{1 + \frac{1}{p_i} + \frac{1}{p_i^2} + \cdots + \frac{1}{p_i^n}} \right).$$

But this expression is bounded by the reciprocal of $\sum_{k=1}^{n} \frac{1}{k}$, since any $k \leq n$ is the product of prime powers $p_i^{a_i}$, with $a_i \leq n$. Sincc the harmonic series grows without bound, the reciprocal of this partial sum approaches 0 as n gets larger.

14. A PERFECT BOX

14.1. Suppose x, y, and z are edge-lengths for a perfect box, and are pairwise relatively prime. Since two odd squares cannot sum to a square (see solution to Exercise 13.2(b)), exactly one of each pair of edge-lengths is odd. Assume x is odd. Note that $y^2 + z^2$ is an odd square. It follows that $x^2 + y^2 + z^2$ is the sum of two odd squares and, therefore, cannot be a square as supposed.

14.2. Assume the lower-left corner of the square is at $(0, 0)$. Then the three equations are easily solved for the point $(45, 24)$.

14.3. Consider a unit segment PQ and let C be the circle of radius 1 centered at Q. Only countably many points have rational distance from each of P and Q (countably many pairs of circles intersecting in at most two points each), and for each such point E only countably many points on C have rational distance from E. Since a union of countably many countable sets is countable, the uncountable set C contains some point R for which no point P has rational distance from each of P, Q, and R. Since Q is at distance 1 from both P and R, $\triangle PQR$ is the desired triangle.

14.4. Following the idea of the hint yields that the triangle can be split into two right triangles, the first of which, by Exercise 13.2, has sides of the form

$$2rsc, \qquad (r^2 - s^2)c, \qquad (r^2 + s^2)c,$$

and the second of which has sides of the form

$$2mnd, \qquad (m^2 - n^2)d, \qquad (m^2 + n^2)d,$$

where one of r, s, and one of m, n is even. So the area of the given triangle is

$$c[rs(r^2 - s^2)] + d[mn(m^2 - n^2)].$$

Now, $rs(r^2 - s^2)$ must be divisible by 3 (if neither r nor s is divisible by 3, then $r^2 - s^2 \equiv (\pm 1)^2 - (\pm 1)^2 \equiv 0 \pmod 3$). Similarly $mn(m^2 - n^2)$ is divisible by 3. It follows that the area is divisible by 6.

14.5. Given $n > 0$, choose a rational-sided triangle whose vertices are within n of three of the given points. Then use Almering's result to get a fourth point within n of the fourth given point having rational distances from the three chosen points.

14.6. It is easy to compute, using the law of cosines, that the three medians of the given triangle have lengths 35, 97, and $4\sqrt{1373}$.

14.7. (a) Place two copies of the triangle adjacent to each other along AB, so that a parallelogram $ACBC'$ is formed. If the triangle is acute then $CC' > AB$ and we may lift C and C' simultaneously, thus rotating the two triangles, until the distance between C and C' equals c. Connecting the moved points C and C' completes the tetrahedron. One can analyze the cross sections of the tetrahedron oriented this way—they are similar parallelograms—to obtain the formula given in the hint to Exercise 3(b).

(b) Heron's formula applied to $\triangle ABC$ yields that the area is 13650. The formula in the hint yields a volume of 611520.

14.8. If a circle of rational radius r is parametrized by

$$\left(r\frac{1-t^2}{1+t^2}, r\frac{2t}{1+t^2}\right), \qquad t \in \mathbb{R} \cup \infty,$$

then rational values of t lead to points with both coordinates rational. This yields the first result of the hint, by using a circle whose diameter is a rational close to the given hypotenuse. Now, given $\triangle ABC$ let D be the foot of the perpendicular from A to BC, and let a rational right-triangle approximating $\triangle ADB$ be placed so that its right angle coincides with $\angle ADB$ and do the same for a rational right-triangle approximating $\triangle ADC$. These two approximating triangles may have different heights, but one of them can be scaled down by a rational so that the heights coincide. The resulting triangle is rational-sided and has rational height, and hence is the desired Heron triangle.

15. EGYPTIAN FRACTIONS

15.1. A sum of odd unit fractions, when combined into a single fraction, will have an odd denominator.

15.2. FALSE. On input $\frac{103}{165}$ the "odd" algorithm yields

$$\frac{1}{3} + \frac{1}{5} + \frac{1}{11}$$

while the ordinary algorithm yields

$$\frac{1}{2} + \frac{1}{9} + \frac{1}{77} + \frac{1}{6930}.$$

A smaller example: $\frac{16}{55}$.

15.3. (a) The greedy algorithm yields

$$\frac{2}{2k+1} = \frac{1}{k+1} + \frac{1}{(k+1)(2k+1)}.$$

(b) $\dfrac{3}{2k} = \dfrac{1}{2k} + \dfrac{1}{k}$.

(c) If $p = 6k - 1$ is the hypothesized prime, then

$$\frac{3}{dp} = \frac{1}{2kd} + \frac{1}{2kd(6k-1)}.$$

(d) If

$$\frac{3}{b} = \frac{1}{m} + \frac{1}{n},$$

let $d = \gcd(m, n)$. Then

$$\frac{3d}{b} = \frac{1}{s} + \frac{1}{t} = \frac{s+t}{st},$$

where s and t are relatively prime. Because b is odd, both s and t must be odd. And since $s + t \equiv 0 \pmod 3$, one of s, t is congruent to $+1 \pmod 3$ and the other to $-1 \pmod 3$. Suppose $s \equiv -1 \pmod 3$. Then s is divisible by some prime $p \equiv -1 \pmod 6$ (all odd primes are congruent to $\pm 1 \pmod 6$). Now, we have that for some integer r, $s + t = 3dr$ and $st = br$. Since p divides s, p must divide b (as desired) or p must divide r. But if p divides r, then p divides both s and $3dr$, whence p divides t, contradicting the fact that s and t are relatively prime.

15.4. If $\frac{a}{b} > 1$ the greedy algorithm will proceed by choosing as large an initial segment of the harmonic series as will fit. If $\frac{1}{m}$ is the first term that doesn't fit, then $m \geq 2$ and the remainder after the first $m-1$ terms have been subtracted from $\frac{a}{b}$ is less than $\frac{1}{m}$, and hence less than 1. By the case discussed in the text, the greedy algorithm when applied to this remainder will terminate, and it will necessarily use unit fractions less than $\frac{1}{m}$. This means the algorithm applied to $\frac{a}{b}$ terminates.

15.5. Assume that the fraction is less than 1. The other case can then be handled by considering the harmonic series as in Exercise 4 (more precisely, use the divergent subseries of the reciprocals of the even integers). Now consider first the case of an even denominator, $\frac{a}{2k}$. Then if the greedy algorithm first produces $\frac{1}{2m}$, the remainder will be

$$\frac{2ma - 2k}{4mk} = \frac{ma - k}{2mk}.$$

The fact that $2m$ rather than $2(m-1)$ was chosen means that this last numerator is less than a and, as for the unrestricted greedy algorithm, this suffices. This proof never uses the assumption that the numerator and denominator of the given fraction are relatively prime. So given a fraction with an odd denominator we may multiply numerator and denominator by 2 and use the preceding case.

15.6. By induction on $i = 2, 3, \ldots, k$. If $m_2 \leq d$, then

$$\frac{1}{m_1} + \frac{1}{m_1(m_1 - 1)} \leq \frac{a}{b},$$

which implies that

$$\frac{1}{(m_1 - 1)} \leq \frac{a}{b},$$

contradicting the choice of m_1. Now, assume inductively that

$$m_i > d^{2^{i-2}},$$

and let $R = \frac{a}{b} - \sum_{j=1}^{i-1} \frac{1}{m_j}$. Then

$$r - \frac{1}{m_i} < \frac{1}{m_i - 1} - \frac{1}{m_i} = \frac{1}{m_i(m_i - 1)} < \frac{1}{(m_i - 1)^2} \leq \frac{1}{\left(d^{2^{i-2}}\right)^2} = \frac{1}{d^{2^{i-1}}},$$

which implies that $m_{i+1} > d^{2^{i-1}}$.

15.7. (a) $\frac{2}{11} = \frac{1}{6} + \frac{1}{66}$ and $\frac{2}{11} = \frac{1}{11} + \frac{1}{22} + \frac{1}{33} + \frac{1}{66}$.

Multiplying these and adding $\frac{1}{121}$ yields

$$\frac{5}{121} = \frac{1}{121} + \frac{1}{66} + \frac{1}{132} + \frac{1}{198} + \frac{1}{396} + \frac{1}{726} + \frac{1}{1452} + \frac{1}{2178} + \frac{1}{4356}.$$

(b) $\frac{5}{11} = \frac{1}{2} + \frac{1}{9} + \frac{1}{99}$, so $\frac{5}{121} = \frac{1}{33} + \frac{1}{99} + \frac{1}{1089}$.

(c) Take $n = 3$ and consider the divisors 1, 3, and 11 of 363, which add up to $15 = 3 \cdot 5$. This yields that

$$\frac{5}{121} = \frac{5 \cdot 3}{121 \cdot 3} = \frac{1}{33} + \frac{1}{121} + \frac{1}{363}.$$

15.8. Because the series of square reciprocals adds up to $\pi^2/6$, the conclusion follows by considering whether or not the unit fraction $1/1$ is used.

16. PERFECT NUMBERS

16.1. If $n = ab$ then $2^a - 1$ divides $2^n - 1$ since the latter equals

$$(2^a - 1)(1 + 2^a + 2^{2a} + \cdots + 2^{(b-1)a}).$$

16.2. The two inequalities follow from the fact that the divisors of ab all have the form $d_1 d_2$ where d_1 divides a and d_2 divides b.

16.3. $220 = 4 \cdot 5 \cdot 11$ so

$$\sigma(220) = \sigma(4)\sigma(5)\sigma(11) = (1 + 2 + 4)(1 + 5)(1 + 11) = 504.$$

The factorization of 284 is $4 \cdot 71$ so

$$\sigma(284) = 7 \cdot 72 = 504.$$

Since

$$\sigma(220) = \sigma(284) = 220 + 284,$$

these two numbers form an amicable pair. The other pair is handled similarly.

16.4. (a) Dividing $2N$ by $22022 = 2 \cdot 7 \cdot 11^2 \cdot 13$ and by $1+3+9 = 13$ reduces it to 1387323 which equals the product of the remaining factors on the left side of (1).

(b) The computation of $\sigma(N)$ uses the fact that the numbers 22021, 3, 7, 11, and 13 are all prime. But $22021 = 192 \cdot 61$.

16.5. If some n divides all the products, then any prime divisor of n does so as well; choose one and call it p. Now, for any integer m making each of the given s functions prime, one of the functions must equal p (since p divides their product). But p cannot equal $a_i + b_i m$ for two values of m. It follows that there are at most r choices of m for which all the functions are prime.

16.6. The given arithmetic progression has 12 prime values. The common difference has the form $2 \cdot 3 \cdot 5 \cdot 7 \cdot 11 \cdot 13 \cdot 17 \cdot 19$, and it is easy to see that any arithmetic progression yielding 12 consecutive prime values must have a common difference that is divisible by $2 \cdot 3 \cdot 5 \cdot 7 \cdot 11$.

16.7. (Sylvester) Suppose $n = p^a q^b$ is perfect (the case of a single prime power is easy). Then

$$\begin{aligned} 2 &= \frac{\sigma(p^a q^b)}{p^a q^b} \\ &= \frac{\sigma(p^a)\sigma(q^b)}{p^a q^b} \\ &= \left(1 + \frac{1}{p} + \frac{1}{p^2} + \cdots + \frac{1}{p^a}\right)\left(1 + \frac{1}{q} + \frac{1}{q^2} + \cdots + \frac{1}{q^b}\right), \end{aligned}$$

using the fact that on prime powers σ is just the sum of the smaller powers of the same prime. Now, since p and q must be odd primes, we may assume that $p \geq 3$ and $q \geq 5$. This implies that the last product is less than the product of

$$1 + \frac{1}{3} + \frac{1}{9} + \frac{1}{27} + \cdots \qquad \text{with} \qquad 1 + \frac{1}{5} + \frac{1}{25} + \frac{1}{125} + \cdots.$$

But this last product is less than $\frac{3}{2} \cdot \frac{5}{4}$ which is less than 2, a contradiction.

17. THE RIEMANN HYPOTHESIS

17.1. It suffices to consider the quotient of $\mathrm{li}(x) - \mathrm{li}(2)$ with $x/\log x$; because $\mathrm{li}(x) - \mathrm{li}(2)$ is a definite integral from 2 to x, we can apply l'Hôpital's rule and the fundamental theorem of calculus. This yields

$$\frac{\log x}{\log x - 1},$$

whose limit is 1 as $x \to \infty$.

17.2. The left-hand side equals

$$\sum \frac{1}{n^s} - 2 \sum \frac{1}{(2n)^s},$$

which equals the right-hand side.

17.3. Since the error is less than the first omitted term, one wants n large enough that $1/\sqrt{n}$ divided by the absolute value of $(1 - 2 \cdot 2^{-s}) = -0.414\ldots$ is less than 0.05. It suffices to take $n \geq 2{,}334$. (There are much more efficient ways to compute $\zeta(s)$!)

17.4. If the limit did not equal 0, then $\pi(x) \cdot \log(x)/x$ would approach ∞, contradicting the prime number theorem which says that this fraction has the limit 1.

18. PRIME FACTORIZATION

18.1. $2^{560} \equiv (2^{16})^{35} \equiv 1 \pmod{17}$, by Fermat's little theorem. The other two cases are similar. Now, if two numbers are congruent modulo relatively prime integers, they are congruent modulo their product.

18.2. If m's base two representation is, for example, 1100101...001, then the algorithm forms, in succession, the numbers $b^{11}, b^{11}, b^{110}, b^{1100}, \ldots, b^{1100101\ldots001}$.

18.3. The hypothesis implies that p divides $a^2 - 1$, which equals $(a+1)(a-1)$. But if p, which is prime, divides a product of two numbers, p must divide one of the two numbers, so p divides one of $a+1$, $a-1$, as required.

18.4. Suppose $a \equiv r^2 \pmod{p}$. Then $a^{(p-1)/2} \equiv r^{p-1} \equiv 1 \pmod{p}$ by Fermat's little theorem. The hypothesis on p implies that $(p+1)/4$ is an integer, so we have

$$(\pm a^{(p+1)/4})^2 = a^{(p+1)/2} = a \cdot a^{(p-1)/2} \equiv a \pmod{p}.$$

Since, by the method of Exercise 3, a has at most two square roots mod p, and since a nonzero number and its negative are unequal mod p (this uses the fact that p is odd), the two square roots (one, in the case that $a \equiv 0$) have been found.

18.5. If b denotes one of the proposed square roots, then $b \equiv \pm r \pmod{p}$ and $b \equiv \pm s \pmod{q}$. Therefore $b^2 \equiv a$ modulo each of p and q, whence $b^2 \equiv a \pmod{pq}$. These four are the only square roots for if $h^2 \equiv a \pmod{pq}$, then $h \equiv \pm r \pmod{p}$ and $h \equiv \pm s \pmod{q}$. It follows that h is congruent to one of the four choices for b modulo both p and q, and hence modulo N.

18.7. First note that the oracle yields a one-step primality test; simply ask whether $(N, N) \in A$. For the rest, it suffices to produce a single nontrivial divisor of a composite number N. First ask whether $(N, \lfloor n/2 \rfloor) \in A$. If it is, then ask whether $(N, \lfloor N/4 \rfloor) \in A$; if it isn't, ask whether $(n, \lfloor 3N/4 \rfloor) \in A$. Continuing this interval-halving search will produce a prime divisor in $O(\log N)$ steps.

19. THE 3n +1 PROBLEM

19.1. Assume all integers greater than 1 have finite stopping time (the other direction is clear). Then the trajectory of 2 ends in the 1-4-2 loop so, by induction, the same is true for every larger starting value.

19.2. After 111 iterations of f, 27 is transformed to 1, entering Figure 19.1 along the branch $\ldots 92, 46, 23, 70, \ldots$.

19.3. There is the nontrivial loop: 13, 66, 33, 166, 83, 416, 208, 104, 52, 26.

19.4. It has been conjectured that all negative trajectories end in one of the loops:

$$(-1, -2), \qquad (-5, -14, -7, -20, -10),$$

or

$$(-17, -50, -25, -74, -37, -110, -55, -164, -82,$$

$$-41, -122, -61, -182, -91, -272, -136, -68, -34).$$

19.5 (a) The case of $32k+11$ is typical. Its trajectory begins with $32k+11$, $48k+17$, $36k+13$, $27k+10$, whence the stopping time is 4.

(b)
$$203 = 64 \cdot 1 + 16 \cdot 2 + 8 \cdot 3 + 8 \cdot 3 + 2 \cdot 4 + 2 \cdot 4 + 2 \cdot 4 + 1 \cdot 5 + 1 \cdot 5 + 1 \cdot 5 + 1 \cdot 5 + 1 \cdot 5 + 1 \cdot 5 + 1 \cdot 5.$$

19.6. (a) $r_4 = 3$ and $r_4 = 7$.

(b) The first binomial coefficient is the total number of sequences satisfying (a) and (b): $i - 1$ $\frac{3}{2}$s may be placed in any position but the first and last. From this overestimate must be subtracted the number of sequences that fail to satisfy (c). Each such has a largest proper initial segment (with k $\frac{3}{2}$s, say) that satisfies (b) and (c), and the number of ways of extending such a segment to one with i $\frac{3}{2}$s by adjoining the appropriate number of $\frac{3}{2}$s and $\frac{1}{2}$s is the coefficient of r_k on the right.

20. DIOPHANTINE EQUATIONS AND COMPUTERS

20.1. What is needed is a systematic way of checking every possible pair of integers and plugging them into P to see if they form a solution. One approach is to first try $(0, 0)$, then try all pairs where the largest number has absolute value 1, then all untried pairs with largest absolute value 2, and so on. Each of these groups is finite, so infinite loops are avoided. Another approach is to go through the positive integers in order. If a number m is divisible by something other than 2 and 3, ignore it. Otherwise, write m as $2^a 3^b$ and substitute each of the four pairs of the form $(\pm a, \pm b)$ for x and y in P.

20.2. (a) Given $P(x, y, \ldots)$, let $Q(x, y, \ldots)$ be $P(x-1, y-1, \ldots)$.
(b) Given $P(x, y, \ldots)$, let $Q(x, y, \ldots)$ be $P(x+1, y+1, \ldots)$.

20.3. Given $P(x, y, \ldots)$, let Q equal $P(x' - x'', y' - y'', \ldots)$. Then $P = 0$ is solvable in $\mathbb{Z}$ if and only if $Q = 0$ is solvable in $\mathbb{N}$.

20.4. First show that for any Fibonacci number F there exists an integer x such that $(F^2 - Fx - x^2)^2 = 1$. Now, $F = p(x, F)$, where p is the polynomial

$$y(2 - (y^2 - yx - x^2)^2).$$

20.5 The sequence starts with 4, 26, 41, 60, 83, 109. After a gigantic number of steps, it starts decreasing, and x_n is first 0 when $n = d \cdot 2^d$ where $d = 3 \times 2^{27}$. This value of n is greater than $10^{120,000,000}$.

20.6. First check that P is not just a constant; if it is, output "YES" only if the constant is 0. Now, a nonconstant polynomial approaches ∞ as x approaches ∞, so there is a number H (which can be determined explicitly in terms of the coefficients) such that if $|x| > H$ then $|P(x)| > 1$. It follows that any integer solution can be found by checking the finitely many integers in $[-H, H]$.

20.7. If P is a homogeneous polynomial with variables $x_1, \ldots, x_n$, let Q_i be the polynomial obtained from P by replacing each occurrence of x_i with 1. Then let Q be the product of the n polynomials Q_i. It is not hard to see that P has a nontrivial solution in $\mathbb{Z}$ if and only if Q has a solution in $\mathbb{Q}$.

HINTS AND SOLUTIONS: INTERESTING REAL NUMBERS

21. PATTERNS IN PI

21.1. There are 66 0s and 177 1s.

21.2. $\log(\frac{1}{18})^{800{,}001} < -1{,}000{,}002$. 284,757 terms for $\arctan \frac{1}{57}$. 210,225 terms for $\arctan \frac{1}{239}$.

21.3. One would want the first omitted term to be less than 10^{-11}. This requires n to be about 50 billion.

21.4. Hint: Observe that $c_{n-1}^2 = 4(a_n^2 - b_n^2)$.

21.5. Hint: It takes $8 \times 19 = 152$ operations to get a_{19}, b_{19}, d_{19}, and

$$1 - \sum_{j=0}^{19} 2^{j+1} d_j.$$

Then, using the relation $4a_{20} = (a_{19} + b_{19})^2$, one addition, one multiplication, and one division yield π_{19}.

22. CONNECTIONS BETWEEN π AND e

22.1. (a) If both were rational then their sum would be too, contradicting the irrationality of π. The same argument shows that at least one of $\pi \pm e$ is transcendental.

(b) If $\pi e = q$ and $\pi + e = r$ where q and r are rational, then $\pi(r - \pi) - q = 0$, contradicting π's transcendence. Exercise 1(c) is similar.

22.2. (a) If this happened, then, for some integer k and rational q, π would equal $q + (e/10^k)$. This means $10^k\pi - e - 10^k q = 0$ which, after multiplication by the denominator of q, becomes an algebraic relation between π and e.

(b) Such a possibility implies that the decimal expansion of π (and of e) is periodic, in contradiction to π's (or e's) irrationality.

22.3. If a is algebraic then so is a^m, since the product of algebraics is algebraics. But if $p(x)$ is a polynomial that has a^m as a root, then $p(x^n)$ is a polynomial having $a^{m/n}$ as a root.

22.4. If a and b are algebraic then so is bi and hence also $a+bi$. Conversely, if $p(x)$ is a polynomial with integer coefficients having $a+bi$ as a root, then $p(a-bi)=0$ too. To see this assume, for example, that $p(x)$ has degree three; multiply everything out and compare real and imaginary parts. Now, if $a-bi$ is algebraic then the sum and difference of $a+bi$ and $a-bi$ are algebraic, yielding that a and b are both algebraic.

22.5. The hypothesis implies that $\log_m n$ is irrational. Since $m^{\log_m n} = n$, Theorem 22.1 implies that $\log_m n$ cannot be algebraic.

22.6. Since $\sqrt{163}i$ is algebraic, this follows from Theorem 22.1 in exactly the same way that e^π's transcendence was proved.

23. COMPUTING ALGEBRAIC NUMBERS

23.1. In all cases the runs of 0s get arbitrarily long. A rational number has an eventually periodic decimal expansion (unless the digits are all 0 beyond a certain point) and so all runs of 0s have length bounded by the longest run occurring prior to the end of the first repeating section.

23.2. x_2 can be handled similarly to Liouville's number. Using an auxiliary row under the main row one can proceed in a way that requires the writing of one auxiliary digit for each of x_2's digits. x_3 is similar since the string of zeros increases by two each time. Write 100 and then write a 0 corresponding to each 0 of the previous run of zeros; this uses fewer than $2n$ steps for n digits. x_5 is similar. For x_4, write the digits an integer at a time. Get the next one by placing a 1 under the units digits of the previous integer and performing the addition. The 1 at the preceding step will serve to mark the left end of the integer. Then the sum can be copied and the process continued. Each integer making up x_4 is written twice, and there are the auxiliary 1s, but in all fewer than $3n$ digits are used to get n digits of x_4.

23.3. Let m/n be as in the hint, i.e,

$$\frac{m}{n} = 10^{-1!} + 10^{-2!} + 10^{-3!} + 10^{-4!} + \cdots + 10^{-s!}.$$

Then $n = 10^{s!}$ and the difference between x_1 and m/n is $10^{-(s+1)} + 10^{-(s+2)} + \cdots$, which is bounded by a geometric series and so is less than $2 \times 10^{-(s+1)!}$. But this last is less than $1/n^d$ because $s > d$.

23.4. One way to prove that x_6 is real-time computable is to begin by writing a 1. Now, to generate more digits look at the first "unread" digit and use the following rules to write two more digits: $0 \to 01$ and $1 \to 10$. Then read the next digit, use it to generate two more digits, and so on. The first unread digit is a 1, so the main sequence becomes 110. Now the next unread digit is 1 so the main sequence becomes 11010, and so on. Since no auxiliary digits are written (except in the table), this shows x_6 is real-time computable.

To prove the irrationality of x_6 observe first that if $k < 2^{m-1}$ then the digits in the positions corresponding to $2^m + 2k$ and $2^m + 2k + 1$ are 1, 0 or 0, 1 according as the value of the $(2^m + k)$th position is 0 or 1. This follows from the definition using the parity of the nunber of 1s in the base2 representation of the position number. Now, this implies nonperiodicity, for suppose the sequence eventually had a repeating part of length k. Choose m so that $2^{m-1} > k$. Then the value of the $(2^m + 2k - 1)$th position equals the value of the $(2^m + k - 1)$th position, which contradicts the rules deduced above.

23.5. (a) $x - (x^2 - 2)/2x = (x^2 + 2)/2x = \frac{1}{2}(x + 2/x)$.

(b) If x_n is the x-coordinate of the current point on the graph, then the equation of the tangent line is

$$y - (x_{n-2}) = 2x_n(x - x_n).$$

The x-intercept of this line (set $y = 0$) agrees with the formula of Newton's method.

24. SUMMING RECIPROCALS OF POWERS

24.1. The error is less than

$$\int_k^\infty \frac{1}{x^n}\, dx = \frac{1}{k^{n-1}(n-1)}.$$

24.2. (a) By Exercise 1,

$$1 < \zeta(n) < \frac{1}{1^{n-1}(n-1)},$$

which implies $\zeta(n) \to 1$.

(b) Use the result of Exercise 1 with $k = 2$ to get upper bounds on $\zeta(n+1) - 1 - 1/2^{n+1}$ and on $\zeta(n) - 1 - 1/2^n$. Use these to bound the quotient of the exercise between expressions that approach $1/2$.

24.3. Since π is transcendental, any power of π is irrational.

24.4. Instead of repeatedly taking derivatives (which is complicated and requires l'Hôpital's rule for evaluation at $x = 0$) just take the series for $\sin x$ and divide through by x. This yields the series

$$1 - \frac{x^2}{3!} + \frac{x^4}{5!} - \frac{x^6}{7!} + \cdots.$$

24.5. (a) This is a consequence of the well-known inequality: $\sin x < x < \tan x$ for $0 < x < \pi/2$.

(b) This follows from (a) by straightforward substitution.

(c) Part (b) implies that

$$\sum_{k=1}^{m} \frac{1}{k^2}$$

is trapped between two quantities, each of which converges to $\pi^2/6$ as $m \to \infty$.

24.6. The first 14 Bernoulli numbers are:

$$1, -\frac{1}{2}, \frac{1}{6}, 0, -\frac{1}{30}, 0, \frac{1}{42}, 0, -\frac{1}{30}, 0, \frac{5}{65}, 0, -\frac{691}{2730}, 0.$$

GLOSSARY

Affine combination. An affine combination is a linear combination in which the sum of the coefficients is 1.

Affinely equivalent. Two sets are affinely equivalent if there is a one-to-one affine transformation that carries one set onto the other.

Affinely independent. A set is affinely indpendent if none of its points is an affine combination of the remaining points.

Affine transformation. An affine transformation is a transformation T that preserves affine combinations. That is, whenever $p_1, \ldots, p_n$ are points of T's domain, and $\lambda_1, \ldots, \lambda_n$ are real numbers whose sum is 1, it is true that

$$T\left(\sum_1^n \lambda_i p_i\right) = \sum_1^n \lambda_i T(p_i).$$

Equivalently, an affine transformation consists of a linear transformation followed by a translation.

Algebraic number. A real or complex number that is a root of some polynomial with integer coefficients. For example, $\sqrt{2}$ is a root of $x^2 - 2$ and $i\ (= \sqrt{-1})$ is a root of $x^2 + 1$. The algebraic numbers form a *field,* that is, the sum, difference, product, and quotient (with nonzero denominator) of two algebraic numbers is again algebraic.

Alternating series test. If $\{a_n\}$ is a decreasing sequence of positive numbers and $a_n \to 0$ as $n \to \infty$ then the infinite series $a_0 - a_1 + a_2 - \cdots$ converges and the approximation to the sum obtained by using a partial sum is in error by less than the first omitted term.

Ball. A ball consists of a sphere plus its boundary. Thus a ball in Euclidean d-space consists of all points whose distance from a given point is less than or equal to a given positive number. When $d = 2$, the balls are called disks.

Base-two notation. The representation of an integer in terms of powers of two. The base-two (or binary) notation of 45, for example, is 101101 because $45 = 2^5 + 2^3 + 2^2 + 2^0$.

Body. See *convex body.*

Boundary. The boundary of a set S consists of all points p such that each neighborhood of p includes both points of S and points not in S.

Bounded set. A bounded subset of $\mathbb{R}^d$ is one that is contained in some ball.

$\mathbb{C}$. The set $\mathbb{C}$ of complex numbers, that is, numbers of the form $x + iy$ where x and y are in $\mathbb{R}$ and i denotes $\sqrt{-1}$.

Centrally symmetric. A subset S of $\mathbb{R}^d$ is centrally symmetric if there is a point c (the center) such that for each point p of S, the point $2c - p$ also belongs to S. (Note that c is the midpoint of the segment joining p and $2c - p$.)

Circle. In the Euclidean plane, a circle is the set of all points at a given positive distance from a given point. (A "solid circle"—a circle together with the points inside it —is here usually called a disk.)

Closed set. A closed set is one that includes all of its boundary points.

Collinear. A set is collinear if it is contained in some line.

Compact set. For a subset S of $\mathbb{R}^d$, the following three conditions are equivalent, and a set satisfying these conditions is said to be compact: S is bounded and closed; for each sequence of points of S there is a subsequence that converges to a point of S; for each collection of open sets that covers S, there is a finite subcollection that also covers S.

Comparison test. If $0 \leq a_n \leq b_n$ for each n, and if $\sum b_n$ converges, then $\sum a_n$ converges.

Computable set. A set for which there is an algorithm (i.e., computer program) that can recognize membership in the set. Thus the even numbers are computable because it is easy to write a computer program that outputs "Yes" when the input integer is even, and "No" when it is not. This definition can be used for sets that do not consist of integers by coding the sets as integers in some way.

Concyclic. A set is concyclic if it is contained in some circle.

Congruent. Two sets are congruent if there is a one-to-one distance-preserving transformation between the points of one and the points of the other.

Congruent mod *m*. Two integers a and b are said to be congruent mod m (written $a \equiv b \bmod m$) if they leave the same remainder after division by m; equivalently, $a \equiv b \bmod m$ if $b - a$ is divisible by m.

Connected. A connected set is one that cannot be expressed as the union of two disjoint closed subsets.

Convex combination. A convex combination of points $p_1, \ldots, p_k$ is a linear combination in which the coefficients are nonnegative and their sum is 1.

Convex body. As the term is used here, a convex body is a convex set that is bounded, closed, and has nonempty interior. Occasionally, when it clear from context that convexity is assumed, the shorter term "body" is used.

Convex hull. The convex hull ($\mathrm{con}\, X$) of a set X is the intersection of all convex sets containing X. Equivalently, it is the set of all convex combinations of points of X. It is known that when $X \subseteq \mathbb{R}^d$, each point of $\mathrm{con}\, X$ can be expressed as a convex combination of $d + 1$ or fewer points of X.

Convex set. A convex set is a set S such that for each pair p and q of points of S, the entire segment pq is contained in S. This is equivalent to saying that S includes all convex combinations of its points.

Countable set. A set that can be put into a one-to-one correspondence with the set of natural numbers. It is not hard to see that the sets of even integers, positive integers, and negative integers are all countable. Somewhat trickier are the proofs that the set of rationals and the set of algebraic numbers are also countable. The set of real numbers, on the other hand, is uncountable.

Dense. A set X is dense in a set Y if $X \subseteq Y$ and each neighborhood of each point of Y includes a point of X.

Disjoint. Two sets are disjoint if they have no common point.

Disk. See *ball* and *circle*.

***e*.** The base of the natural logarithms. One way to define e is as the unique number a such that the derivative of a^x is a^x. It can be shown (see a calculus text) that 2 is too small and 3 is too big. Further calculations show that $e = 2.71828\ldots$.

Edge. For a Jordan polygon, the terms *edge* and *side* are used interchangeably.

Euclidean *d*-space. This is the space $\mathbb{R}^d$ with the norm of a point $x = (x_1, \ldots, x_d)$ defined as

$$\left(\sum_1^d x_i^2\right)^{1/2}$$

and the distance between x and a point $y = (y_1, \ldots, y_d)$ defined as

$$\left(\sum_1^n (x_i - y_i)^2\right)^{1/2}.$$

Euler's ϕ-function. The function $\phi(n)$ is defined on positive integers n, for which it is the number of integers in the interval $[1, n]$ that are relatively prime to n. A formula for ϕ can be given in terms of n's prime factorization: if $n = \prod p_i^{a_i}$, then

$$\phi(n) = n \prod \left(1 - \frac{1}{p_i}\right).$$

Fibonacci numbers. The sequence of numbers obtained by starting with 0, 1, and obtaining each succeeding number by adding the previous two. Thus the first few Fibonacci numbers are 0, 1, 1, 2, 3, 5, 8, 13, 21, 34, 55, 89.

General position. Points in the plane are in general position if no three of them are collinear.

Geometric series. An infinite series where each term is obtained by multiplying the preceding term by a fixed number r. These series therefore have the form $a + ar + ar^2 + ar^3 + \cdots$. A geometric series with nonzero first term converges if and only if $|r| < 1$, and the sum is $a/(1 - r)$.

Graph. A graph consists of a finite set V of points (called vertices or nodes) together with a set of unordered pairs of points of V (called edges).

Half-line. See *ray.*

Harmonic series. The infinite series

$$1 + \frac{1}{2} + \frac{1}{3} + \frac{1}{4} + \frac{1}{5} + \cdots,$$

which (see any calculus text) diverges to infinity.

Incidence. Two sets are incident if one is contained in the other. In particular, a point and a line are incident if the point lies on the line or, equivalently, the line passes through the point.

Infinite series. Infinitely many numbers that want to be added together. If the partial sums (the sums of the first n terms) converge to the limit S as $n \to \infty$ then the series is said to converge to the sum S. Otherwise the series is said to diverge.

Inner product. The inner product of two points

$$x = (x_1, \ldots, x_d) \quad \text{and} \quad y = (y_1, \ldots, y_d)$$

of $\mathbb{R}^d$ is the number

$$\sum_1^d x_i y_i.$$

Recall that this is equal to the product $\|x\| \, \|y\| \cos\theta$, where $\|\cdot\|$ denotes the Euclidean norm and θ is the angle between the rays that issue from the origin and pass through the points x and y, respectively.

Integral test. Suppose $f(x)$ is a continuous and decreasing function on $[m, \infty]$. Then:
(a) If $0 \le a_n \le f(n+1)$ for $n \ge m$ and $\int_m^\infty f(x)\,dx$ is finite, then the series $a_m + a_{m+1} + \cdots$ converges to a sum that is no greater than the integral.
(b) If $0 \le f(n) \le a_n$ for $n \ge m$ then $a_m + a_{m+1} + \cdots$ is at least as large as $\int_m^\infty f(x)\,dx$, and the series diverges if the integral diverges.

Interior point. A point p is an interior point of a set S if some neighborhood of p is contained in S.

Intersection. The intersection of two sets X and Y is the set of all points that belong to both X and Y.

Irrational number. A real number that is not rational. Historically, the first example is $\sqrt{2}$, which can be seen to be irrational as follows. If $\sqrt{2} = \frac{m}{n}$ then $2n^2 = m^2$. But the base-two representation of m^2 ends in an even number of zeros while that of $2n^2$ ends in an odd number of zeros.

Jordan curve. A Jordan curve is a simple closed curve in the plane.

Jordan polygon. A Jordan polygon is a Jordan curve that is formed from a finite number of segments.

Jordan region. When a Jordan curve is removed from the plane, what's left is the union of two disjoint connected sets, only one of which is bounded. The union of the curve and that bounded component is a Jordan region.

Line. The line determined by two distinct points x and y is the set of all points of the form $\lambda x + (1-\lambda)y$, where λ is an arbitrary real number.

Line-segment. The line-segment (or simply segment) determined by two distinct points x and y is the set of all points of the form $\lambda x + (1-\lambda)y$, where $0 \le \lambda \le 1$. (This is the "closed segment." For the "open segment," use $0 < \lambda < 1$.)

Linear combination. A linear combination of points $p_1, \ldots, p_n$ is a point expressible in the form $\sum_1^n \lambda_i p_i$ where the λ_i are real numbers.

Linear transformation. A linear transformation L is one that preserves linear combinations, so that always $L(\sum_1^n \lambda_i p_i) = \sum_1^n \lambda_i L(p_i)$. The linear transformations of $\mathbb{R}^n$ into $\mathbb{R}^m$ are precisely the transformations that can be represented in the usual way by an $m \times n$ matrix A, so that $L(x) = Ax$ for each point $x \in \mathbb{R}^n$.

Maclaurin series. Many functions can be expressed as an infinite series of the form $\sum_{n=1}^{\infty} a_n x^n$. The Maclaurin series of an infinitely differentiable function $f(x)$ is the series $\sum_{n=1}^{\infty} a_n x^n$ where each a_n equals $f^{(n)}(0)/n!$. Many familiar functions ($\sin x$, $\cos x$, e^x, polynomials, etc.) are represented by their Maclaurin series, but this is not true for all infinitely differentiable functions. For example, the function $e^{-1/|x|}$ is infinitely differentiable but all the coefficients in its Maclaurin series are zero.

$\mathbb{N}$. The set of natural numbers: $0, 1, 2, 3, 4, 5, \ldots$.

Neighborhood. A neighborhood of a point p is a set that contains some ball centered at p.

Nondeterministic polynomial-time algorithm. This is a complicated concept and the reader is referred to the book *Computers and Intractability: A Guide to the Theory of NP-Completeness* by M. Garey and D. Johnson (Freeman, San Francisco, 1979) for a more detailed discussion. In short, a nondeterministic algorithm to recognize a set A is one that can be implemented by a computer program that includes instructions of the form "`GOTO 100 OR 200 OR 300`" where the line number chosen depends on, say, a coin-flip. If, for any input in A, at least one sequence of coin-flips leads to the ouput "Yes," and if for any input not in A all sequences of coin-flips lead to a "No," and if the program runs in polynomial time for all inputs, then the algorithm is said to be a nondeterministic polynomial-time algorithm, and the set A is said to be in the class $\mathbb{NP}$. An equivalent definition is the following: A set A is in $\mathbb{NP}$ if there is a polynomial-time algorithm that accepts as input potential members of A and an additional string called a *certificate,* and such that for each input in A there is at least one certificate for which the algorithm halts with a "Yes," while for inputs not in A, all certificates lead to a "No." For example, the set of composite numbers is in $\mathbb{NP}$ because the program that accepts two integers n and c ($< n$) as input and checks whether c divides n and $1 < c$ runs in polynomial-time, and if n is composite then there is a certificate c leading to a "Yes," namely any proper divisor of n. Many problems for which the naive algorithm works in exponential time lie in $\mathbb{NP}$. It has not been proved that the class $\mathbb{NP}$ is, as expected, different than $\mathbb{P}$; indeed, this problem, called the $\mathbb{P} = \mathbb{NP}$ problem, is the most important problem of theoretical computer science.

Normed linear space. A normed linear space consists of a real vector space together with a function $\|\cdot\|$ (the norm) that satisfies the following conditions for all points x and y of the space and all real numbers λ : $\|x\| \geq 0$, with equality if and only x is the origin; $\|\lambda x\| = |\lambda| \, \|x\|$; $\|x+y\| \leq \|x\| + \|y\|$. The distance between two points x and y of a normed linear space is defined as $\|x - y\|$.

Open set. A set is open if it contains a neighborhood of each of its points.

Parallelotope. A parallelotope in $\mathbb{R}^d$ is a set that is the image, under some affine transformation of $\mathbb{R}^d$ onto itself, of the cube $\{x = (x_1, \ldots, x_d) : |x_i| \leq 1 \text{ for all } i\}$. Equivalently, a parallelotope is a set of the form

$$\{c + \sum_1^d \epsilon_i p_i : |\epsilon_i| < 1 \text{ for all } i\},$$

where c is a point of $\mathbb{R}^d$ and $p_1, \ldots, p_d$ are linearly independent points of $\mathbb{R}^d$.

Partition. A partition of a set S is a collection of pairwise disjoint sets whose union is S.

Pi (π). The ratio of the circumference of a circle to its diameter. Equivalently, the ratio of the area of a circle to the square of its radius.

Plane. See $\mathbb{R}^2$.

Polygon. This is used ambiguously here. Sometimes it means a simple closed curve that is formed from a finite number of segments. In other cases, it means a polygonal Jordan region.

Polynomial-time algorithm. An algorithm that runs in time (by which is meant the number of fundamental bit operations) that is bounded above by a polynomial function of the length of the input. Thus there are constants c and k such that for any input, the algorithm halts in cn^k steps where n is the length of the input. The class of problems solvable by a polynomial-time algorithm is denoted by $\mathbb{P}$.

$\mathbb{Q}$. The set of rational numbers.

$\mathbb{R}$. The set of real numbers, that is, numbers corresponding to a point on an unbounded (in both directions) line. The standard representation for a real number is by its decimal expansion.

$\mathbb{R}^2$. See $\mathbb{R}^d$.

$\mathbb{R}^d$. The points of $\mathbb{R}^d$ are d-tuples of real numbers. The space is made into a real vector space by defining addition and scalar multiplication in the usual coordinatewise fashion. Thus if $x = (x_1, \ldots, x_d)$ and $y = (y_1, \ldots, y_d)$, we have $x + y = (x_1 + y_1, \ldots, x_d + y_d)$ and $\lambda(x_1, \ldots, x_d) = (\lambda x_1, \ldots, \lambda x_d)$. Unless some other distance function is specified, it is assumed that the Euclidean distance is used (see *Euclidean d-space*).

Rational number. A number of the form $\frac{m}{n}$ where m and n are integers. The rationals are distinguished among the real numbers by having decimal expansions that are eventually periodic; that is, after some point the decimal expansion consists of a finite block of digits that repeats.

Ray. Except in Section 1, this means a subset of $\mathbb{R}^d$ that is of the form $\{q + \lambda p : \lambda \geq 0\}$ for some $q \in \mathbb{R}^d$ and some nonzero $p \in \mathbb{R}^d$. (This is the "closed ray" that issues from q in the direction p. For "open ray," use $\lambda > 0$.) In Section 1, rays in this sense are called "half-lines," because the word "ray" is used to denote the polygonal path followed by a light ray under the usual reflection rule.

Recursive set. This is a synonym for *computable set*.

Region. A region is a connected set that has interior points.

Relatively prime integers. Two integers such that the only positive integer that divides both is 1.

Rigid motion of the plane. A function from the plane $\mathbb{R}^2$ to itself that preserves distance and preserves orientation (a consequence of this definition is that the function is one-to-one and onto). The rigid motions of the plane consist of translations and rotations; these mappings can be carried out rigidly in the plane, that is, continuously and without leaving the plane. A reflection preserves distance, but is not a rigid motion since a triangle, say, would have to be torn, or taken out of the plane, in order to be reflected in a line.

Segment. See *line-segment.*

Side. For a Jordan polygon, the terms *side* and *edge* are used interchangeably.

Simple closed curve. A simple closed curve is the image of a circle under a one-to-one continuous transformation. A subset S of $\mathbb{R}^d$ is a simple closed curve if and only if there are continuous real-valued functions $f_1, \ldots, f_d$ on $[0, 1]$ such that, with $p(\tau) = (f_1(\tau), \ldots, f_d(\tau)) \in \mathbb{R}^d$, the following conditions are satisfied: $p(0) = p(1)$; for $0 \leq \sigma < \tau < 1$, $p(\sigma) \neq p(\tau)$; $S = \{p(\tau) : 0 \leq \tau \leq 1\}$.

Simplex. A d-simplex is the convex hull of $d + 1$ affinely independent points. The 2-simplices are the triangles, and the 3-simplices are the tetrahedra.

Sphere. A sphere is the set of all points at given distance from a given point. In a normed linear space, each sphere has the form $\{x : \|x - c\| = \rho\}$, where c is the sphere's center and ρ is its radius.

Square. This is used ambiguously here. Sometimes it means just the boundary, and sometimes its means the entire "solid square."

Symmetric. See *centrally symmetric.*

Tetrahedron. See *simplex.*

Transcendental number. A real or complex number that is not algebraic.

Translation. A translation is a transformation that adds a fixed point to each point in its domain. Thus if T is a translation of $\mathbb{R}^d$ there is a point $q \in \mathbb{R}^d$ such that $T(x) = x + q$ for all $x \in \mathbb{R}^d$.

Union. The union of two sets X and Y is the set of all points that belong to X or to Y (or to both).

$\mathbb{Z}$. The set of all integers (positive, negative, or zero).

INDEX OF NAMES

Adleman, L. M. 200, 221–222
al-Banna, I. 180
Alexander, R. 87
Almering, J. H. 132, 204–205, 302
Alon, N. 109
Amman, R. 114
Anderson, C. W. 213
Ang, D. 132
Anning, N. H. 132–133, 135
Apéry, R. 249, 263
Appel, K. xii, 120
Archimedes 1
Augustine, Saint 180
Avis, D. 87, 98

Bailey, D. 253, 256
Bálint, V. 95, 106
Bálintova, A. 106
Banach, S. 128, 130
Barnette, D. 120
Beck, J. 103, 107
Beckman, F. S. 123
Bellamy, D. P. 148
Benda, M. 124, 127
Berger, M. 76
Besicovitch, A. S. 92, 132
Beukers, F. 263
Bezdek, A. 117
Bezdek, K. 78, 92
Bhascara 132
Bhattacharya, B. K. 87
Bialostocki, A. 101
Biggs, N. L. 120
Bing, R. H. 145, 147–149
Birkhoff, G. D. 73
Bistriczky, K. 97
Blaschke, W. 80
Bleicher, M. N. 206–207
Bohl, P. 145
Boldrighini, C. 71
Bollobás, B. 87
Bondy, J. A. 146
Borsuk, K. 93
Bouligand, G. 87
Brahmegupta 132
Brent, R. P. 213, 252
Breusch, R. 207
Brocard, H. 174, 203
Brouwer, L. E. J. 145
Brown, R. F. 148
Buchholz, R. H. 204–205
Buhler, J. P. 170, 221
Butler, G. J. 80

Campbell, P. 207
Cantor, G. 258
Capoyleas, V. 87
Cayford, A. H. 90
Chakerian, G. D. 92,94
Champernowne, D. 251
Chilakamarri, K. B. 124, 127
Chudnovsky, D. 253
Chudnovsky, G. 253
Chvátal, V. 77
Clarkson, K. L. 103
Cobham, A. 258–259
Cohen, G. L. 213
Collatz, L. 225
Condict, J. 213
Connelly, R. 92
Connett, J. E. 71
Conway, J. 226
Coppersmith, D. 201
Corrádi, K. 118
Coxeter, H. S. M. 105–106
Crandall, R. E. 170, 200, 225
Craveiro de Carvalho, F. J. 84
Croft, H. T. 3, 73
Crotty, J. M. 81
Csima, J. 105
Cunningham, F. 93

Dantzig, G. B. 288
Danzer, L. 88, 117
Davies, R. O. 92
Davis, M. 230
Daykin, D. E. 132–133
De Bruijn, N. G. 94, 105, 124, 282
Debrunner, H. 126
Dehn, M. 129
Dekster, B. V. 124
DeTemple, D. W. 75
Descartes, R. 180–181, 214

Dickson, L. E. 132, 181, 211
Dierker, P. 101
Diophantus 168
Dirac, G. A. 80, 103
Dirichlet, P. G. L. 169
Dobkin, D. P. 98
Douady, R. 74
Dubins, L. 130
Duff, G. F. D. 91

Eberstark, H. 241
Edelsbrunner, H. 98, 103
Edwards, H. M. 201
Eggleston, H. G. 92, 204, 295
Ehrhart, E. 80
Elkies, N. 171
Elliott, P. D. T. A. 105–106, 281
Emch, A. 137
Engel, A. 52
Engel, P. 118
Erdős, P. 3, 94–95, 98, 103, 105, 107–108, 124, 132–133, 135, 168, 176, 180, 192, 206–207, 212, 282
Euclid 1, 179
Euler, L. 169–171, 179, 212, 217, 240, 248–249, 261–262
Everett, C. J. 225
Ewald, G. 78

Falconer, K. J. 3, 122
Faltings, G. 169–170, 201–202, 300
Fejes Tóth, G. 97, 111
Fejes Tóth, L. 3, 116
Fenn, R. 142
Ferguson, H. 256, 263–264
Fermat, P. de 168, 171
Fibonacci 175, 206
Filaseta, M. 201–202
Fischer, K. G. 127
Fisk, S. 77
Forcade, R. W. 256
Fouvry, E. 200
Franel, J. 251
Frankl, P. 121–122
Fredman, M. L. 87
Frye, R. 171
Fujiwara, M. 80

Gale, D. 87, 287
Galperin, G. A. 73
Gardner, M. 111, 120
Gardner, R. 80, 84, 130
Garner, L. 226
Gauss, C. F. 182–183, 252
Gelfond, A. O. 255–256
Germain, S. 200
Gillies, D. B. 209–210
Girault-Beauquier, D. 115
Goldwasser, S. 222
Goodman, J. E. 99
Graham, R. L. 93, 136, 146, 168, 207
Granville, A. 170, 200
Greenwell, D. 123
Groemer, H. 94
Gromov, M. L. 87, 139
Gruber, P. 3, 74–75
Grünbaum B. 3, 88–89, 92, 94, 103, 106, 111–112, 114, 117, 139, 286
Grünert, J. A. 172
Guibas, L. J. 103
Guillemin, V. 74
Guy, R. K. 3, 71, 133, 168

Habicht, W. 275
Hadwiger, H. 3, 87, 120, 122, 126
Hagis, P. 212–213
Hagopian, C. L. 148–149
Hajos, G. 118
Haken, W. xi–xii, 120
Hales, A. W. 116
Hallstrom, A. 80
Hammer, J. 3
Hansen, H. C. 91
Hansen, S. 105–107
Harary, F. 124
Harborth, H. 98, 132–133
Hardy, G. H. 72
Hausdorff, F. 128
Hayashi, T. 80
Head, A. K. 222
Heath-Brown, D. R. 200
Helly, E. 90
Hermite, C. 214, 255
Hilbert, D. 195, 255–256
Hirsch, M. 130
Hirschhorn, M. D. 112
Hoffman, A. J. 288
Holt, F. 71
Hopf, H. 140
Horton, J. D. 98
Hu, T. C. 288
Huang, M. 222
Hunt, D. C. 112
Hunter, J. A. H. 203

Imai, H. 87
Inkeri, K. 169–170
Isbell, J. R. 132

Ishihata, K. 225

James, R. 111
Jamison, R. E. 108
Johnson, P. D. 123
Johnson, S. 95
Jones, J. 230

Kakutani, S. 124
Kanada, Y. 253
Kannan, R. 247, 256
Karush, J. 130
Kasimatis, E. A. 115–116
Katok, A. 71
Keane, M. 71
Kelly, J. B. 84
Kelly, L. M. 105, 139
Kelly, P. J. 85
Kemnitz, A. 133
Kershner, R. B. 111
Kilian, J. 222
Kind, B. 93
Kirszbraun, M. 88
Klamkin, M. S. 92
Klee, V. 88, 71, 126
Kleinschmidt. P. 93
Kneser, M. 86–87, 89, 275
König, D. 72
Korec, I. 203
Kranakis, E. 103
Kreinovič, V. 75
Kreisel, G. 196
Kronecker, L. 72
Kronheimer, E. H. 138, 142
Kronheimer, P. B. 138, 142
Kuiper, N. H. 75
Kummer, E. 132, 169, 199–200
Kuperberg, W. 111, 117
Kupitz, Y. 109

Laczkovich, M. xii–xiii, 50, 287
Lagrange, J. 132
Lamé, G. 169
Lander, L. J. 171
Larman, D. G. 78, 80, 121–122, 135
Lassak, M. 78, 94
Lawrence, J. 118
Lazutkin, V. F. 74
Lebesgue, H. 91
Leech, J. 132, 205
Legendre, A. M. 169
Lehman, R. S. 218
Lehmer, D. H. 169, 200
Lehmer, E. 169, 200

Lenstra, A. K. 221, 247
Lenstra, H. 221
Levi, F. W. 78
Lévy, P. 140
Lieb, E. 87
Lindemann, F. 50, 214, 255, 293
Liouville, J. 258
Littlewood, J. E. 184, 218
Lloyd, E. K. 120
Lovász, L. 95, 247, 256
Loxton, J. H. 258
Lubiw, A. 287
Lucas, E. 209

McGeoch, L. 256
McMullen, P. 117
Mahler, K. 258–259
Manasse, M. 221
Marchetti, M. 71
Masur, H. 71
Mather, J. N. 74
Matijasevič, Y. 195–196, 230, 232
Mead, D. G. 116
Melchior, E. 105
Melrose, R. 74
Mendés-France, M. 259
Mertens, F. 219
Meyerson, M. D. 137, 142
Michelacci, G. 80
Miller, G. 189, 221–222, 224
Minty, G. J. 89
Monagan, M. B. 201
Monier, L. 189
Monsky, P. xii, 115
Mordell, L. J. 132, 204
Moser, L. 3, 121
Moser, W. 3, 95, 105, 121
Motzkin, T. S. 98, 105, 107, 282
Murty, U. S. R. 146

Nagell, E. 132
Nakayama, N. 208
Nickel, L. 209
Nivat, M. 115
Niven, I. 252, 258
Nocco, G. 214
Noll, C. 209

Odlyzko, A. 219
O'Neill, P. E. 98
O'Rourke, J. 77
Ogilvy, C. S. 137
Overmars, M. 98

Pach, J. 3, 87, 89–90, 95, 106, 276

Paganini, N. 181
Pál, J. 91–92
Parkin, T. R. 171
Penrose, L. 71
Penrose, R. 113
Perles, M. 109, 124, 127
Perrin, R. 99
Perron, O. 118
Petty, C. M. 81
Pintz, J. 219
Pocchiola, M. 103
Pollack, R. 99
Pollard, J. 221
Pomerance, C. 213, 221
Pratt, V. 222
Purdy, C. 124
Purdy, G. 3, 107–108, 124
Putnam, H. 230

Quarles, F. S. 123

Rabin, M. 189, 221–223
Radon, J. 277
Raiskii, D. E. 120, 122, 126
Ramsey, F. P. 26
Rappaport, D. 98
Rathbun, R. L. 204–205
Rauch, J. 71
Rehder, W. 87
Reinhardt, K. 93, 111
Renegar, J. 116
Rennie, B. C. 92
Ribenboim, P. 170–171
Rice, M. 111
Richman, F. 115
Riemann, B. 182, 216
Riesel, H. 221
Robertson, J. 75
Robertson, S. A. 84
Robinson, J. 230, 232
Robinson, R. M. 232
Rogers, C. A. 78, 80, 121–122, 130, 135
Rogers, J. 80
Root, S. C. 214
Rosenbaum, J. 84
Rothe, H. 80
Rothschild, B. L. 136
Rumely, R. S. 221

Salamin, E. 252
Sawyer, E. T. 105
Schäfke, R. 80
Schattschneider, D. 111
Schinzel, A. 212
Schneider, T. 255–256
Schoenberg, I. J. 88, 93, 132
Scholten, B. 98
Schönhage, A. 260
Schramm, O. 78
Schubert, H. 204
Scott, P. R. 108
Seidel, J. J. 135
Senechal, M. 117
Sharir, M. 103
Sheng, T. K. 132–133
Shephard, G. C. 3, 111–112, 114, 117
Shermer, T. 77
Shtogrin, M. I. 111
Siegel, C. L. 256
Simmons, G. J. 124
Simon, B. 87
Sine, R. 75
Skewes, S. 218
Skolem, T. 232
Slowinski, D. 209
Smillie, J. 72
Sompolski, R. W. 170, 200
Spaltenstein 81
Sperner, E. 146
Spieker, T. 123
Spirakis, P. G. 87
Sprague, R. 91
Steiger, F. 135
Stein, R. 112
Stein, S. K. 115–116, 118
Steuerwald, R. 212, 214
Stewart, B. M. 207
Stoneham, R. G. 251
Straus, E. G. 116, 136, 176
Stromquist, W. 137
Sudakov, V. N. 87
Sudan, G. 72
Suryanarayana, D. 212
Süss, W. 80
Swinnerton-Dyer, H. P. F. 73
Sydler, J. P. 129
Sylvester, J. J. 105, 214, 306
Szabó, S. 118
Székely, L. A. 122
Szekeres, G. 95
Szemerédi, E. 103, 107, 212
Szemerédi, G. 212
Szucs, A. 72

Tamvakis, N. K. 80
Tanner, J. 170, 199–201

Tarski, A. 50–51, 116, 124, 128–129, 131
Taylor, M. 71
te Riele, H. J. J. 213, 217–219
Terras, A. 225–226
Thomas, J. 115
Thue Poulsen, E. 86
Thwaites, B. 225
Toeplitz, O. 137
Trotter, W. T. 103, 107
Turan, P. 212
Turner, P. H. 76
Tutte, W. T. 124

Ulam, S. 93, 132
Ungar, P. 108

van de Lune, J. 216
van der Poorten, A. J. 258–259
Vandiver, H. S. 169, 200
Van Leeuwen, J. 87
Van Stigt, W. P. 145
Vaughan, H. E. 138
Venkov, B. A. 117
Vincent, I. 98
Volčič, A. 80
Volkmer, H. 80
von Neumann, J. 130
Vose, M. 206
Voxman, B. 101

Wagstaff, S. S. 169–170, 199–201, 210
Webber, W. 105
Wegner, G. 123, 127
Weide, B. 87
Weil, A. 171
Weitzenbock, W. 80
Welzl, E. 103
Wilbour, D. 109
Wilson, P. R. 106
Wilson, R. M. 120–122
Winter, D. T. 217
Wirsing, E. 80, 213
Wiseman, J. A. 106
Wood, D. 87
Woodall, D. R. 120, 124, 126, 289
Wormald, N. C. 122
Wright, E. M. 72

Yaglom, A. M. 261
Yaglom, I. M. 261
Yanagihara, K. 84
Yoneda, N. 225
Yorinaga, M. 219
Young, G. S. 149

Zaks, J. 124, 142
Zamfirescu, T. 75, 137
Zemlyakov, A. 71
Zuccheri, L. 84

SUBJECT INDEX

acute triangle 7, 10, 270
alcoholic drinks 242
algebraic curve 106
algebraic numbers 243, 246, 256
allowable sequence 100–102, 108, 280–281
amicable numbers 180–181
anisohedral 40
aperiodic tiling 38, 112–115
arithmetic-geometric mean 252–254, 260
arc 67, 69, 148–149, 299
arcwise connected 69, 148–149
area 16–17, 19, 21–24, 86–87, 91–94, 279–280
art gallery theorem 9–11, 76–77, 271

balanced 109–110
ball *See* sphere
Banach–Tarski paradox 128–129
barycentric coordinates 147
barycentric subdivision 150, 298
billiard path 6–8, 10–11, 269–272
Bernoulli numbers 199, 262, 264, 314
Borsuk's problem 93–94
Brent–Salamin formula 252, 253
Brouwer fixed-point theorem 66, 70, 146–147, 297–298

caustic 7, 74–76
Champernowne's number 239, 245, 251, 253
chord 12
chromatic number 45–49, 120–127, 289–291
circle 7, 10–12, 15, 21, 69, 116, 270–271 *See also* ordinary circle, inversion
circle-squaring 50–52, 129, 200, 240
circumscribed figures 60–61
collinear 56
coloring 9, 11, 45, 120 *See also* painting
computable set 213, 226
concyclic 34, 56
congruent by dissection 51–52
conic 106
connecting line 29–35, 103–110, 281–285
constant width 22, 24, 75, 272, 295–296
continuous shrinking 16, 19, 86, 88, 277
continuum 66, 68–70, 138, 141, 145–150
convex hull 87, 276–278
convex polygon 23, 25–28, 95–102, 280–281 *See also* names of particular sorts of polygons
convex set 1, 12
cover 21–24, 91–94
cryptography 222–223
crystallographic group 115
cube 94, 118, 280
curvature 273
cyclic *See* concyclic

degree 145–146
Δ-set 48, 120
de Moivre's formula 131, 264
de Moivre–Stirling approximation to $n!$ 28
dense billiard path 6, 71–73
dense set 54, 57, 132–135
diameter 21–24, 91–94, 279–280
Diophantine equations 174, 195–197, 204, 230–232
Dirichlet's theorem 211
disk *See* circle
dissection xii, 42, 44, 51–52, 119, 287–288
double normal 272
drilling square holes 22

e 210, 215, 239–240, 243, 250, 255–257, 293
eccentricity 14, 273
ellipse 5, 10, 15, 74, 79, 81–83, 270, 272–273
ellipsoid 75–76
equichordal point 12, 15, 80–85, 272
equidecomposable sets 50, 128–130
equidissection 42–43, 119, 286
equipower point 84
equiproduct point 12, 15, 84, 273
equireciprocal point 12, 15, 84, 273
Euclid–Euler formula 179, 213
Euler ϕ-function 74

Fermat numbers 210–211, 221
Fermat's last theorem xi, 45, 167–172, 196, 199–202, 231
Fermat's little theorem 187–188, 256, 307–308
Fibonacci numbers 230–231, 233, 245
Fisher's inequality 283
fixed-point property 66–70, 145–150, 297–299
fixed-point theorem *See* Brouwer fixed-point theorem
focus 5, 10, 15, 270
four color theorem xii, 45, 120

γ 210, 219, 249–250, 264
general position 25–26, 29, 100–101, 110
God 180
Goodstein's theorem 231, 233
Gödel's theorem 195–196, 231
graph 109, 145–149
greedy algorithm 175–177
Guinness Book of Records 242

half-line 10
Hartmanis–Stearns Conjecture 247, 258
Helly's theorem 90, 278
heptagon 44, 286
Heron triangle 174, 204–205, 302
Heron's formula 174, 302
hexagon 23, 36–37, 40, 43, 46, 94, 111, 114, 132, 139
Hilbert's tenth problem 195–197, 230–232
homeomorphic 66–67
homothet 57–60, 65, 139

illumination 3–6, 9–11, 71, 76–79, 269–273
incidence 29–35, 103–107, 109–110, 145–146, 149, 281–285, 298
inscribed figures 58–65, 137–144, 295–297
inversion 34, 281–282
irregular prime 199–200
isohedral 39–40, 111

Jordan curve 4, 58–65, 69–70, 137–138, 272
Jordan curve theorem 58–59
Jordan polygon 9–11, 114
Jordan region 12–15, 112

Kakeya set 92–93
Kummer's criterion 169

labeling 146–147, 150, 298
Lagrange's four squares theorem 127, 197, 289
lattice 36–37
Lebesgue tiling theorem 122
level set 141–142, 144
lighting *See* illumination
line *See* ordinary line, connecting line, pseudoline, line at infinity
line at infinity 103–104
Liouville's number 239, 245–246, 312
listable 230–231
logarithmic integral, $\mathrm{li}(x)$ 182–185, 218–219

measure 86
Mersenne primes 178–179, 209–210, 224
midpoint-free 27
Möbius function, $\mu(n)$ 184, 219
monohedral 36, 39–39, 41, 43–44, 117
monotone subsequence 27–28, 95–101
morphine 258

near-pencil 33–34, 284
noncomputable set 226
nondeterministic polynomial time ($\mathbb{NP}$) 222–224, 232
nonseparating continuum 66, 69
normal numbers 251
normed linear space 83,85

octahedron 140
ordinary circle 34, 106, 282
ordinary line 29, 33–35, 109
ordinary plane 107, 109

painting 45-49, 120–127, 289–291
parabola 106
parallelogram 36, 57, 132, 294
parallel body 276
parallel pencil 103–104
parallelotope (= parallepiped) 140, 276, 296
pencil *See* near-pencil, parallel pencil
Penrose rhombs 113–114
pentagon 38, 44, 111–112, 114, 286
perfect box 173, 203
perfect numbers 167, 178–181, 209–213
perimeter 19, 22, 87, 276, 279
periodic *See* periodic billiard path, periodic tiling
periodic billiard path 7–8, 71–75, 271
periodic tiling 36–37, 40, 111–112
phase space 73
p-hexagon 39, 44, 118, 286
ϕ-function *See* Euler ϕ-function
pi (π) 50, 239–243, 248–257, 261
$\pi(x)$ 182–183, 218–219
picnic 141–142
pigeonhole principle 26
point at infinity 103–104
polygon *See* names of special types
polynomial-time algorithm 186, 189, 221–224
prime number theorem 183, 218
prime numbers 182, 186–189, 196, 209–212, 216–223, 231, 262
product 147, 148, 299
projective plane 103–104
prototile 36–40, 44
pseudoline 101
Putnam competition 205
Pythagorean triples 168, 171–174, 201

quadrilateral 65, 132, 137, 139, 286
quantum mechanics 242

Radon's theorem 277–278

Ramsey's theorem 26, 142, 280
random polynomial time ($\mathbb{RP}$) 222
rationally approximable set 54–57, 132–136, 294
rational polygon 6, 71–73, 79, 271–272
rational set 54–57, 132–136, 294
real-time computable 244–246, 259
reflection 3–8, 10–11, 71–76, 79, 269–272
regular n-gon 23–24, 65, 75, 109
regular prime 170, 199–200
retraction 68–70, 297–299
Reuleaux triangle 21–22, 24
Rhind papyrus 176
Riemann hypothesis 168, 182–184, 188, 215–219, 256
Riemann ζ-function 182, 215–218, 249, 261–264
right triangle 7
rigid cover 23–24, 91–92
ringworm problem 92
Rolle's theorem 140
Rosetta stone 176
RSA cryptography scheme 223

separating continuum 138
Sierpiński–Mazurkiewicz paradox 131
$\sigma(n)$ 179
simple closed curve *See* Jordan curve
simplicial subdivision 146
smooth 5–6, 8, 11–14, 74–76, 79, 137, 271
snakelike continuum 148
Sperner's lemma 146
sphere 81, 86, 116, 119, 136, 289, 294
square xii, 22, 36–37, 43, 58–65, 72, 79, 133, 137–138, 142
starshaped 3, 93
stopping time 225–229
string construction 76, 79, 113, 272
symmetric 14–15, 36, 40, 42–43

tetrahedra 204–205
$3n+1$ conjecture 191–194, 225–229
tiling 36–44, 46, 111–119, 285–289
trajectory 191–194, 225
transcendental numbers 50, 239, 243–244, 246, 251, 255–256
translation cover 21–22
tree 108–109, 148, 150
treelike continuum 148–150
triangle 22, 24, 43, 57, 61–62, 139, 142, 294
triangulation 9, 271
twin prime conjecture 211

universal cover *See* rigid cover

Wallace–Bolyai–Gerwien Theorem 50–51, 128–129
wallpaper group 115
Wolfskehl prize 199

$\zeta(s)$ *See* Riemann zeta function